Frontier Orbitals
and
Organic Chemical Reactions

Frontier Orbitals
and
Organic Chemical Reactions

Ian Fleming

*Assistant Director of Research in the Department
of Organic and Inorganic Chemistry, Cambridge
Fellow of Pembroke College, Cambridge*

A Wiley–Interscience Publication

JOHN WILEY & SONS
London · New York · Sydney · Toronto

Copyright © 1976, by John Wiley & Sons, Ltd.

Reprinted with corrections May 1978
Reprinted December 1978
Reprinted September 1980

Library of Congress Cataloging in Publication Data:

Fleming, Ian, 1935–
  Frontier orbitals and organic chemical reactions.

  'A Wiley–Interscience publication.'
  Bibliography: p.
  Includes index.
  1. Molecular orbitals. 2. Chemistry, Physical
organic. I. Title.
QD461.F53 1977   541'.28   76–3800

ISBN 0 471 01820 1 (Cloth)
ISBN 0 471 01819 8 (Pbk)

Printed and bound in Great Britain
at The Pitman Press, Bath

Acknowledgements

In writing a book about frontier orbital theory I am conscious of a very considerable debt to the pioneers of the subject. It is they, not I, who devised the theory and demonstrated its wide applicability, and it is therefore with a sense of great gratitude that I here list the names of the people who have provided chemists with this important new insight into the mechanics of chemical reactions. In alphabetical order, they are: N. T. Anh, K. Fukui, W. C. Herndon, R. Hoffmann, K. N. Houk, R. F. Hudson, G. Klopman, L. Salem and R. Sustmann. I have done little more than bring their material together, find a few more illustrative examples and simplify some of their arguments. They are innocent of any breaches of the discipline of theoretical chemistry that I may have made, by way of simplification, in my efforts to reach a wide audience. Finally, I should like to thank Dr. W. Carruthers, Professor R. F. Hudson, Professor A. R. Katritzky and Dr. A. J. Stone, each of whom made several helpful suggestions.

IAN FLEMING

Contents

CHAPTER 1

Introduction

Molecular orbital theory is a powerful and versatile asset to the practice of organic chemistry. As a theory of bonding it has almost completely superseded the valence bond theory: it has proved, in the long run, to be just as amenable to pictorial, non-mathematical expression; it has given the right answers to some decisive questions; it is the theory most theoreticians prefer; and quite importantly, organic chemists find it less misleading in everyday use. It is widely known today, not only as a theory of bonding but also as a theory capable of giving some insight into the forces involved in the making and breaking of chemical bonds. Most conspicuously, it was used by Woodward and Hoffmann[1] to explain the pattern of reactivity in pericyclic reactions; indeed, most of the other theories explaining the Woodward–Hoffmann rules are also based on molecular orbital theory.

More recently, molecular orbital theory has provided a basis for explaining many other aspects of chemical reactivity besides the allowedness or otherwise of pericyclic reactions. The new work is based on the perturbation treatment of molecular orbital theory, introduced by Coulson and Longuet–Higgins,[2] and is most familiar to organic chemists as the frontier orbital theory of Fukui.[3] Earlier molecular orbital theories of reactivity concentrated on the product-like character of transition states: the concept of localization energy in aromatic substitution is a well-known example. The perturbation theory concentrates instead on the other side of the reaction coordinate. It looks at how the interaction of the molecular orbitals of the starting materials influences the transition state. Both influences on the transition state are obviously important, and it is therefore important to know about both of them, not just the one, if we want a better understanding of transition states, and hence of chemical reactivity.

In this book, I have presented the theory in a much simplified, and, in particular, an entirely non-mathematical language. I shall assume only that the reader is familiar with the concept of a molecular orbital and its expression as a linear combination of atomic orbitals. I have simplified the treatment in order to make it accessible to every practising organic chemist, whether student or research worker, whether mathematically competent or not. In order to reach such a wide audience, I have frequently used very simple arguments and assumed relatively little knowledge. I hope that experienced organic chemists can recognize and skip these sections. Also, an undergraduate reader may, in many other places, wonder why so much space is devoted to a particular topic. The general importance of some of these topics lies in the history of the subject and the importance the problem had at one time. In other cases it may simply represent a minor facet of chemistry which just happened to be outstandingly puzzling until frontier orbital theory came along. Pericyclic reactions, for example, although hardly a minor facet, are nevertheless not quite as important as the size of Chapter 4 might lead one to think: the disproportionate emphasis they receive is a consequence of the effectiveness of frontier orbital theory in this field. I hope the less experienced reader will persevere, recognizing that I have tried simultaneously to produce an introduction to the subject and a review of it. A perfectly uniform level is hard to maintain with both these objectives in mind, but all organic chemists ought to know the theoretical basis and the wide and growing applications of this important theory.

I can perhaps best show its importance and usefulness by posing a number of familiar questions from a wide range of organic chemistry, to all of which—and to many more—the frontier orbital theory provides a satisfying answer.

(i) Why do enolate ions (1) react more rapidly with protons on oxygen, but with primary alkyl halides on carbon?

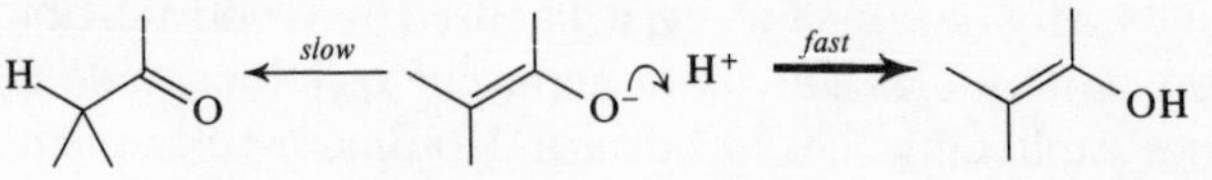

(1)

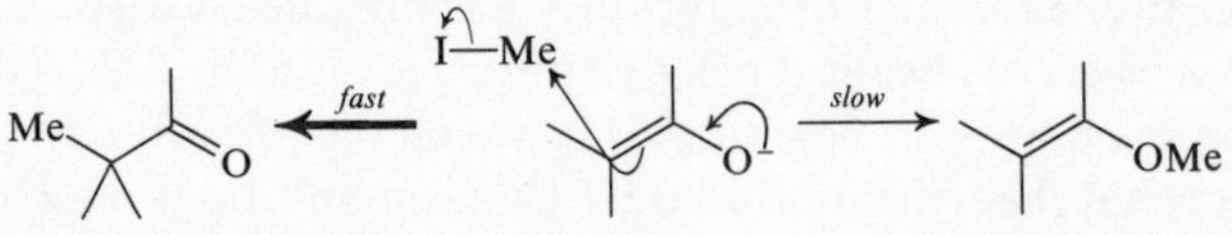

(1)

(ii) Why is pyrrole (2) attacked by electrophiles faster in the 2-position than in the 3-position?

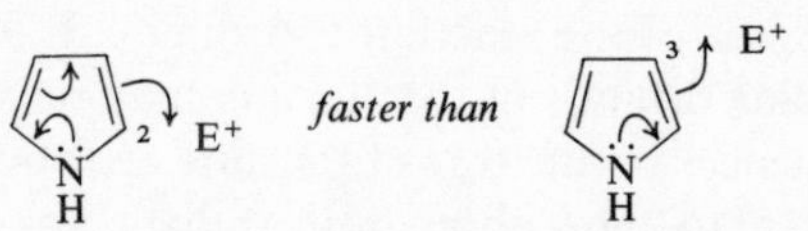

faster than

(2a)   (2b)

(iii) Why do some electrophiles attack pyridine *N*-oxide (3) at the 2-position, others at the 3-position and yet others at the 4-position?

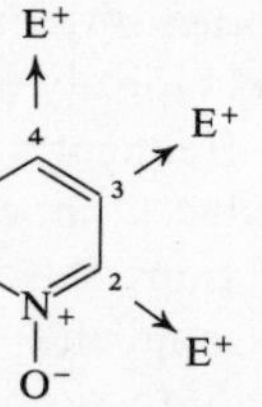

(3)

(iv) Hydroxide ion is much more basic than hydroperoxide ion. Why, then, is it so much less nucleophilic?

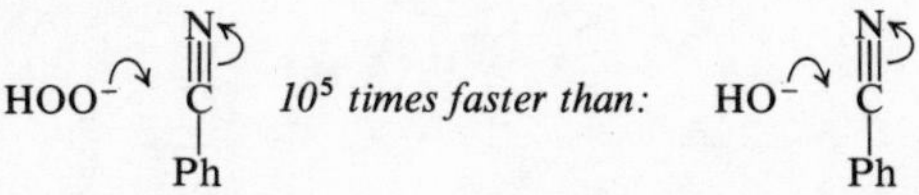

$$HOO^- \quad 10^5 \text{ times faster than:} \quad HO^-$$

(v) Why does maleic anhydride (5) react easily with butadiene (4), but not at all easily with ethylene (6)?

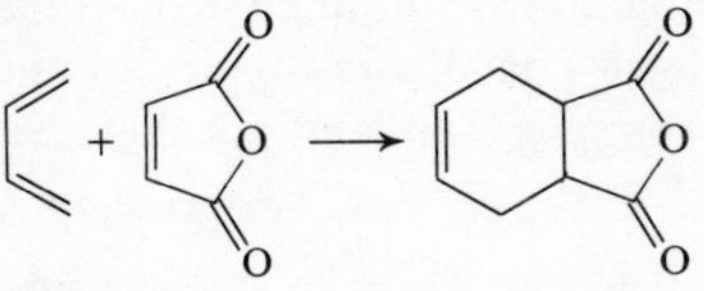

(4) (5)

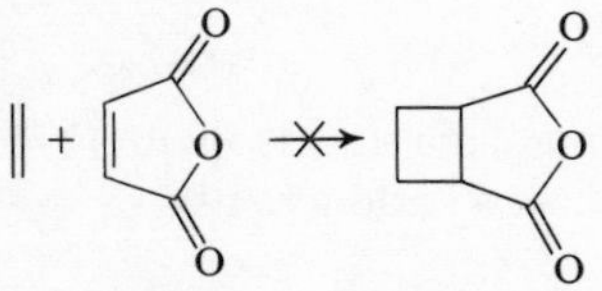

(6) (5)

(vi) Why does the Diels–Alder reaction give *endo* adducts such as (7)?

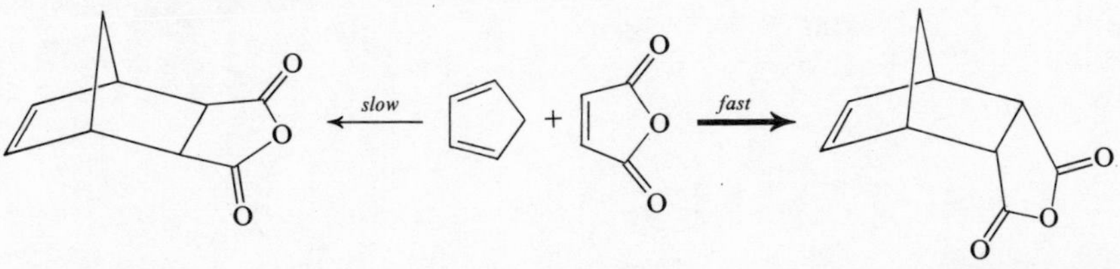

slow ← + → fast

(7)

(vii) Why do Diels–Alder reactions usually go faster when there is an electron-withdrawing group on the dienophile? And why does Lewis-acid catalysis speed them up even more?

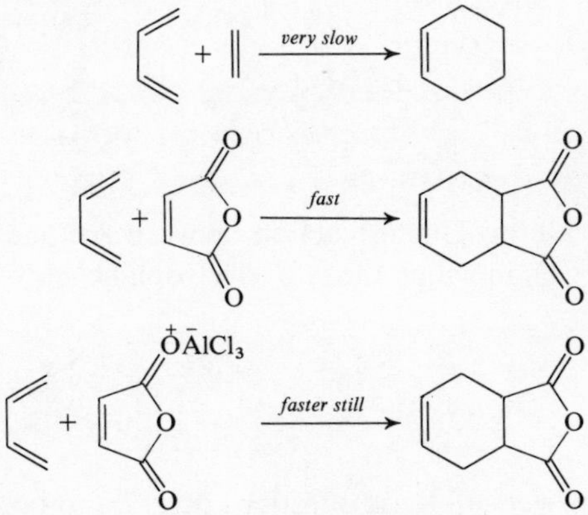

+ || —*very slow*→

+ —*fast*→

$\overset{+}{O}\overline{A}lCl_3$

+ —*faster still*→

4

(viii) Why does 1-methoxybutadiene (9) react with acrolein (10) to give only the 'ortho' isomer (11) and not the 'meta' isomer (8)?

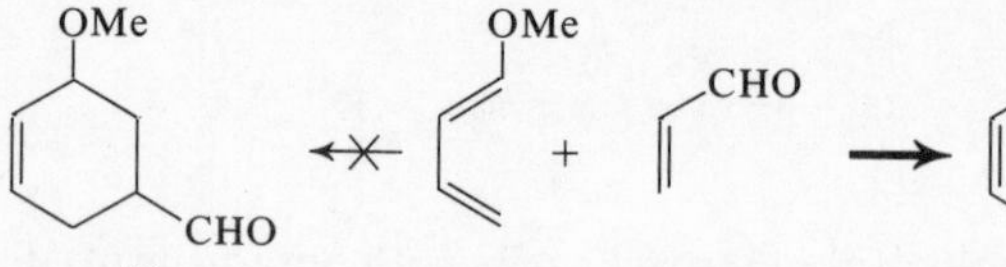

(8)   (9)   (10)   (11)

(ix) Why does diazomethane (13) add to methyl acrylate (14) to give the isomer (15) in which the nitrogen end of the dipole is bonded to the carbon atom bearing the methoxycarbonyl group, and not the other way round (12)?

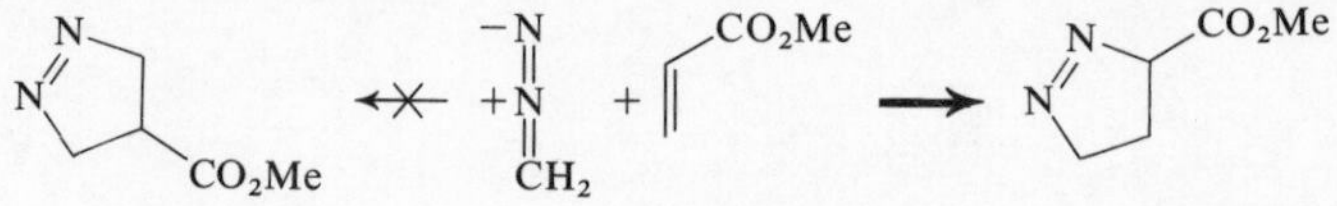

(12)   (13)   (14)   (15)

(x) Why do cyclopentadiene (16) and tropone (17) react to give the [4 + 6] adduct (18) and not the [2 + 4] adduct (19)?

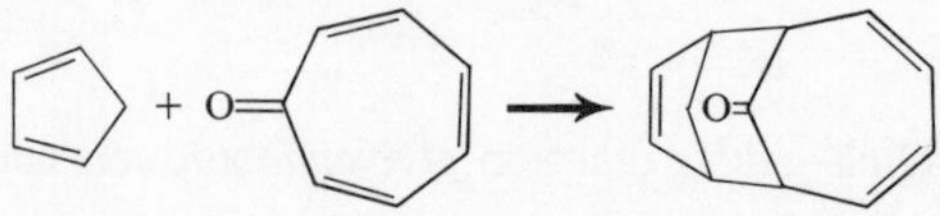

(16)   (17)   (18)

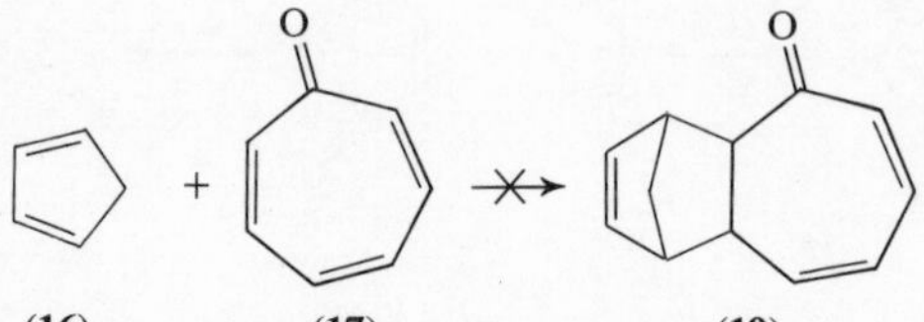

(16)   (17)   (19)

(xi) When methyl fumarate (20) and vinyl acetate (21) are copolymerized with a radical intitiator, why does the polymer consist largely of alternating units (22)?

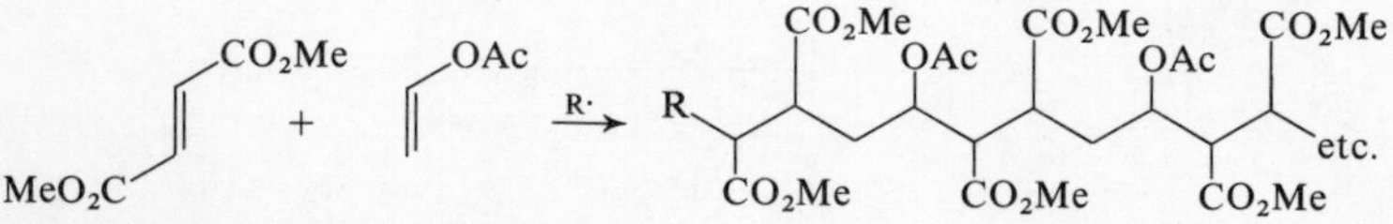

(20)   (21)   (22)

(xii) Why does the Paterno–Büchi reaction between acetone and acrylonitrile give only the isomer (23) in which the two 'electrophilic' carbon atoms become bonded?

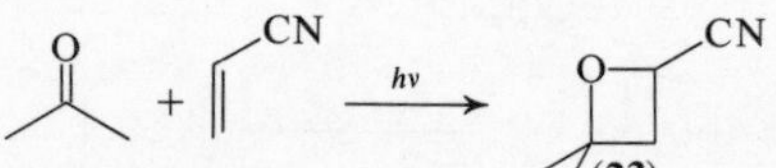

A theory which answers all these questions deserves to be widely known.

CHAPTER 2

Molecular Orbitals and Frontier Orbitals

2.1 Chemical Bonds

2.1.1 Homonuclear Bonds

2.1.1.1 The Hydrogen Molecule. We imagine a bond to be made by bringing two atoms from infinity to within bonding distance. If we are making a bond between two hydrogen atoms, we use the 1s orbitals from each atom. When the atoms arrive at bonding distance, we find that we can combine the orbitals in two ways (Fig. 2-1). The first is a bonding way, where the orbitals of the same sign are placed next to each other; the electron density *between* the two atoms is increased (hatched area), and hence the negative charge which these electrons carry *attracts* the two positively charged nuclei. This results in a lowering in overall energy and is illustrated in Fig. 2-1, where the line next to the drawing of this orbital is placed low on the diagram. The second way in which we can combine the orbitals is called antibonding. The signs of the function which describes the electron-distribution are opposite on each nucleus, and, if there were any electrons in this orbital, there would be a very low electron density in this volume of space, since the function is changing sign between the nuclei.

6

We represent the sign change by filling in one of the orbitals, and we call the plane which divides the function at the sign change a node. The low electron density would lead to strong repulsion between the nuclei; thus, if we wanted to have electrons in this orbital and still keep the nuclei reasonably close, energy would have to be put into the system.

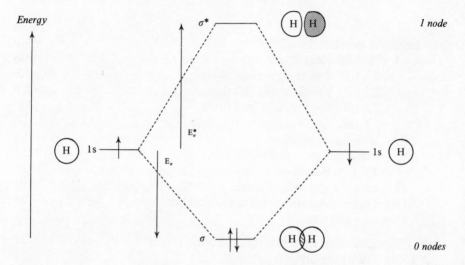

Fig. 2-1 The orbitals of hydrogen

The simplest mathematical description of this situation is shown in equation 2-1, where the function which describes the new electron distribution is called σ and the functions which describe the electron distribution in the atomic orbitals are called ϕ_1 and ϕ_2 (the subscripts 1 and 2 refer to the two atoms). The co-

$$\sigma = c_1\phi_1 + c_2\phi_2 \qquad\qquad 2\text{-}1$$

efficients, c, are a measure of the contribution which the atomic orbital is making to the molecular orbital, and we shall return to their numerical values shortly. In summary, by making a bond between two hydrogen atoms, we create two new orbitals, σ and σ^*, which we call the *molecular orbitals*; the former is bonding and the latter antibonding (the asterisk generally signifies an antibonding orbital). In the ground state of the molecule, the two electrons will be in the orbital labelled σ. There is, therefore, when we make a bond, a lowering of energy equal to twice the value of E_σ in Fig. 2-1 (*twice* the value, because there are two electrons in the bond).

In fact, putting *two* electrons into this orbital does not achieve twice the energy-lowering of putting *one* electron into it. We are *allowed* to put two electrons into the one orbital if they have opposite spins, but they still repel each other, because they have to share the same space; consequently, in forcing a second electron into the σ-orbital, we lose some of the bonding we might otherwise have gained. For this reason, the value of E_σ in Fig. 2-1 is smaller than that of E_σ^*. This is why two helium atoms do not combine to form an He_2 molecule. There are four electrons in two helium atoms, two of which would go into the σ-bonding orbital and two into the σ^*-antibonding orbital. Since $2E_\sigma^*$ is greater than $2E_\sigma$, we would need extra energy to keep the two helium atoms together.

We must now look at the coefficients, c, of equation 2-1. When there are electrons in the orbital, the squares of the c-values are a measure of the electron population in the neighbourhood of the atom in question. Thus *in each orbital the sum of the squares of all the c-values must equal one*, since only one electron in each spin state can be in the orbital. Now the orbitals in hydrogen are symmetric about the mid-point of the H—H bond; in other words $|c_1|$ must equal $|c_2|$. Thus we have defined what the values of c_1 and c_2 in the bonding orbital must be, namely $1/\sqrt{2} = 0.707$. If all molecular orbitals were filled, then there would have to be one electron in each spin state on each atom, and this gives rise to a second criterion for c-values, namely that the sum of the squares of all the c-values *on any one atom* in *all* the molecular orbitals must also equal one. Thus the antibonding orbital of hydrogen, σ^*, will have c-values of 0.707 and -0.707, because these values make the whole set fit both criteria:

$$
\begin{array}{cccc}
 & c_1 & c_2 & \\
\sigma^* & 0.707 \bigcirc\!\!\bullet & -0.707 & \Sigma c^2 = 1.000 \\
\sigma & 0.707 \bigcirc\bigcirc & 0.707 & \Sigma c^2 = 1.000 \\
 & \Sigma c^2 = 1.000 & \Sigma c^2 = 1.000 &
\end{array}
$$

Of course, we could have taken c_1 and c_2 in the antibonding orbital the other way round, giving c_1 the negative sign and c_2 the positive.

2.1.1.2   C—H Bonds: Methane. Carbon atoms have six electrons, of which the 1s are the core electrons, not involved in bonding. In methane, therefore, there are eight valence electrons, for which we need four molecular orbitals. These are easily provided by allowing the 1s orbitals of the hydrogen atoms to combine successively with the 2s, $2p_x$, $2p_y$ and $2p_z$ orbitals of the carbon atom. What is not perhaps so obvious is where in space to put the hydrogen atoms. They will, of course, repel each other, and the furthest apart they can get is the tetrahedral arrangement. In this arrangement, it is still possible to retain bonding interactions between the hydrogen atoms and the carbon atoms in all four orbitals (Fig. 2-2). In fact the maximum amount of total bonding is obtained this way. These four orbitals are therefore the bonding molecular orbitals, and there will be, higher in energy, a corresponding set of antibonding orbitals, with which we shall not be concerned now.

One difference between this situation and that of the hydrogen molecule is immediately apparent: there is no *single* orbital in this description which we can

8

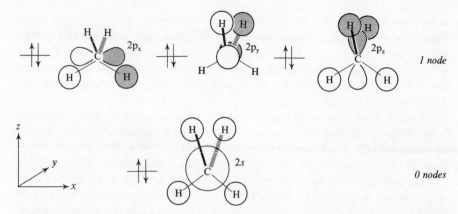

z
↑
|
|____→ y
↙
x

Fig. 2-2   The occupied orbitals of methane. (The lobes are drawn small and thin for clarity rather than accuracy.)

equate with the C—H *bond*. To get round this difficulty, chemists used the idea of hybridization; that is, they took all these orbitals and mixed them together to produce a set of *hybrids*, each of which retained only those parts of the total bonding in methane which were involved in the bonding to an individual hydrogen atom. The result, an sp^3-hybrid, looked like an unsymmetrical p orbital, and one hybrid for each of the hydrogen atoms led to the familiar picture for the bonds in methane shown in Fig. 2-3a. This picture has the advantage over that in Fig. 2-2 that it does bear some resemblance to the lines

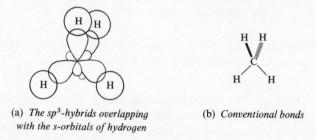

(a) *The sp^3-hybrids overlapping with the s-orbitals of hydrogen*

(b) *Conventional bonds*

Fig. 2-3   Methane represented by sp^3-hybridized orbitals

drawn on the conventional structure (Fig. 2-3b), and that the C—H bonds are now localized. The bonds drawn on Fig. 2-2 do not represent anything actually there; but without them the picture would be hard to interpret. The two descriptions of the overall wave function for methane are in fact identical; hybridization involves the same approximations as those used to arrive at Fig. 2-2. However, for some purposes we actually want to avoid localizing the electrons in the bonds (see pp. 79–80, for example), and in any case, hybridization is an extra concept, which has to be learned. It should only be used when it offers a considerable simplification. It is not difficult or inconvenient simply to remember that there is no single orbital representing the C—H bond, but rather several orbitals contributing to the bonding between these two atoms.

2.1.1.3   C—C σ-Bonds: Ethane. With a total of fourteen valence electrons to accommodate in molecular orbitals, ethane presents a more complicated picture. Again, various combinations of the 1s orbitals on the hydrogen atoms and the 2s, $2p_x$, $2p_y$ and $2p_z$ orbitals on the two carbon atoms (Fig. 2-4) give a set of

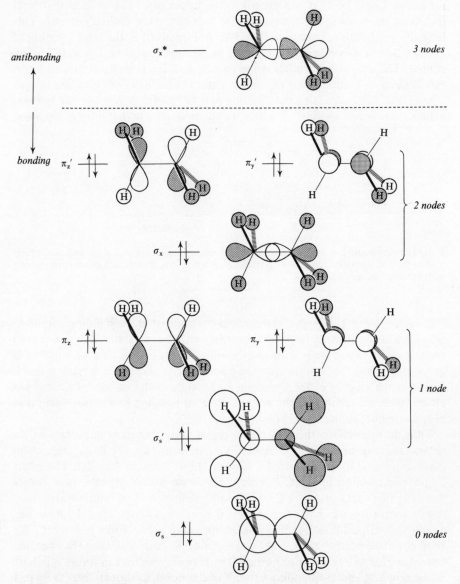

Fig. 2-4   The bonding orbitals and one antibonding orbital of ethane

seven molecular orbitals, each of which is bonding as a whole. We shall concentrate for the moment on those orbitals which give rise to the force holding the two carbon atoms together; between them they make up the C—C bond. The molecular orbitals (σ_s and σ_s'), made up from 2s orbitals on carbon, are very like the orbitals in hydrogen, in that the region of overlap is directly on a line between the carbon nuclei; as before, they are called σ-orbitals. The bonding in the lower one is very strong, but it is somewhat offset by the antibonding (as far as the C—C bond is concerned) in the upper one. The molecular orbital (σ_x) using the p_x orbital of carbon is also σ in character, and very strong. This time, its antibonding counterpart (σ_x^*) is not involved in the total bonding of ethane, nor is it bonding overall. It is in fact the lowest-energy antibonding orbital. The molecular orbitals using the $2p_y$ and $2p_z$ orbitals of carbon again fall in pairs, a bonding pair (π_y and π_z) and (as far as C—C bonding is concerned, but not overall) an antibonding pair (π_y' and π_z'). The overlap in these orbitals is redrawn in Fig. 2-5, where we see more clearly that it is not head-on,

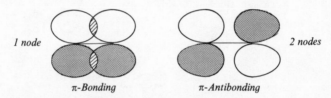

1 node   *2 nodes*

π-*Bonding* π-*Antibonding*

Fig. 2-5 π-Bonding. (The p orbitals on carbon are drawn somewhat more realistically here, to show the overlap (shaded). In most diagrams, the p orbitals are drawn much thinner in order that everything can be seen more clearly.)

but sideways-on. The extra electron density (shaded) in the bonding combination is no longer directly between the nuclei, with the result that it does not hold them together as strongly as in a σ-bonding orbital. This kind of bonding is called π-bonding. In fact, the electron population in these four orbitals is much higher in the vicinity of the hydrogen atoms than in the vicinity of the carbon atoms, with the result that the amount both of bonding and antibonding that they contribute to the C—C bond is small.

Thus, in total effect, the orbital (σ_x) is just about the most important single orbital making up the C—C bond. We can construct for it an interaction diagram (Fig. 2-6), just as we did for the H—H bond in Fig. 2-1. The other major contribution to C—C bonding, which we cannot go into now, comes from the fact that σ_s is more C—C bonding than σ_s' is C—C antibonding.

For simplicity, we shall often discuss the orbitals of σ-bonds as though they could be localized into bonding and antibonding orbitals like σ_x and σ_x^*. We shall not often need to refer to the full set of orbitals, except when they become important for one reason or another. Any property we may in future attribute to the bonding and antibonding orbitals of a σ-bond, as though there were just one such pair, can always be found in the full set of all the bonding orbitals.

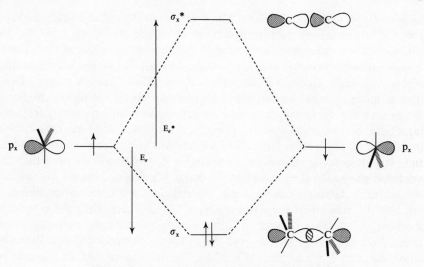

Fig. 2-6 A major part of the C—C σ-bond

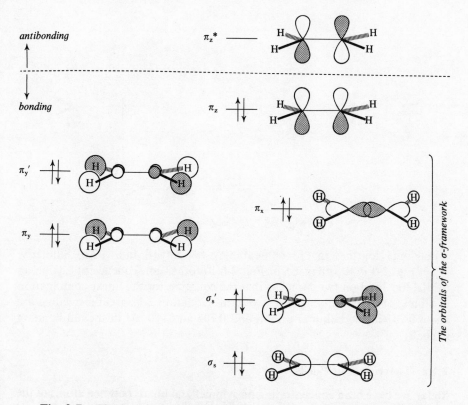

Fig. 2-7 The bonding orbitals and one antibonding orbital of ethylene

2.1.1.4 C=C π-Bonds: Ethylene. The orbitals of ethylene are again made up from the 1s orbitals of the four hydrogen atoms and the 2s, $2p_x$, $2p_y$ and $2p_z$ orbitals of the two carbon atoms (Fig. 2-7). One group, made up from the 1s orbitals on hydrogen and the 2s, $2p_x$ and $2p_y$ orbitals on carbon, is substantially σ-bonding, which causes the orbitals to be relatively low in energy. These orbitals make up what we call the 'σ-framework'. But standing out, higher in energy than the σ-framework orbitals, is an orbital made up entirely from the $2p_z$ orbitals of the carbon atom overlapping in a π-bond. This time, the π-orbital is localized on the carbon atoms, and we run into no problems in treating it as such. It gives much greater strength to the C—C bonding in ethylene than the π-orbitals give to the C—C bonding in ethane, which is one reason why we talk of ethylene as having a *double* bond. Nevertheless, the C—C σ-bonding provided by the orbitals of the σ-framework is much greater than the π-bonding provided by $π_z$. This is because, other things being equal, π-overlap is inherently less effective in lowering the energy than is σ-overlap. Thus in the interaction diagram for a π-bond (Fig. 2-8), the drop in energy, $E_π$, caused by

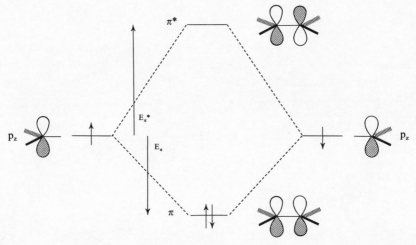

Fig. 2-8   A C=C π-Bond

π-bonding is less than that for comparable σ-bonding ($E_σ$ in Fig. 2-6). Similarly, $E_π^*$ in Fig. 2-8 is less than $E_σ^*$ in Fig. 2-6. Another consequence of having an orbital localized on two atoms is that the equation for the linear combination of atomic orbitals contains only two terms (equation 2-1), and the c-values are again 0·707 in the bonding orbital and 0·707 and −0·707 in the antibonding orbital.

2.1.2   Heteronuclear Bonds

So far, we have been concentrating on symmetrical bonds between atoms of the same kind (homonuclear bonds). The interaction diagrams (Figs. 2-1, 2-6 and

2-8) were constructed by combining atomic orbitals of *equal* energy, and the coefficients, c_1 and c_2, in the molecular orbitals were equal to each other. It is true that C—H bonds were created by the molecular orbitals of Figs. 2-2, 2-4 and 2-7, but we were not concerned there with the consequences of the fact that the two atoms in the bond were different. To deal with this problem, we shall take a C—O σ-bond. The C—O bond in a molecule such as methanol, like the C—C bond in ethane (Fig. 2-4), has several orbitals contributing to the force which keeps the two atoms bonded to each other; but, just as we could, in the case of ethane, abstract one of the important molecular orbitals and make a typical interaction diagram (Fig. 2-6) for it, so can we now take the corresponding orbital from the set making up a C—O σ-bond. The important thing is the *comparison* between the C—C orbital and the corresponding C—O orbital. What we learn about the properties of C—O bonds by looking at this one orbital will be the same as we would have learned, at much greater length, from the set as a whole.

In a C—O σ-bond, one element, the oxygen atom, is more electronegative than the other. Other things being equal, the energy of an electron in an atomic orbital on an electronegative element is lower than that of an electron on a less electronegative element. In making a covalent bond between carbon and oxygen from the $2p_x$ orbitals on each atom, we shall have an interaction between orbitals of *unequal* energy, as in Fig. 2-9. Also, because of the loss of symmetry,

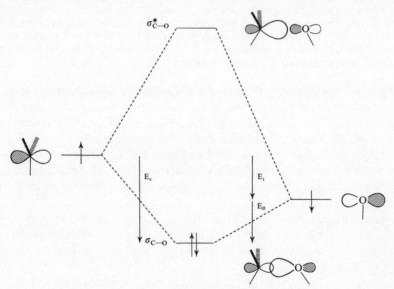

Fig. 2-9   A major part of the C—O σ-bond

the coefficients are no longer equal: the oxygen atom keeps a larger share of the total electron population. In other words, the coefficient on oxygen is larger than that on carbon for the major bonding orbital, σ_{C-O}. It follows that the coefficients in the corresponding antibonding orbital, σ^*_{C-O}, must reverse this situation: the one on carbon will have to be larger than the one on oxygen.

The mathematical form of the relationship between the various contributions to the overall energy is, very approximately (and when $\alpha_1 \approx \alpha_2$) a pair of simultaneous equations:[4–6]

$$\frac{c_2}{c_1} = -\frac{\alpha_1 - E}{\beta - SE} \qquad 2\text{-}2$$

$$\frac{c_2}{c_1} = -\frac{\beta - SE}{\alpha_2 - E} \qquad 2\text{-}3$$

where c is the coefficient of the atomic orbital in the molecular orbital; α, the Coulomb integral, is the energy associated with having the electron localized on the atom (i.e., the energy levels for the isolated orbitals on the left and right of Fig. 2-9); E is the overall energy; β, the resonance integral, is the energy associated with having the electron shared by the atoms in the form of the covalent bond (i.e. related to E_0 in Fig. 2-9); and S is the overlap integral, which gives a measure of how effective the overlap is in lowering the energy. Subtracting 2-2 from 2-3 gives an equation which is quadratic in E, and solutions are:

$$E = \frac{\alpha_1 + \beta}{1 + S} \quad \text{and} \quad E = \frac{\alpha_2 - \beta}{1 - S} \qquad 2\text{-}4$$

The former is the bonding combination and the latter the antibonding combination. It is a property of equations 2-2 and 2-3 that the larger the difference between the α-values (i.e. the larger the value of E_i in Fig. 2-9), the more unequal the coefficients, c, will be. It is also a property of equations 2-2 and 2-3 that, in the bonding orbital, the larger value of c will be on the atom with the lower value of α. (Incidentally, equation 2-4 also shows that, since the denominator in the bonding combination is $1 + S$ and the denominator in the antibonding combination is $1 - S$, the bonding orbital is not as much lowered in energy as the antibonding is raised.)

What we have done is to take the lower-energy atomic orbital and, in a bonding way, mix in with it some of the character of the higher-energy orbital. This creates the new bonding molecular orbital, which naturally resembles the atomic orbital nearer it in energy more than the one further away. We have also taken the higher-energy orbital and mixed in with it, in an antibonding way, some of the character of the lower-energy orbital. This produces the antibonding molecular orbital, which naturally more resembles the atomic orbital nearer it in energy.

When the coefficients are unequal, the overlap of a small lobe with a larger lobe does not lower the energy of the bonding molecular orbital as much as the overlap of two atomic orbitals of more equal size. $2E_0$, in Fig. 2-9, is not as large as $2E_\sigma$ in Fig. 2-6.

We might be tempted to say at this stage that we have a weaker bond than we had before, but we must be careful in defining what we mean by a weaker bond in this context. If you look up tables of bond strengths, you will find that the C—O bond strength is given as something like 85·5 kcals/mole (358 kJ mol^{-1}), whereas the C—C bond is something like 83 kcals/mole (347 kJ mol^{-1}). Only part of the bond strength represented by these numbers comes from the purely covalent bonding given by $2E_0$ in Fig. 2-9. The other part of the strength of the C—O bond comes from the electrostatic attraction between the high electron population concentrated on the oxygen atom and the relatively exposed carbon nucleus. We usually say that the bond is polarized, or that it has ionic character. This energy is related to the value E_i on Fig. 2-9, as we can readily see by using an extreme example: suppose that the energies of the interacting orbitals are very far apart (Fig. 2-10, where the isolated orbitals could be Na· and F·, for example); the overlap will be negligible, and the new molecule will now have almost entirely isolated orbitals in which the higher-energy orbital has given up its electron to the lower-energy orbital. In other

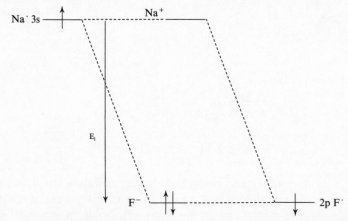

Fig. 2-10 An ionic bond. (This picture is very much oversimplified. It ignores the repulsion between the two electrons in F⁻, and also the electrostatic attraction between the Na⁺ and the F⁻. The two effects oppose each other, which is why the picture is useful, though misleading.)

words, we shall have a pair of ions, where formerly we had a pair of (imaginary) radicals. There will be no covalent bonding to speak of, and the drop in energy in going from the pair of radicals to the cation plus anion is now E_i in Fig. 2-10, which, we can see, is indeed related to E_i in Fig. 2-9.

The C—O bond *is* strong, if we try to break it homolytically to get a pair of radicals, and the C—C bond *is* easier to break this way. This is what the numbers 85·5 and 83 kcals/mole refer to. In other words, $E_c + E_o$ in Fig. 2-9 is evidently greater than $2E_\sigma$ in Fig. 2-6. But it is very much easier to break a C—O bond heterolytically to the cation (on carbon) and the anion (on oxygen) than to cleave a C—C bond this way. In other words, $2E_0$ in Fig. 2-9 is less than $2E_\sigma$ in Fig. 2-6.

The important thing to get out of all this argument, because we are going to use it later when we consider real rather than imaginary reactions, is that *when two orbitals of unequal energy interact, the gain in covalent bonding is less than when two orbitals of very similar energy interact.* Conversely, when it comes to transferring an electron, the ideal situation is one in which the electron is in a very high-energy orbital and the 'hole' is in a very low-energy orbital.

Exactly the same arguments apply to making a C=O π-bond (Fig. 2-11), except that, as with the π-bond of ethylene, the raising and lowering of the

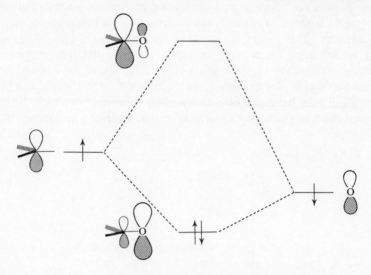

Fig. 2-11   A C=O π-bond

molecular orbitals above and below the atomic orbitals is less than it was for the corresponding σ-bond. Once again it is easier to break a C=O bond heterolytically and a C=C bond homolytically. Some examples of chemical reactions may perhaps bring a sense of reality to what must seem, so far, a very abstract discussion. Nucleophiles readily attack a carbonyl group but not an

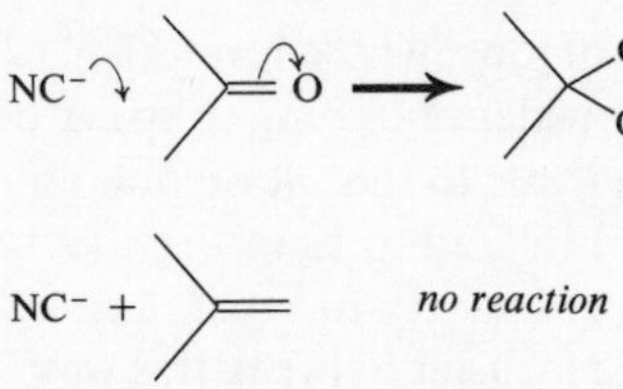

isolated C=C double bond. On the other hand, radicals readily attack C=C double bonds; but only rarely do they add to carbonyl groups.

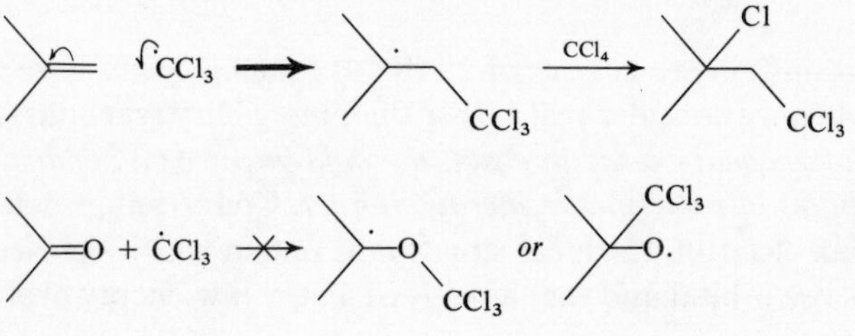

2.1.3 Conjugation—Hückel Theory

2.1.3.1 Butadiene. A further aspect of bonding is also easily described by this simple version of molecular orbital theory, namely that of conjugation. Again we imagine a conjugated system, such as that of butadiene, as being set up by bringing two isolated π-bonds from infinity to within bonding distance. As usual, we get a pattern of raised and lowered energy levels as in Fig. 2-12, and

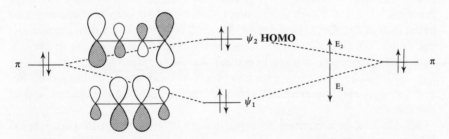

Fig. 2-12 Energies of the π-molecular orbitals of ethylene and butadiene

we get a new set of orbitals, ψ_1, ψ_2, ψ_3^*, and ψ_4^*, each described by an equation of the form:

$$\psi = c_1\phi_1 + c_2\phi_2 + c_3\phi_3 + c_4\phi_4 \qquad 2\text{-}5$$

The lowest-energy orbital has all the c-values positive, and hence bonding is at its best. The next-highest energy level has one node, between C-2 and C-3; in other words, c_1 and c_2 are positive and c_3 and c_4 are negative. There is therefore bonding between C-1 and C-2 and between C-3 and C-4, but not between C-2 and C-3. But what are the c-values for butadiene? Are they all equal—in other words, $\pm\frac{1}{2}$? They are not; and, to explain why, we draw an analogy between the electron in these orbitals and the quantum-mechanical situation of the electron in the box.

In the case of the electron in the box, we look at each energy level as a series of sine waves. We now do the same for conjugated systems, but this time the wave is seen in the coefficients, c. Thus the lowest-energy orbital of butadiene, ψ_1, reasonably enough, has a high concentration of electrons in the middle,

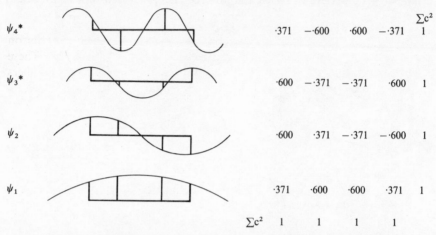

					Σc^2
$\psi_4{}^*$	·371	−·600	·600	−·371	1
$\psi_3{}^*$	·600	−·371	−·371	·600	1
ψ_2	·600	·371	−·371	−·600	1
ψ_1	·371	·600	·600	·371	1
Σc^2	1	1	1	1	

Fig. 2-13   Coefficients of the π-molecular orbitals of butadiene

but in the next orbital up, ψ_2, because of the repulsion between the wave functions of opposite sign on C-2 and C-3, the electron density is concentrated at the ends of the conjugated system. We shall have to take on trust the numerical values[7] which are generally used for the coefficients, if we are to avoid a very long and necessarily mathematical digression. But they can be seen, in Fig. 2-13, to obey the pattern we can expect of them from the electron-in-the-box analogy, and, squared and summed, they do add up in columns and in rows to one.

We are now in a position to explain the well-known fact that conjugated systems are often, but not always, more stable than unconjugated systems. It comes about because ψ_1 is lowered in energy more than ψ_2 is raised (E_1 in Fig. 2-12 is larger than E_2). The energy (E_1) given out in forming ψ_1 comes from the overlap between the atomic orbitals on C-2 and C-3; this overlap did not exist in the isolated π-bonds. It is particularly effective in lowering the energy of ψ_1, because the coefficients on C-2 and C-3 are large. By contrast, the increase in energy of ψ_2, caused by the repulsion between the orbitals on C-2 and C-3, is not as great, because the coefficients on these atoms are smaller in ψ_2. Thus the energy lost from the system in forming ψ_1 is greater than the energy needed to form ψ_2, and the overall energy of the ground state of the system ($\psi_1{}^2\psi_2{}^2$) is lower.

In this example, we have for the first time seen more than one filled and more than one empty orbital in the same molecule. In fact, of course, the σ-framework, with its strong σ-bonds, has several other filled orbitals lying lower in energy than either ψ_1 or ψ_2, but we do not usually pay much attention to them, simply because they lie so much lower in energy. In fact, we shall be paying very special attention to the filled orbital which is highest in energy: we call it the highest occupied molecular orbital (HOMO). We shall also be paying special attention to the unoccupied molecular orbital of lowest energy (LUMO). The HOMO and LUMO are the *frontier orbitals*.

2.1.3.2   The Allyl System. Another conjugated system we shall need later on is that of the allyl cation (**24**), allyl radical (**25**), and allyl anion (**26**). These three reactive intermediates all have the same orbitals, but different numbers

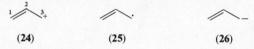

(24)   (25)   (26)

of electrons in the orbitals. To derive the orbitals of the allyl system (Fig. 2-14), we imagine a p-orbital being brought into conjugation with a π-bond. The result is again to split the original π and π* orbitals; but the lowering of the π* and the raising of the π orbital are equal, and a single orbital ψ_2 is created. The

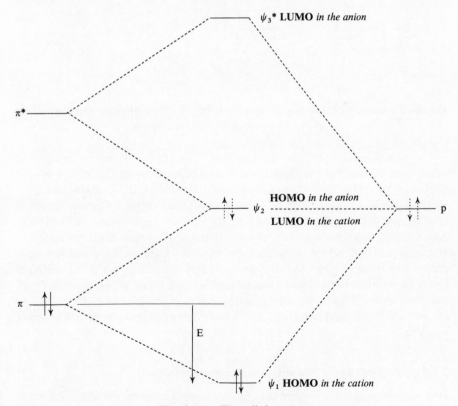

Fig. 2-14   The allyl system

20

lowest-energy orbital, ψ_1, has bonding across the whole conjugated system, with the electrons concentrated in the middle. The next orbital up in energy, ψ_2, is different from those we have met so far. As with ψ_2 in butadiene, its symmetry demands that the node be in the middle; but this time the centre of the conjugated system is occupied by an *atom* and not by a *bond*. Having a node in the middle means having a zero coefficient on C-2, and hence the coefficients on C-1 and C-3 in this orbital must be, if, squared and summed, they are to equal one, $\pm 1/\sqrt{2}$: in other words $+0.707$ and -0.707 respectively. Again, because of symmetry, $|c|$ in ψ_3^* must mirror $|c|$ in ψ_1, and hence we get the unique set of c-values shown in Fig. 2-15.

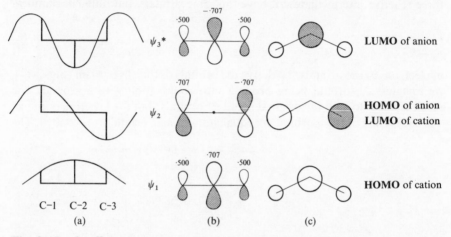

Fig. 2-15 Coefficients in the molecular orbitals of the allyl system (a) as sine curves, (b) in elevation and (c) in plan

The atomic orbitals of opposite sign on C-1 and C-3 in ψ_2 are so far apart in space that their repulsive interaction does not much raise the energy of this molecular orbital relative to that of an isolated p orbital; this explains why ψ_2 is on the same level as p in Fig. 2-14. Such an orbital is often called *non*-bonding (NBMO), as distinct from a bonding (ψ_1) or an antibonding (ψ_3^*) orbital. Again we see that there has been a gain in bonding as a result of the conjugation. In the allyl cation, where ψ_1 is filled and ψ_2 and ψ_3^* are empty, we have simply gained the energy 2E. In the allyl anion, where ψ_1 and ψ_2 are filled, we have again gained the energy 2E because of the fact that p and ψ_2 are essentially on the same level. (A 'gain' in energy, in this sense, is understood to be to 'us', or the outside world, and hence means a loss of energy in the system and stronger bonding.)

2.1.4 Thermodynamic Stability and Chemical Reactivity

It is very important to realize that having conjugation *may* make a molecule *thermodynamically more stable* than an unconjugated one, for the reason we have just seen, but it does not follow that conjugated systems are less reactive. Indeed, they are very often more reactive—or, we might say, *kinetically less stable*. The distinction between thermodynamic and kinetic stability must always be borne in mind: organic chemists use 'stable' and 'stability' without always identifying which meaning they are assuming.

The reason why dienes are often more reactive than simple alkenes is an example of what this book is about: *reactivity* is determined by a number of factors, but one of them is the energy of the HOMO, as we shall see later in this chapter. The HOMO of butadiene is ψ_2 (Fig. 2-12), and that of ethylene is π (Figs. 2-8 and 2-12). The former is higher in energy than the latter, and this leads to its greater reactivity. On the other hand, thermodynamic stability is determined, as we saw in the last section, by the energies of *all* the filled orbitals. *Thus thermodynamic stability is associated with both ψ_1 and ψ_2, but kinetic stability is quite largely (though not exclusively) determined by ψ_2 alone.*

2.1.5 PES and ESR

Two recent techniques have greatly helped in our understanding of molecular orbital theory, and we shall use evidence from them in the frontier orbital analysis of chemical reactivity. One of these is photoelectron spectroscopy[8] (PES), which measures, in a rather direct way, the energies of filled orbitals. The values obtained by this technique for the energies of the HOMO of some simple molecules are collected in Table 2-1. Here we can see how the change from a simple double bond (entry 6) to a conjugated double bond (entry 10) *raises* the energy of the HOMO. Similarly, we can see how the change from a simple carbonyl group (entry 8) to an amide (entry 14) also raises the HOMO energy, just as it ought to, by analogy with the allyl anion (Fig. 2-15), with which an amide is isoelectronic. We can also see that the interaction between a C=C bond (π-energy $-10\cdot5$ eV) and a C=O bond (π-energy $-14\cdot1$ eV) gives rise to an HOMO of lower energy ($-10\cdot9$ eV, entry 16) than when two C=C bonds are conjugated ($-9\cdot1$ eV, entry 10). Finally, we can see that the more electronegative an atom is, the lower is the energy of its HOMO (entries 1 to 5). All these observations confirm that the theoretical treatment we have been using is supported by some experimental evidence. PES has also proved useful because theory, although it usually gets the order of energy levels right, has proved, except in the most advanced and elaborate treatments, to be inadequate for getting the correct absolute values for the energies of molecular orbitals.

The second technique which both confirms some of our deductions and provides useful quantitative data for frontier orbital analysis is electron spin resonance spectroscopy[9, 10] (ESR). This technique detects the odd electron in radicals; the interaction of the spin of the electron with the magnetic nuclei (^{1}H, ^{13}C, etc.) gives rise to splitting of the resonance signal, and the degree of splitting is proportional to the electron population at the nucleus. Since we already know that the coefficients of the atomic orbitals, c, are directly related

Table 2-1 Energies of HOMOs of some simple molecules from PES

1 eV = 23 kcal = 96·5 kJ

	Molecule	Type of orbital	Energy eV
1	$:PH_3$	n	−9·9
2	$:SH_2$	n	−10·48
3	$:NH_3$	n	−10·85
4	$:OH_2$	n	−12·6
5	$:ClH$	n	−12·8
6	$CH_2{=}CH_2$	π	−10·51
7	$HC{\equiv}CH$	π	−11·4
8	$:O{=}CH_2$	π	−14·09
9		n	−10·88
10	$CH_2{=}CH{-}CH{=}CH_2$	ψ_2	−9·1
11		ψ_1	−11·4 or −12·2
13	$HC{\equiv}C{-}C{\equiv}CH$	π	−10·17
14	$H_2N{-}\underset{\dot{H}}{C}{=}O:$	π	−10·5
15		n	−10·13
16	$CH_2{=}CH{-}\underset{\dot{H}}{C}{=}O:$	π	−10·9
17		n	−10·1
18		π	−8·9
19		π	−9·25
20	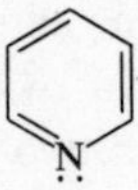	π	−9·3
21		n	−10·5

to the electron population we can expect there to be a simple relationship between these coefficients and the observed coupling constants. This proves to be quite a good approximation. The nucleus most often used is ^1H, and the atomic orbital whose coefficient is measured in this way is that on the carbon atom to which the hydrogen atom in question is bonded. The McConnell equation:

$$a_H = Q_{CH}^H \rho_C \qquad\qquad 2\text{-}6$$

shows the relationship of the observed coupling constant (a_H) to the unpaired spin population on the adjacent carbon atom (ρ_C). The constant Q is different from one situation to another, but when an electron in a p_z orbital on a trigonal carbon atom couples to an adjacent hydrogen, it is about −24 gauss.

However, the relationship between coupling and electron population is not quite as simple as this. Thus, though p orbitals on carbon have zero electron density at the nucleus, coupling is nevertheless observed; similarly, in the allyl radical, which ought to have zero odd-electron population at the central carbon atom, coupling to a neighbouring hydrogen nucleus is again observed. This latter coupling turns out to be opposite in sign to the usual coupling, and hence has given rise to the concept of negative spin density. Nevertheless the technique has provided some evidence that our deductions about the coefficients of certain molecular orbitals have some basis in fact as well as in theory: the allyl radical does have most of its odd-electron density at C-1 and C-3; and several other examples will come up later in this book. We merely have to remember to be cautious with evidence of this kind; at the very least, the observation of negative spin density should remind us that the Hückel theory of conjugated systems (the theory we have been using) *is* a simplification of the truth.

The standard ways of generating radicals for ESR measurements involve adding an electron to a molecule or taking one away. In the former case the odd electron is fed into what was the LUMO, and in the latter case the odd electron is left in the HOMO. Since these are the orbitals which are most important in determining chemical reactivity, it is particularly fortunate that ESR spectroscopy should occasionally give us access to their coefficients. It is now time to see why it is these orbitals which are so important.

2.2 Frontier Orbitals: HOMO and LUMO

2.2.1 Transition States

In order to approach the problem of chemical reactivity, let us imagine two molecules which are about to combine with each other in a simple, one-step reaction. We can assume that we often know fairly well what are the energies of the starting materials and the possible products (Fig. 2-16). Our problem in assessing the mutual reactivity of the two molecules is much more likely to be estimating the energy for the *transition states*. Perturbation theory,[11] introduced by Coulson and Longuet-Higgins,[2] can be adapted to this end: we treat the interaction of the molecular orbitals of the two components as a perturbation on each other. The perturbation leads, as we shall find, to the very same kind of bonding and antibonding interactions that we have just seen when two separate orbitals are brought together to give a bond. (The difference now is that the two components really are coming from far apart to within bonding distance, instead of our simply imagining it.) However, as the perturbation increases, it ceases to be merely a perturbation, and the theory fails to be able to accommodate so large a change. We are therefore unlikely to have direct access to a good picture of the transition state; nevertheless, we do get an estimate of the *slope* of an early part of the path along the reaction coordinate leading up to the transition state (labelled path A and path B on Fig. 2-16). Unless something unusual happens nearer the transition state, the slopes will probably predict—for instance—which of two transition states is the easier to get to: on the whole, the steeper path is likely to lead to a higher-energy transition state.

The situation shown in Fig. 2-16 is a common one: the higher-energy transition state leads to the higher-energy product. However, there are many situations where this is not so, or where we do not know which is the higher-energy product. In these cases perturbation theory, which looks at the reactant side of the reaction coordinate, can help us out. Even in those cases where we do know the relative energies of the products A and B, the orbital-interaction effect on the slopes of path A and path B ought to be taken into consideration in explaining the order of the two transition states. Influences from both sides of the reaction coordinate affect the transition state. Hitherto organic chemists have more often concentrated on the product side, but now, obviously, we have a useful, and in some situations unique, new tool for examining the reactant side of the reaction coordinate. The Hammond postulate says that transition states for exothermic reactions are reactant-like, and for endothermic reactions product-like.[12] We can therefore expect that frontier orbital effects will be particularly strong in exothermic reactions.

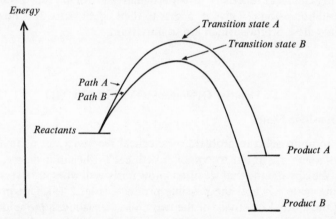

Fig. 2-16   The energy along two possible reaction coordinates

2.2.2   The Perturbation Theory of Reactivity

Now let us look at the perturbation which the reacting molecules exert upon each other's orbitals. Let the two reacting molecules have orbitals, filled and unfilled, as shown in Fig. 2-17. As the two molecules approach each other, the orbitals interact. Thus we can take, let us say, the highest occupied orbital of the molecule on the left and the highest occupied orbital of the molecule on the right and combine them in a bonding and an antibonding sense, just as we did, following Hund and Mulliken, when making a π-bond from two isolated p orbitals. The new molecular orbitals, in the centre of Fig. 2-17 will then be an approximation to two of the orbitals of the transition state.

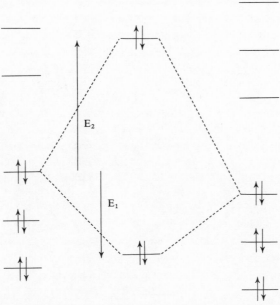

Fig. 2-17 The interaction of the HOMO of one molecule with the HOMO of another

The formation of the bonding orbital is, as usual, exothermic (E_1), but the formation of the antibonding orbital is endothermic (E_2), because there are two electrons which must go into it. As in the attempt to form a helium molecule from two helium atoms, the energy needed to force the molecules together in an antibonding combination is greater than that gained from the bonding combination. This situation will be true for *all* combinations of fully occupied orbitals. Combinations of unfilled orbitals with other unfilled orbitals will have no effect on the energy of the system, because without the electrons there is no way of gaining or losing energy.

The interactions which do have an important energy-lowering effect are the combinations of filled orbitals with unfilled ones. Thus, in Fig. 2-18 and Fig. 2-19, we have such combinations, and in each case we see that the energy-lowering in the bonding combination is the usual one, and that the rise in energy of the antibonding combination is without effect on the actual energy of the system, because there are no electrons to go into that orbital.

We can also see on Fig. 2-18 that it is the interaction of the HOMO of the left hand molecule and the LUMO of the right hand molecule that leads to the largest drop in energy ($2E_A > 2E_B$). The interaction of other occupied orbitals with other unoccupied orbitals—as in Fig. 2-19—is less effective, because the closer the interacting orbitals are in energy, the greater is the splitting of the levels (see p. 15). Now we can see why it is the HOMO/LUMO interaction

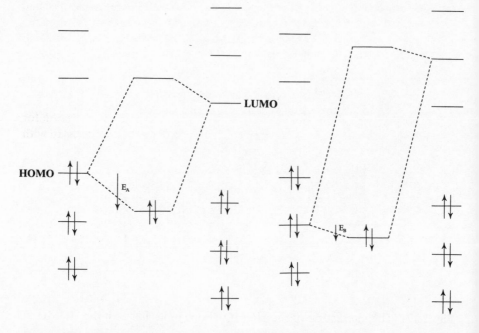

Fig. 2-18 The interaction of the HOMO of one molecule with the LUMO of another

Fig. 2-19 The interaction of a lower filled MO of one molecule with a higher unfilled MO of another

which we look at, and why these orbitals, the *frontier orbitals*, as Fukui named them,[13] are so important. The other occupied orbital/unoccupied orbital interactions contribute to the energy of the interaction and hence to lowering the energy of the transition state, but the effect is usually less than that of the HOMO/LUMO interactions.

The HOMO/HOMO interactions (Fig. 2-17) are large compared with the HOMO/LUMO interactions (Fig. 2-18): both E_1 and E_2 in Fig. 2-17 are much larger than E_A in Fig. 2-18. This is because HOMO/HOMO interactions will usually be between orbitals of comparable energy, whereas the HOMO of one molecule and the LUMO of another are usually well separated in energy. (In the mathematical form of perturbation theory, the former are first-order interactions, whereas the latter are usually second-order.) This will also be true of many of the interactions of the other occupied orbitals on one compound with the occupied orbitals on the other. Although the bonding (E_1) and antibonding (E_2) interactions cancel one another out to some extent, the net antibonding interaction between two molecules will be large: many such orbitals interact in this way, and their interactions are first-order. These interactions give rise to a large part of the activation energy for many reactions. The second-order interactions, like those of Figs. 2-18 and 2-19, even though they are entirely bonding in character and reduce the activation energy, are relatively small. The HOMO/LUMO interaction is merely the largest of a lot of small interactions. We shall discuss this matter further in the next section, where we meet a formidable-looking equation, from which the strength of these interactions can be estimated quantitatively.

2.2.3 The Equation for Estimating Chemical Reactivity

Using perturbation theory, Klopman[14] and Salem[15] derived an expression for the energy (ΔE) gained and lost when the orbitals of one reactant overlap with those of another. Their equation has the following form:

$$\Delta E = \underbrace{-\sum_{ab} (q_a + q_b)\beta_{ab}S_{ab}}_{\textit{first term}} + \underbrace{\sum_{k<l} \frac{Q_kQ_l}{\varepsilon R_{kl}}}_{\substack{\textit{second} \\ \textit{term}}}$$

$$+ \underbrace{\sum_r^{\text{occ.}} \sum_s^{\text{unocc.}} - \sum_s^{\text{occ.}} \sum_r^{\text{unocc.}} \frac{2(\sum_{ab} c_{ra}c_{sb}\beta_{ab})^2}{E_r - E_s}}_{\textit{third term}} \qquad \text{2-7}$$

q_a and q_b are the electron populations (often loosely called electron densities) in the atomic orbitals a and b.

β and S are the resonance and overlap integrals mentioned on p. 14.

Q_k and Q_l are the total charges on atoms k and l.

ε is the local dielectric constant.

R_{kl} is the distance between the atoms k and l.

c_{ra} is the coefficient of atomic orbital a in molecular orbital r, where r refers to the molecular orbitals on one molecule and s refers to those on the other.

E_r is the energy of molecular orbital r.

The derivation of this equation involves, as one might expect, many approximations and assumptions, which we shall not go into. It is valid only because S will always be small for the overlap of orbitals of p character. The integral S has the form shown in Fig. 2-20: as the orbitals approach, there is an initial gain in bonding; then the gain slows up, as the front lobe of one orbital begins to overlap with the back lobe of the other; it reaches a maximum value of 0·27 at a distance of 1·74 Å (for a C—C pσ-bond) and then rapidly falls off. Thus any reasonable estimate of the distance apart of the atoms at the transition state cannot fail to make S small. The integral β is roughly proportional to S, so the third term of equation 2-7 above is the second-order term. With S always small, the higher-order terms are naturally very small indeed, and we can neglect them. This is why a *second-order* perturbation treatment works. Let us now look at each of the three terms of equation 2-7.

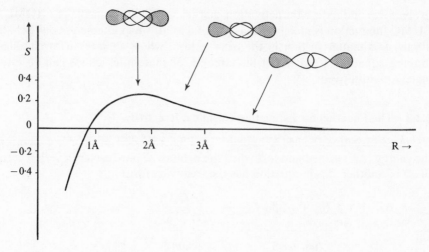

Fig. 2-20 The Function S with distance R for a $2p\sigma$-$2p\sigma$ C—Cbond

(i) *The first term* is the first-order closed-shell repulsion term, and comes from the interaction of the filled orbitals of the one molecule with the filled orbitals of the other (as in Fig. 2-17). Overall it is antibonding in effect, and can be compared to the interaction of two helium atoms.

This term will usually be large relative to the other two terms: it represents a good deal of the enthalpy of activation for many reactions. Apart from this, its main effect on chemical reactivity can probably be identified with the well-known observation that, on the whole, the smaller the number of bonds to be made or broken at a time, in a chemical reaction, the better. If a reaction can take place in several, not too difficult stages, it will probably go in stages, rather than in one concerted process. The concerted process, whatever it is, must involve the making (or breaking) of more than one bond, and for every bond to be made (or broken), we must have an antibonding contribution from the first term of equation 2-7. Another important reason for the general preference for stepwise reaction is, of course, the much more favourable entropy term when a relatively small number of events happen at once.

The important observation about the first term, in any particular case, is that it is usually very similar for each of two possible pathways. Thus, if a molecule can be attacked at two possible sites, the first term will often be nearly the same for attack at each site. Similarly, if there are two possible orientations in a cyclo-addition, the first term will not be very different in either orientation. This is not so for the other two terms, and it is therefore with them that we shall mainly be concerned in explaining differential reactivity of this kind.

We shall, in fact, be ignoring the first term from now on, because frontier orbital theory is mainly used to explain features of *differential* reactivity. We are on somewhat weak ground in doing so, and we should not forget it. The interaction of a filled orbital with a filled orbital, as in Fig. 2-17, leads to a small antibonding effect, but there are *many* filled orbitals interacting with *many* filled orbitals, and the total effect is the sum of many small ones. The overall effect of the first term of equation 2-7 ought, therefore, to be rather unpredictable, but it seems that adding up a lot of small items very often averages out the total effect.

(ii) *The second term* is simply the Coulombic repulsion or attraction. This term, which contains the total charge, Q, on each atom, is obviously important when ions or polar molecules are reacting together.

(iii) *The third term* represents the interaction of all the filled orbitals with all the unfilled of correct symmetry (Figs. 2-18 and 2-19 etc.). It is the second-order perturbation term and is only true if $E_r \neq E_s$. (When $E_r \approx E_s$, the interaction is better described in charge-transfer terms, and the perturbation is then a first-order one of the form $\sum_{ab} 2c_{r_a}c_{s_b}\beta_{ab}$.) Here we can see again, this time in simple arithmetical terms, that it is the HOMO and the LUMO which are most important: they are the ones with the smallest value of $E_r - E_s$, and hence they make the largest contribution to the third term of equation 2-7.

In summary:

As two molecules collide, three major forces operate:

(i) *The occupied orbitals of one repel the occupied orbitals of the other.*
(ii) *Any positive charge on one attracts any negative charge on the other (and repels any positive).*
(iii) *The occupied orbitals (especially the HOMOs) of each interact with the unoccupied orbitals (especially the LUMOs) of the other, causing an attraction between the molecules.*

We are now in a position to apply these ideas to the components of a chemical reaction. Let us begin with a negatively charged, conjugated system, like the allyl anion, reacting with a positively charged, conjugated system, like the allyl cation. In this imaginary reaction, the major contributions to bond-making will be the very powerful charge-charge interaction (the second term of equation 2-7) *and* the very strong interaction from the HOMO of the anion and the LUMO of the cation (E_1 in Fig. 2-21). By contrast, the interaction of the HOMO of the cation and the LUMO of the anion is much less effective (E_2 in Fig. 2-21), because $E_r - E_s$ is relatively so large. (The allyl anion + allyl cation reaction is most unusual because $E_r \approx E_s$ for the important interaction of the HOMO of the anion with the LUMO of the cation. For this reason, the important frontier orbital interaction is both very strong and first-order, not second-order like the third term of equation 2-7. Nevertheless, it provides a simple illustration of how the ideas behind equation 2-7 work, and it also shows how it comes about that in general *the important frontier orbitals for a nucleophile reacting with an electrophile are HOMO (nucleophile)/LUMO (electrophile)* and not the other way round.

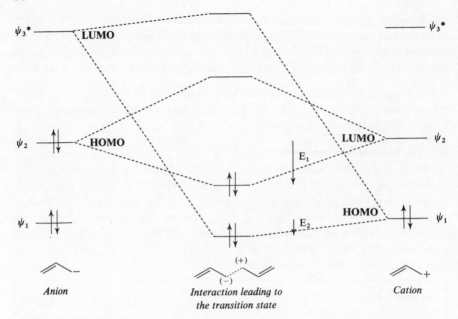

Fig. 2-21 Orbital interactions in the reaction of the allyl anion with the allyl cation

Having identified the causes for the ease of a reaction like this one, we must next use the ideas behind equation 2-7 to identify the *sites* of reactivity in each of the reacting species. To find the contribution of the Coulombic forces, we need the total electron population on each atom. For the allyl cation this is easy; it is given by squaring the c-values of the only filled orbital: the overall electron *deficiency* is seen to be concentrated on C-1 and C-3, where the electron population is lowest, and it is therefore here that charged nucleophiles will attack. For the allyl anion, we sum the squares of the c-values of both ψ_1 and

Fig. 2-22 Total π-electron population (and excess-charge distribution in brackets) in the allyl anion and the allyl cation

(The numbers in this diagram are easily derived from the coefficients shown on Fig. 2-15. For the total π-electron population: square the c-values, multiply the results for each atomic orbital by the number of electrons in the molecular orbital of which it is part, and add up the total on each atom for all the molecular orbitals. For the excess-charge distribution: change the sign, to convert the electron population to charge, and, if only π-orbitals are under consideration, as here, add one to each to allow for the nuclear charge.)

ψ_2 and get the values shown in Fig. 2-22. The overall excess of electrons is again concentrated on C-1 and C-3, and it is therefore here that charged electrophiles will attack. Thus when the reaction takes place, the charge-charge attraction represented by the second term of equation 2-7 will lead C-1 ($=$C-3) of the allyl anion to react with C-1 ($=$C-3) of the allyl cation, and C-2 will have little nucleophilic or electrophilic character.

When we add the contribution from the frontier orbitals, the picture is even more striking. The HOMO of the anion has coefficients at C-1 and C-3 of ±0.707, and similarly the LUMO of the cation has coefficients at C-1 and C-3 of ±0.707. In both frontier orbitals the coefficient on C-2 is zero. Thus the frontier orbital term is overwhelmingly in favour of reaction of C-1 (C-3) of the anion with C-1 (C-3) of the cation. Only the relatively ineffectual HOMO (cation)/LUMO (anion) interaction shows any profit in bonding at C-2 of either component.

We have now seen how the attraction of charges *and* the interaction of frontier orbitals combine to make a reaction between two such species as the allyl anion and allyl cation both fast and site-specific. We should remind ourselves that this is not the whole story: another reason for both observations is that the reaction is very exothermic when a bond is made between C-1 and C-1: we gain the energy of a full σ-bond with cancellation of charge, which we could not easily

do if reaction were to take place at C-2 on either component. In other words, we find, as we often shall, that we are in the situation of Fig. 2-16, path B: the Coulombic forces and the frontier orbital interaction on one side, and the stability of the product on the other, combine to make transition state B a low-energy one.

Indeed, you may well feel that there was little point in looking at the frontier orbitals in a reaction like this, where bonding between C-2 of the anion and C-2 of the cation would be absurd:

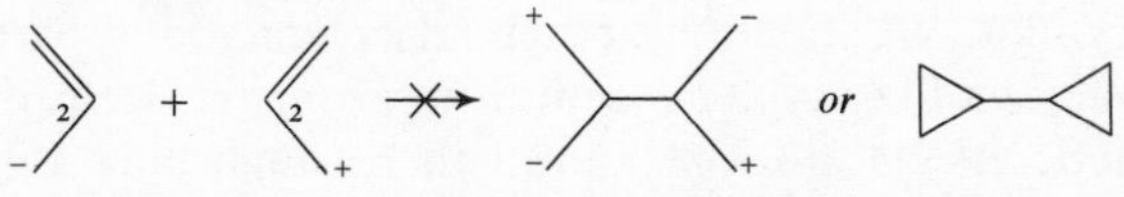

The purpose of doing so was twofold: in the first place, it did show that we get the same answer by considering the frontier orbitals as we do from the product-development argument, and secondly it showed how the allyl anion and allyl cation are nucleophilic and electrophilic respectively at both C-1 and C-3 *without our having to draw canonical structures.*

One of the arguments for retaining valence bond theory has been the ease with which things like the nucleophilicity of the allyl anion at C-1 and C-3 are explained by drawing the canonical structures. Even as simple a version of MO theory as the one presented here does the job just as well. The drawings chemists use for their structures will inevitably be crude representations: we shall always have to make some kind of localized drawing, whether it be of a benzene ring, or an enolate ion, or whatever. At the same time, we shall continue to make, as we already do, considerable mental reservations about how accurately such drawings represent the truth. If our mental reservations are made within the framework of the molecular orbital theory, we shall have a better and more useful picture of organic chemistry at our disposal; furthermore, molecular orbital theory *is* capable of being made as simple as valence bond theory.

2.2.4 Other Factors Affecting Chemical Reactivity

When chemical reactivity has been discussed in the past a number of factors have usually been identified, some of which are obviously involved in the derivation of equation 2-7, and some of which are not. Thus we are including the Coulombic factors which lead ions to react faster with polar or oppositely charged molecules than with non-polar or uncharged ones. We are also, at least in part, including factors such as the strength of the bond being made (it affects β) and the strength of any bond being broken (it affects E_r and/or E_s). The Woodward–Hoffmann rules are included in a sense, in that we have to evaluate whether the overlap integral (S and hence β) is bonding or antibonding; however, it would be easy to overlook this in a calculation. The loss of conjugation—for example, the loss of aromaticity in the first step of aromatic electrophilic substitution—is partly taken account of. Thus, the very low energy of ψ_1 in benzene leads the value of $E_r - E_s$ for that orbital to be much larger than if the aromaticity were not present. The simpler HOMO/LUMO approach, however, makes no such allowance.

But a number of other factors are either being ignored in this treatment or at least being underestimated. *Strain* in the σ-framework, whether gained or lost, is not directly included, except insofar as it affects the energies of those orbitals which are involved. Factors which affect the *entropy* of activation are not included. Finally, *steric effects* are ignored.

We cannot, then, expect this approach to understanding chemical reactivity to explain everything. We should bear in mind its limitations, particularly when dealing with subjects like *ortho/para* ratios in aromatic electrophilic substitution, where steric effects are well known to be important. Likewise *solvent effects* (which usually make themselves felt in the entropy of activation term) are also well known to be part of the explanation of the principal of hard and soft acids and bases. Some mention of all these factors will be made again in the course of this book. Arguments based on the interaction of frontier orbitals are powerful, as we shall see, but they must not be taken so far that we forget these very important limitations.

CHAPTER 3

Ionic Reactions

3.1   The Principle of Hard and Soft Acids and Bases (HSAB)[16]

3.1.1   Inorganic Acids and Bases

Some years ago, Pearson[17] introduced, at first only into inorganic chemistry but later into organic chemistry as well,[18] the concept of hard and soft acids and bases (HSAB). He pointed out that Lewis acids and bases (including H^+ and OH^-) could be classified as belonging, more or less, to one of two groups. One kind he called hard and the other he called soft; they are listed in Table 3-1. The striking observation was, and this was the basis of the classification, that, on the whole, hard acids formed stronger bonds (and reacted faster) with hard bases, and soft acids formed stronger bonds (and reacted faster) with soft bases. For example, a hard acid like the proton is a stronger acid than the silver cation, Ag^+, when a hard base like a hydroxide ion is used as the reference point; but if a softer base like ammonia, NH_3, had been used, we would have come to the opposite conclusion. This situation is summarized in the *rule*: hard-likes-hard and soft-likes-soft.

Pearson's classification was intended to simplify and illuminate the problem, but it did not, and was not intended to, explain it. We are now in a position to point out one of the ways in which molecular orbital theory does explain it.[19] First of all let us look at *thermodynamic* acidity and basicity, namely the way in which the equilibrium

$$\text{acid} + \text{base} \rightleftharpoons \text{salt}$$

is affected by orbital interactions. In outline, it seems that a hard acid bonds strongly to a hard base because the orbitals involved are far apart in energy. As we saw in Fig. 2-10, this leads essentially to an *ionic* bond, and we can associate the strength of the bond with the value E_i on that diagram (p. 15). On the other hand, a soft acid bonds strongly to a soft base because the orbitals involved are close in energy. As we saw in Figs. 2-1 and 2-6 (p. 6 and 11), we get the maximum gain in *covalent* bonding when the interaction is between levels of similar energy, and we can associate the strength of the bond with the value E_σ in Fig. 2-1. We can also see that in practice we shall not often have pure hardness and pure softness in our acids and bases; rather, there is a continuum, and Fig. 2-9 is a case where the bond strength comes from both types of interaction. In summary, the hard acids have high-energy LUMOs and the hard bases have low-energy HOMOs. The higher the energy of the LUMO of an acid, the harder it is as an acid. Similarly, the lower the energy of the HOMO of a base, the harder it is as a base.

Table 3-1 Some hard and soft acids (electrophiles) and bases (nucleophiles)

Bases (Nucleophiles)	Acids (Electrophiles)
Hard	*Hard*
H_2O, OH^-, F^-	H^+, Li^+, Na^+, K^+
$CH_3CO_2^-$, PO_4^{3-}, SO_4^{2-}	Be^{2+}, Mg^{2+}, Ca^{2+}
Cl^-, CO_3^{2-}, ClO_4^-, NO_3^-	Al^{3+}, Ga^{3+}
ROH, RO^-, R_2O	Cr^{3+}, Co^{3+}, Fe^{3+}
NH_3, RNH_2, N_2H_4	CH_3Sn^{3+}
	Si^{4+}, Ti^{4+}
	Ce^{3+}, Sn^{4+}
	$(CH_3)_2Sn^{2+}$
	$BeMe_2$, BF_3, $B(OR)_3$
	$Al(CH_3)_3$, $AlCl_3$, AlH_3
	RPO_2^+, $ROPO_2^+$
	RSO_2^+, $ROSO_2^+$, SO_3
	I^{7+}, I^{5+}, Cl^{7+}, Cr^{6+}
	RCO^+, CO_2, NC^+
	HX (hydrogen bonding molecules)
Borderline	*Borderline*
$C_6H_5NH_2$, C_5H_5N, N_3^-, Br^-, NO_2^-,	Fe^{2+}, Co^{2+}, Ni^{2+}, Cu^{2+}, Zn^{2+}, Pb^{2+},
SO_3^{2-}, N_2	Sn^{2+}, $B(CH_3)_3$, SO_2, NO^+, R_3C^+,
	$C_6H_5^+$
Soft	*Soft*
R_2S, RSH, RS^-	Cu^+, Ag^+, Au^+, Tl^+, Hg^+
I^-, SCN^-, $S_2O_3^{2-}$	Pd^{2+}, Cd^{2+}, Pt^{2+}, Hg^{2+}, CH_3Hg^+,
R_3P, R_3As, $(RO)_3P$	$Co(CN)_5^{2-}$
CN^-, RNC, CO	Tl^{3+}, $Tl(CH_3)_3$, BH_3
C_2H_4, C_6H_6	RS^+, RSe^+, RTe^+
H^-, R^-	I^+, Br^+, HO^+, RO^+
	I_2, Br_2, ICN, etc.
	trinitrobenzene, etc.
	chloranil, quinones, etc.
	tetracyanoethylene, etc.
	O, Cl, Br, I, N, $RO^·$, $RO_2^·$
	M^0 (metal atoms)
	bulk metals
	CH_2, carbenes

3.1.2 Hard and Soft Nucleophiles and Electrophiles

The principles of hard and soft acids and bases has also been applied to kinetic phenomena.[20,21] In this connection, organic chemistry has provided most of the examples, because the reactions of organic chemistry are often slow enough for their rates to be easily measured. In organic chemistry, and in ionic organic chemistry in particular, we are generally interested in the reactions of electrophiles with nucleophiles. These reactions are a particular kind of the general acid-with-base type of reaction, and so the principle of hard and soft acids and bases applies equally to the reactions of electrophiles with nucleophiles.

The striking observation in this field is that the rates with which nucleophiles attack one kind of electrophile are not necessarily a good guide to the rates with which the same nucleophiles will attack other electrophiles. Following the principle of hard and soft acids and bases, we categorize nucleophiles as being hard or soft, and electrophiles as being hard or soft, and again the data fall into the pattern that hard nucleophiles react faster with hard electrophiles and soft nucleophiles with soft electrophiles.

We can now return to equation 2-7 (p. 27), which deals with the application of molecular orbital theory to reaction rates. Essentially, the second term of equation 2-7 represents the bonding gained from a hard-hard interaction, and the third term represents the bonding gained from a soft-soft interaction. We saw from our consideration of the imaginary reaction of the allyl anion (the base or nucleophile) with the allyl cation (the acid or electrophile) that the important frontier orbital of a nucleophile is the relatively high-energy HOMO, and the important frontier orbital of an electrophile is the relatively low-energy LUMO. Hard nucleophiles (Table 3-1) are generally those which are negatively charged and have relatively low-energy HOMOs (in other words, they are the anions centred on the electronegative elements). Hard electrophiles (Table 3-1) are generally those which are positively charged and have relatively high-energy LUMOs (in other words the cations of the more electropositive elements). Thus their reactions with each other are fast because each makes a large contribution to the second term of equation 2-7. On the other hand, soft nucleophiles have high-energy HOMOs and soft electrophiles have low-energy LUMOs, and their reactions with each other are fast because each makes a large contribution to the third term of equation 2-7.

To take a very simple example, a nucleophile like the hydroxide ion is hard at least partly because it has a charge, and because it is based on a small and electronegative element. Accordingly, it reacts much faster with a hard electrophile like a proton than with a soft electrophile like bromine. On the other hand, an alkene is a very soft nucleophile, at least partly because it is uncharged and has a high-energy HOMO. Thus it reacts much faster with an electrophile which

$$HO^- \quad H{-}\overset{+}{O}H_2 \quad \textit{faster than} \quad HO^- \quad Br{-}Br$$

$$CH_2{=}CH_2 \quad Br{-}Br \quad \textit{faster than} \quad CH_2{=}CH_2 \quad H{-}\overset{+}{O}H_2$$

has a low energy LUMO, like bromine or the silver cation, than it does with a proton.

The rates of most reactions are affected by contributions from both terms of equation 2-7, with one often being more important than the other. It is important to realize, for example, that a hard nucleophile *may* react faster with a soft electrophile than a soft nucleophile with the same soft electrophile. Thus hydroxide ion almost certainly reacts faster with silver ion than does ethylene. This is because hydroxide ion is, for several reasons, more generally reactive than ethylene.

In summary:

Hard nucleophiles have a low-energy HOMO and usually have a negative charge.

Soft nucleophiles have a high-energy HOMO but do not necessarily have a negative charge.

Hard electrophiles have a high-energy LUMO and usually have a positive charge.

Soft electrophiles have a low-energy LUMO but do not necessarily have a positive charge.

A hard-hard reaction is fast because of a large Coulombic attraction.

A soft-soft reaction is fast because of a large interaction between the HOMO of the nucleophile and the LUMO of the electrophile.

The larger the coefficient in the appropriate frontier orbital (of the atomic orbital at the reaction centre), the softer the reagent.

3.1.3 Nucleophilicity of Inorganic Ions towards Organic Electrophiles

By using only the HOMO of a nucleophile and the LUMO of an electrophile, we can simplify equation 2-7 to equation 3-1. This will be a good approximation, because the interactions of the other orbitals all have much larger $E_r - E_s$

$$\Delta E = \underbrace{-\frac{Q_{nuc.}\,Q_{elec.}}{\varepsilon R}}_{\substack{\text{The} \\ \text{Coulombic} \\ \text{term}}} + \underbrace{\frac{2(c_{nuc.}\,c_{elec.}\,\beta)^2}{E_{HOMO\,(nuc.)} - E_{LUMO\,(elec.)}}}_{\text{The frontier orbital term}} \qquad 3\text{-}1$$

values and thus make little contribution to the third term of equation 2-7. Klopman[14] has worked out the contribution of frontier orbital and solvation terms to the nucleophilicities and electrophilicities of a range of inorganic bases and acids. From the known ionization potentials and electron affinities, and correcting for the effect of solvation, he calculated values ($E^{\neq}$, Table 3-2) for the effective energy of the HOMO of the nucleophiles and the LUMO of the electrophiles. *The results agree extremely well with Pearson's empirically derived order of softness.* The higher the value of $E^{\neq}$ for the HOMO of a nucleophile, the softer it is, and the higher the value of $E^{\neq}$ for the LUMO of an electrophile, the harder it is.

Klopman[14] has also used equation 3-1 to estimate the nucleophilicities of a range of anionic nucleophiles: I^-, Br^-, Cl^-, F^-, HS^-, CN^-, and HO^-. He

Table 3-2 Calculated softness character for inorganic
nucleophiles and electrophiles

Nucleophile	HOMO $E^{\neq}$ (eV)	Electrophile	LUMO $E^{\neq}$ (eV)
H^-	-7.37 ↑	Al^{3+}	6·01 ↑
I^-	-8.31	La^{3+}	4·51
HS^-	-8.59 Soft	Ti^{4+}	4·35 Hard
CN^-	-8.78	Be^{2+}	3·75
Br^-	-9.22	Mg^{2+}	2·42
Cl^-	-9.94 Hard	Ca^{2+}	2·33
HO^-	-10.45	Fe^{3+}	2·22
H_2O	$-(10.73)$	Sr^{2+}	2·21
F^-	-12.18 ↓	Cr^{3+}	2·06
		Ba^{2+}	1·89
		Ga^{3+}	1·45
		Cr^{2+}	0·91
		Fe^{2+}	0·69
		Li^+	0·49
		H^+	0·42
		Ni^{2+}	0·29
		Na^+	0
		Cu^{2+}	-0.55
		Tl^+	-1.88
		Cd^{2+}	-2.04
		Cu^+	-2.30
		Ag^+	-2.82 Soft
		Tl^{3+}	-3.37
		Au^+	-4.35
		Hg^{2+}	-4.64 ↓

assumed unit charges and unit values for the coefficients, c. For the $E_{HOMO(nuc.)}$
term, he used the value of $E^{\neq}$ which he had already calculated (Table 3-2). The
value of β changes, depending on the nature of the bond being made, but it can
readily be calculated.[22] This left only the energy of the LUMO of the electrophile, $E_{LUMO(elec.)}$, as an unknown on the right-hand side of the equation.
Klopman therefore calculated ΔE values for a series of imaginary electrophiles
with different values for the energy of the lowest unoccupied orbital. What he
found is very striking. (i) Setting $E_{LUMO(elec.)}$ at -7 eV (i.e. at a very low value)
made the $E_{HOMO} - E_{LUMO}$ term small, and hence the frontier orbital term, the
second term of equation 3-1, made a large contribution to ΔE. The order of the
values ΔE was $HS^- > I^- > CN^- > Br^- > Cl^- > HO^- > F^-$, which is the
order of nucleophilicities which has been observed for the attack of these ions
on peroxide oxygen:[23]

$$Nu^- \quad HO{-}OH$$

(ii) Setting $E_{LUMO(elec.)}$ at -5 eV, the order of nucleophilicities is slightly changed,
because the frontier orbital term makes a slightly smaller contribution to ΔE.
The order of ΔE values is now $HS^- > CN^- > I^- > HO^- > Br^- > Cl^- > F^-$,

which parallels the Edwards E values[24] for the nucleophilicities of these ions towards saturated carbon:

$$Nu^- \quad \overset{\curvearrowright}{} \quad \overset{\displaystyle \diagdown}{\underset{\diagup}{C}} \overset{\curvearrowright}{-} X$$

Table 3-3[14]   Nucleophilicity of some inorganic ions towards various electrophiles as a function of $E_{HOMO} - E_{LUMO}$

Nucleo- phile	E_{HOMO} in eV[a]	ΔE Calculated for:			Found:		
		$E_{LUMO} = $ -7 eV	$E_{LUMO} = $ -5 eV	$E_{LUMO} = $ $+1$ eV	$k \times 10^{4b}$	Edwards' E^c	pK_a
HS⁻	−8·59	2·64	1·25	0·55	too fast	too fast	7·1
I⁻	−8·31	2·52	1·07	0·45	6900	2·06	−7·3
CN⁻	−8·78	2·30	1·17	0·56	10	2·79	9·1
Br⁻	−9·22	1·75	0·98	0·48	0·23	1·51	−4·3
Cl⁻	−9·94	1·54	0·97	0·52	0·001	1·24	—
HO⁻	−10·45	1·49	1·01	0·58	0	1·65	15·7
F⁻	−12·18	1·06	0·82	0·54	0	1·0	3·2

[a] From Table 3-2.
[b] For the reaction of the nucleophiles with peroxide oxygen.
[c] A measure of nucleophilicity towards saturated carbon.

(iii) Finally, setting $E_{LUMO(elec.)}$ very high at $+1$ eV, the frontier orbital term is made relatively very unimportant, and the order of ΔE values is governed almost entirely by the Coulombic term of equation 3-1: $HO^- > CN^- > HS^- > F^- > Cl^- > Br^- > I^-$. This is the order of the pK_a's of these ions, in other words, of the extent to which the equilibrium:

$$Nucleophile^- + H_3O^+ \rightleftharpoons Nucleophile\text{-}H + H_2O$$

lies to the right.

Thus, simply by adjusting the relative importance of the two terms of equation 3-1, we can duplicate the otherwise puzzling changes of nucleophilic orders as the electrophile is changed. The proton is a very hard electrophile because it is charged, and especially because it is very small. Hence, a nucleophile can get very close to it in the transition state and R in equation 3-1 is made small. The oxygen-oxygen bond, on the other hand, has no charge, and, being a weak σ-bond, it has a relatively low-lying σ^* LUMO; so it is a very soft electrophile. Similarly, with nucleophiles such as F^-, Cl^-, Br^-, and I^-, the energy of the HOMO will rise as we go down the periodic table, and with nucleophiles like Cl^-, HS^-, and R_2P^-, the energy of the HOMO will rise as we move to the left in the periodic table. This *explains*, therefore, the well-known observation, in the reactions above and in many others, that the 'softness' of a nucleophile increases in these two directions.

Quantitative support for the importance of frontier orbitals comes from the study[25] of a reaction in which the Coulombic term was kept small and relatively constant. Twelve thiocarbonyl compounds of the general formula (27) were treated with methyl iodide.

(27)   (28)   (29)

The rate of the S_N2 reaction (27 + 28 → 29) correlated well with the ionization potential of the lone pairs on the sulphur atom. The ionization potential was measured by photo-electron spectroscopy and is a direct measure of the energy of the HOMO.

So far, we have allowed for the effect of the solvent in a chemical reaction either by ignoring it or, as Klopman did, by correcting for it in some way. Even if a solvent is not involved in a reaction with an organic electrophile, its orbitals may interact with those of the electrophile. This interaction is part of the power of a solvent to solvate ions, and it should be amenable to treatment by perturbation theory. If the orbitals are close enough in energy for a first-order treatment to be appropriate, reaction would occur; so solvation is a second-order interaction. The second term of equation 3-1 will therefore be a good approximation, and the major interactions will be between the HOMO of the solvent and the LUMO of an electrophile, and between the LUMO of the solvent and the HOMO of a nucleophile. Using this idea (and hence the second term of equation 3-1), and using ionization potentials and electron affinities as measures of the energies of the HOMO and the LUMO of a range of solvents, Dougherty[26] has been able to explain some otherwise puzzling changes of solvating power. Thus no single scale of solvating power works for all reactions, just as there is no single scale of nucleophilicity. We can now see that—amongst other things, no doubt—a balance between both sets of frontier orbital interactions ($HOMO_{solvent}/LUMO_{reagent}$ and $LUMO_{solvent}/HOMO_{reagent}$) may help to account for this.

3.2   Ambident Nucleophiles

3.2.1   Charged Nucleophiles

3.2.1.1   Cyanide ion. One of the most useful aspects of the principle of hard and soft acids and bases is the way in which it sorts out our ideas on ambident reactivity. A cyanide ion, for example, can react with an alkyl halide, depending on the conditions, to give either a nitrile (30)[27] or an isonitrile (31):[28]

$$KCN + EtI \longrightarrow Et—C≡N$$

(30)

$$AgCN + EtI \longrightarrow Et—N=C:$$

(31)

In other words, it is sometimes nucleophilic on carbon and sometimes on nitrogen; nucleophiles like this are called ambident. The principle of hard and soft acids and bases helps us to *classify* and *remember* this pattern of reactivity. It tells us that the carbon atom will be the softer and the nitrogen atom the harder end of the nucleophile (since, other things being equal, as we have seen in the last section, hard nucleophiles are on the right in the periodic table).

Thus the electrophiles which attack cyanide ion on nitrogen will be the harder ones, and the ones which attack on carbon will be the softer ones. This fits with the reactions illustrated above. As already seen (pp. 38–39) alkyl halides in simple S_N2 reaction are soft electrophiles; thus it is appropriate for cyanide to react from the soft end of the ion (**32**). When a silver ion is present (other Lewis acids like zinc and mercuric ion behave similarly), the halide ion is assisted in leaving the carbon atom, and in the transition state there is now a greater development of charge on the carbon atom undergoing substitution (**33**). Carbonium ions are hard electrophiles, and therefore it is again appropriate that on this occasion cyanide ion should react from the harder end of the ion.

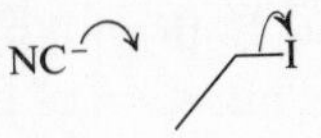

(**32**)

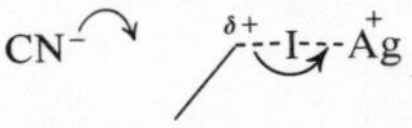

(**33**)

We can also see that this behaviour is *explained* by the different contributions made in the two cases to equation 3-1. The contribution from the first term is greater in the silver-catalyzed reaction because of the greater charge on the carbon atom; in the cyanide ion itself, the greater total charge will be on nitrogen, the more electronegative atom, and hence this is the site of nucleophilic reactivity with the harder electrophile. With the softer electrophile, it is not the overall charge distribution that counts, because the contribution from the first term of equation 3-1 is now much smaller. For the contribution from the second term of equation 3-1, it is the distribution of the electron density in the HOMO of the cyanide ion that we need to look at. It is not easy simply to guess what this is. We can, however, guess that the lowest-energy filled π-orbital will have a very high degree of polarization towards the nitrogen atom; it is quite likely that this will lead the next orbital up in energy, namely the HOMO, to be polarized the other way round; so it *is* reasonable that the carbon atom should be the more nucleophilic site when the second term of equation 3-1 is the more important contributor to ΔE. Furthermore, the β-value for C—C bond formation is higher than for N—C bond formation; when the frontier orbital term is dominant, this will enhance the tendency for bond formation to carbon, because it is in this term that β appears.

3.2.1.2   The Nitrite Ion (and the Nitronium Ion). We might, at first sight, be tempted to think that the HOMO of a nitrite ion would resemble that of the allyl anion; if it did, it would react with soft (and with hard) electrophiles on oxygen, because both the HOMO and the overall charge distribution of the allyl anion make the two ends of the conjugated system the nucleophilic sites. It is well known that the nitrite ion does not behave this way. Although silver **nitrite** does react with alkyl halides to give nitrites, sodium nitrite gives more

42

nitroalkane than alkyl nitrite. As explained above, an alkyl halide is a hard electrophile in the presence of silver ion and a soft electrophile in its absence.

$$AgNO_2 + RBr \longrightarrow RONO$$

$$NaNO_2 + RBr \longrightarrow RNO_2$$

Perhaps solvation is important here: the nitrogen, with so little of the charge, must be less crowded by solvent molecules and is therefore more accessible. But this is not the whole story, as we can tell from the complementary case of nitration. In nitration, the important frontier orbital will be the LUMO of NO_2^+, and this is a similar orbital to the HOMO of NO_2^-. The nitronium ion, NO_2^+, always reacts on nitrogen, both with soft nucleophiles like benzene and with hard ones like water. In the nitration of benzene, the solvent is often non-polar; thus differential solvation is not likely to be responsible for the fact that the nitrogen atom is the electrophilic site.

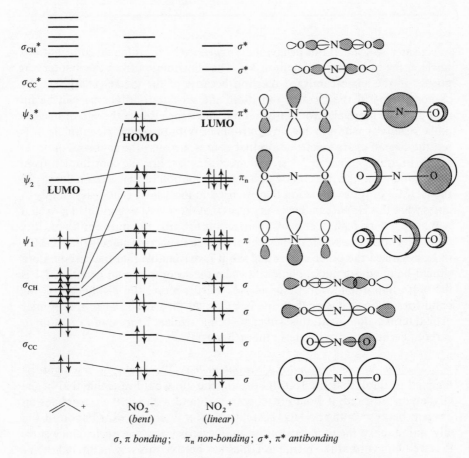

σ, π bonding; π_n non-bonding; σ^*, π^* antibonding

Fig. 3-1 Molecular orbitals of the nitronium ion

In the allyl cation, there are sixteen valence electrons in all, of which fourteen are used in the σ-framework (two C—C bonds and five C—H bonds).[29] Consequently there are two electrons left to go into the lowest π-orbital, ψ_1, as we have seen in Fig. 2-14. If we imagine the protons of the CH$_2$ and CH groups amalgamated with the carbon nuclei, we get oxygen and nitrogen nuclei respectively, and the number of valence electrons is the same. This is why we can expect some relationship between the allyl system and the NO$_2$ system. However, only four orbitals are required for σ-bonding in the NO$_2$ system, and the other orbitals, although similar, are not quite the same. They are shown in Fig. 3-1, where we can see that the last two electrons of the sixteen in NO$_2{}^+$ will go into an orbital, π_n, which resembles ψ_2 of the allyl system and not ψ_1. Thus the LUMO of NO$_2{}^+$ is the orbital labelled π^*, which resembles ψ_3^* in the allyl system. The coefficients of this orbital are large on the central (nitrogen) atom and small on the other two (oxygen) atoms, just as they are for the allyl system. Thus we see that, when the frontier orbital term is dominant, the nitronium ion will be electrophilic at the nitrogen atom. Similarly, when we accommodate two more electrons, to give the nitrite ion, NO$_2{}^-$, they will go into an orbital similar to that shown as π^* in Fig. 3-1. In fact, NO$_2{}^-$ and NO$_2{}^.$, are non-linear, and the orbitals π, π_n and π^*, are no longer degenerate. The actual shape of the HOMO of NO$_2{}^-$ is shown in Fig. 3-2. Once again, the large coefficient is on nitrogen, and nitrogen is the soft centre when the nitrite ion is a nucleophile.

Fig. 3-2   Plan, and perspective drawing of the HOMO of the nitrite anion

Furthermore, ESR studies[30] of nitrogen dioxide (NO$_2{}^.$) have shown that the site of highest odd-electron population is indeed on nitrogen (about 53% of the electron, with the oxygens sharing the other 47%). This confirms our deduction that the nitrogen atom bears the larger coefficient in the HOMO of the nitrite ion and in the LUMO of the nitronium ion, and hence that it is the softer site.

3.2.1.3   Enolate Ions. The most important ambident nucleophile is the enolate ion (**34**). Why does an enolate ion react with some electrophiles at carbon and

$$\text{Me} \diagdown \!\!\!\diagup\!\!\!\diagdown\!\!\!=\!\!\!\text{O} \quad \xleftarrow{\text{MeI}} \quad \diagup\!\!\!=\!\!\!\text{O}^- \quad \xrightarrow{\text{H}^+} \quad \diagup\!\!\!=\!\!\!\text{OH}$$

(**34**)

with others at oxygen? We can now use the explanation based on the relative importance of the Coulombic and frontier orbital terms to account for this well-known observation.[31] The π-orbitals of the enolate ion are shown in

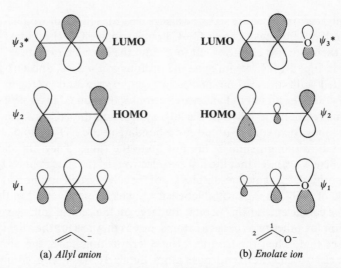

(a) *Allyl anion*   (b) *Enolate ion*

Fig. 3-3 π-Molecular orbitals of the allyl anion and the enolate ion

Fig. 3-3b. The size of the lobes can be taken as roughly representing the size of the c-values (or c^2 values) of the atomic orbitals which make up the molecular orbitals. The system is closely related to that of the allyl anion, which is shown in Fig. 3-3a, but the effect of the oxygen atom is to polarize the electron distribution. Thus the lowest-energy orbital, ψ_1, is, as we would expect, strongly polarized towards oxygen. The next orbital up in energy, ψ_2, however, is polarized away from oxygen. This follows because of the way the coefficients on the atoms, squared, have to add up to one, both in columns and in rows (p. 7). Thus, with a very large value of c on oxygen in the lowest orbital, the values of c on oxygen in the other two (ψ_2 and ψ_3^*) must be relatively small. Likewise, with a very small value of c on C-2 in the lowest orbital, the values of c on the other two orbitals must be relatively large. The effective charge on each atom in the ion is proportional to the sum of the squares of the c-values for the filled orbitals, namely ψ_1 and ψ_2. The result is that more of the total charge is on the oxygen atom than on the carbon atom C-2, because the c-value on oxygen in ψ_1 is so very large and the c-value on carbon so small. However, the c-values in the HOMO are the other way round, although not strongly so. With charged electrophiles, then, the site of attack will be oxygen, as is indeed the case, kinetically, with protons and carbonium ions. With electrophiles having little charge and relatively low-lying LUMOs, the reaction will take place at carbon. In other words, hard electrophiles react at oxygen and soft electrophiles at carbon. Once again, the fact that β-values for bonds to carbon are usually higher than β-values for bonds to oxygen enhances the tendency for the frontier orbital term to encourage reaction at carbon.

We can also explain why the nature of the leaving group on an alkyl halide (or tosylate, for example) affects the proportion of C- to O-alkylation. The observation is that the harder the leaving group (i.e. the more acidic the conjugate acid of the leaving group), the lower the proportion of *C*-alkylation (Table 3-4). Plainly, the harder the leaving group, the more polarized is its bond to carbon (p. 13), and hence the more charge there will be on carbon in the transition state. As a result, the Coulombic term of equation 3-1 will grow in importance with the hardness of the leaving group, and *O*-alkylation will become easier.

Table 3-4 The Proportion of *C*- to *O*-Alkylation as a Function of the Leaving Group[32]

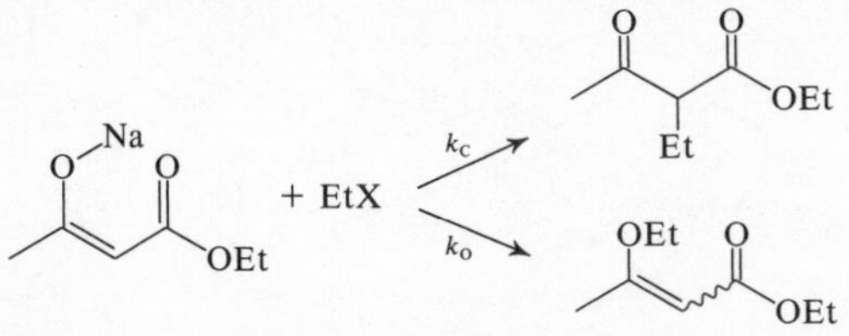

Leaving group, X^-	I^-	Br^-	TsO^-	$EtSO_4^-$	$CF_3SO_3^-$
k_C/k_O	>100	60	6·6	4·8	3·7

3.2.1.4 Pentadienyl Anions and Dienolate Ions. It is well known that the dienyl anion (**35**) is protonated most rapidly at the central carbon atom. This gives the energetically less favourable product. Furthermore, it appears at first sight to be anomalous, when we consider the contribution from the molecular orbitals of the starting material. The sums of the squares of the coefficients of the filled orbitals (these are listed on p. 123) are equal on C-1 and on C-3 (and, of course, on C-5). Also, the frontier orbital coefficients are equal on C-1 and C-3. However, the Hückel calculation which gave these values neglected the fact that C-3 is flanked by two trigonal carbon atoms, but that C-1 has only one trigonal carbon adjacent to it. Naturally, this perturbs the system. This is borne out

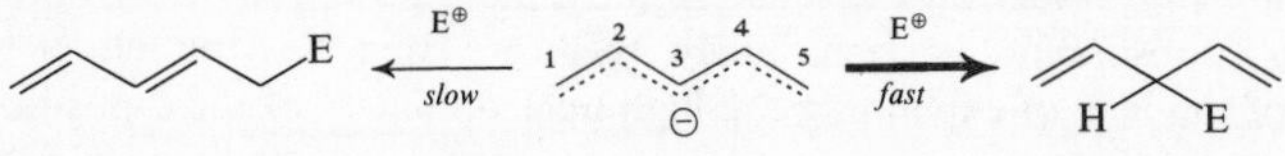

(**35**)

by ESR measurements[33] on the cyclohexadienyl radical, which clearly show a larger coupling to the hydrogen on C-3 than to that on C-1 (Fig. 3-4). HOMO coefficients and total π-electron population have also been calculated[34] by including a term for the overlap of the C—H bonds at C-6 with the π-system (i.e. hyperconjugation). Both the experiment and the calculation have a greater coefficient at C-3 than at C-1. Thus it appears that this should be the more reactive site. It is, of course, a very exothermic reaction, and it is therefore just the kind of reaction which should show the influence

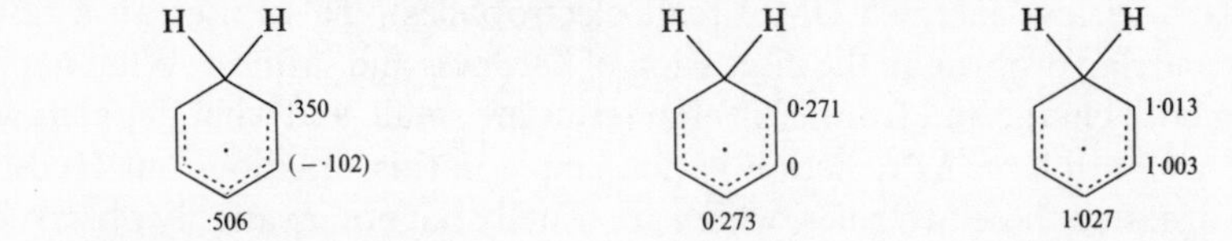

Spin densities (c^2 values) obtained from ESR measurements by applying the McConnell equation (p. 22)

Calculated c^2 values for the HOMO

Calculated total π-electron population

Fig. 3-4 Electron distribution in the cyclohexadienyl radical

from the interaction of the orbitals of the starting materials, rather than the influence from the relative energies of the two possible products.

The same factors may well account for the fact that protonation takes place faster at the α-carbon of a dienolate ion (36) than at the γ-carbon. Again, a calculation[35] on a

(36)

related system, the ether (37), indicates that the total π-electron population is higher at the α- than at the γ-carbon atom. However, another calculation,[36] on the same ether, gives HOMO-coefficients (38) which suggest that the γ-carbon ought to be the more nucleophilic site towards soft electrophiles. This is not the case: methyl iodide, like the proton, reacts faster at the α-carbon. The difference in the coefficients (38) is small,

MeO $\overset{\alpha}{\diagup}\overset{\gamma}{\diagup}$

1·071 1·053

π-Electron population[35]

(37)

MeO $\overset{\alpha}{\diagup}\overset{\gamma}{\diagup}$

−0·492 0·498

HOMO coefficients[36]

(38)

and an oxyanion substituent on the diene is much more powerfully electron donating than a methoxy substituent. It seems likely, therefore, that it is the model which is inadequate, not the idea of explaining the high nucleophilicity of the α-position along these lines.

3.2.1.5 Summary. We can extend these ideas to a large number of other ambident nucleophiles and hence account for many of the well-known examples of hard and soft behaviour. In summary: the site where most of the charge is (the hard centre) will be the site of attack by charged, or relatively charged, electrophiles (hard electrophiles), and the site of the largest coefficient in the HOMO of the nucleophile (the soft centre) will be the site of attack by electrophiles with a relatively low energy LUMO (soft electrophiles). This raises an awkward but recurring problem in the discussion of hardness and softness. What happens when both charge and frontier orbital terms are small, and what happens when they are both large? Prediction is not simple in this situation, but Hudson[37] has suggested these two rules, which are usually but not invariably observed:

(a) When both charge and orbital terms are *small*: nucleophiles and electrophiles will be *soft* (that is, orbital control is more important).

(b) When both charge and orbital terms are *large*: nucleophiles and electrophiles will be *hard* (that is, charge control is more important).

The former is the situation in S_N2 reactions, where the bond being broken is a σ-bond (and hence the LUMO is high in energy) but the charge on the carbon atom is very small. Carbon atoms undergoing S_N2 reactions are well known, as we have seen earlier (pp. 38–39), to be soft centres, responsive to the presence of a high-energy HOMO in the nucleophile.

The latter is the situation in acylation and phosphorylation, where the electrophilic carbon or phosphorus atoms have a considerable ground-state electron deficiency, and hence charge, and also, because it is a π-bond that is being broken, a relatively low-energy LUMO. Carbonyl and phosphoryl groups are well known to be hard centres, responsive to the basicity of the nucleophile. However, carbonyl groups *are* responsive to the frontier orbital terms too; thus a sulphur nucleophile is about 100 times more nucleophilic towards a carbonyl group than is an oxygen nucleophile of the same basicity.[38] We shall see a further example of the responsiveness of carbonyl groups to frontier orbital effects when we come to the α-effect (p. 77).

3.2.2 Aromatic Electrophilic Substitution

A monosubstituted benzene ring is an ambient nucleophile, but it is special among ambident nucleophiles because, along with a number of polycyclic aromatic compounds, it has received so much attention from the theorists of organic chemistry. Molecular orbital theory found some of its earliest and most considerable applications in the realm of aromaticity and in explaining the reactivity of aromatic molecules. Not all these explanations involve frontier orbital theory: arguments based on the product side of the reaction coordinate work quite well, as we shall see, in explaining many of the observations in this field. But, because of the usefulness of MO theory in general, we shall, in this long section of the chapter, look carefully at both sides of the reaction co-ordinate. We shall find that both product-development explanations and frontier orbital explanations give the same answer. We shall, of course, give special attention to those occasions when only the frontier orbital theory provides some kind of an explanation for an otherwise puzzling observation.

3.2.2.1 A Notation for Substituents. Before we begin to discuss aromatic electrophilic substitution and the effects on it of having a substituent already present in the benzene ring, it is convenient to have at our disposal a notation for the various kinds of substituents which we may come across. There are three common types of substituents, each of which modifies the reactivity of conjugated systems in a different way. They are (a) simple conjugated systems, like vinyl or phenyl, which we shall designate with the letter C; (b) conjugated systems which are also electron-withdrawing, like formyl, acetyl, cyano, nitro, and carboxy, which we shall designate with the letter Z; and (c) heteroatoms which carry a lone pair of electrons capable of overlap with the benzene ring; these we shall designate with the letter $\ddot{X}$. We usually include simple alkyl groups in this category, because they are able, by overlap of the C—H (or

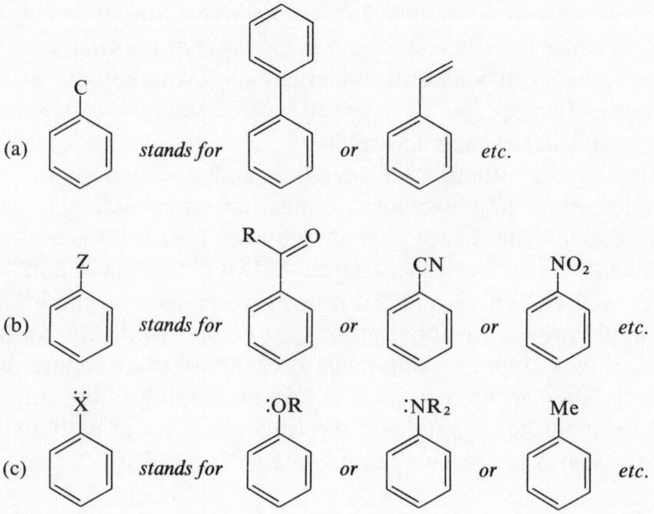

(a) [C-substituted benzene] *stands for* [biphenyl] *or* [styrene] *etc.*

(b) [Z-substituted benzene] *stands for* [R-C(=O)-benzene] *or* [CN-benzene] *or* [NO₂-benzene] *etc.*

(c) [Ẍ-substituted benzene] *stands for* [:OR-benzene] *or* [:NR₂-benzene] *or* [Me-benzene] *etc.*

C—C) bonds with the π-system (hyperconjugation, see p. 80) to supply electrons to the conjugated system.[39] The effect is like that of the lone pairs; it is usually much smaller, but quite noticeable. We shall be using this classification again in Chapter 4, on pericyclic reactions, the subject into which it was first introduced by Houk.[40]

3.2.2.2   Product-Development Control. The rate-determining step in most aromatic electrophilic substitutions is well known to be the attack of the electrophile on the aromatic ring. The rate of such a reaction (like that of all reactions)

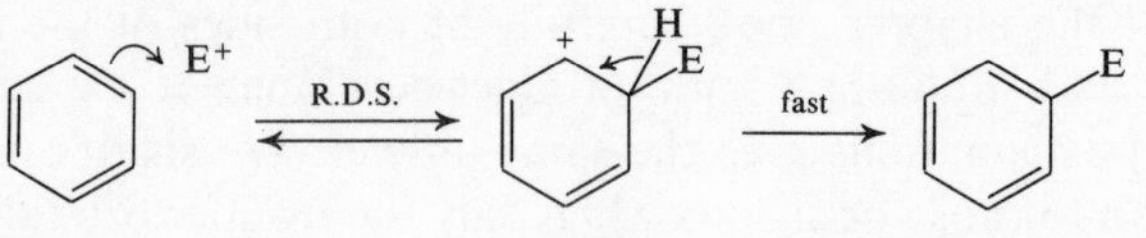

(39)

will be affected on one side of the reaction coordinate by the energy of the 'product', in this case the intermediate (39).[41] Until recently, this has been the contribution to the energy of the transition state which has most often been considered.

Most chemists, and most chemistry text-books, still use what is essentially a valence-bond description of this reaction, based on the 'product'-stability argument. In its simplest form it goes something like this:

With anisole (40) (an Ẍ-substituted benzene), for example, substitution takes place in the *ortho* and *para* positions, rather than in the *meta* position, because the intermediates produced by *ortho* and *para* attack (41 and 43) are lower in energy than the intermediate (42) produced by *meta* attack. The energies of the former intermediates are lower because of the coherent overlap of the lone pair of electrons on the oxygen atom with the orbital containing the positive charge. This overlap is not possible with

(40)   (41)   (42)   (43)

the *meta* intermediate. The overlap is easy to illustrate with curly arrows, and we can see that arrows cannot be drawn in the same way on the *meta* intermediate (42).

However, we ought to be clear that this is a superficial argument (which fortunately works). *Curly arrows*, when used with a molecular orbital description of bonding, work as well as they do simply because they *illustrate the electron distribution in the frontier orbital*, and for reaction kinetics it is the frontier orbital that is most important. But in the present case, we are using a thermodynamic argument, for which we need to know the energy of each of the filled orbitals, and not just one of them.

To assess the energies of the three possible intermediates (41, 42 and 43), we shall, for simplicity, ignore the fact that one of the atoms is an oxygen atom, and we shall use instead the simple hydrocarbon-conjugated systems (45, 46 and 47) which are isoelec-

(44)   (45)   (46)   (47)

tronic with them. We are using, in other words, the benzyl anion (44) as a model for an $\ddot{X}$-substituted benzene ring. The energies and the coefficients of the orbitals of these intermediates have been calculated[4] and are shown in Fig. 3-5. Although the calculations will not give good absolute values for the energies, they will probably get the relative energies about right, which is all we need be concerned with. We can see in Fig. 3-5 that the main reason why the total energy of 46 is higher (-3.08β) than that of 45 (-3.50) and 47 (-3.45) is that the highest filled orbital, ψ_3 in 46, is not lowered in energy (as ψ_3 is in the intermediates 45 and 47), because there are no π-bonding interactions between any of the adjacent atoms. It is, in fact, a nonbonding molecular orbital. This is, of course, the same point that the curly arrows were making, but we should be sure that the two lower-energy orbitals do not compensate for the high energy of ψ_3. In fact, they do to some extent, but not much: we can see that ψ_1 of 47 is actually lower in energy than ψ_1 of 46, and ψ_2 of 45 is lower than ψ_2 of 46. Indeed, the sum of ψ_1 and ψ_2 for 46 is more negative than ($\psi_1 + \psi_2$) for both 45 and 47. We can also see on Fig. 3-5 the reason for both these results. The circles drawn round the atoms are very roughly in proportion to the c^2-values—in other words, the electron population; the clear and darkened circles serve to identify changes of sign in the wave-function. If we look at ψ_1 of 46, we see four atoms with high coefficients (two of 0.316 and one of 0.512, each flanking one of 0.602) close together and all of the same sign. This leads to strong π-bonding and a low energy for this orbital. Such qualitative arguments can also be applied to ψ_2 in each case, and they serve to give us some confidence in the general rightness of the calculated values of the energies of the orbitals in Fig. 3-5. These small effects on ψ_1 and ψ_2 clearly do not compensate for the effect of having no π-bonding in ψ_3 in 46, but we do now have a more thorough version of the original explanation for *ortho/para* substitution in $\ddot{X}$-substituted benzenes.

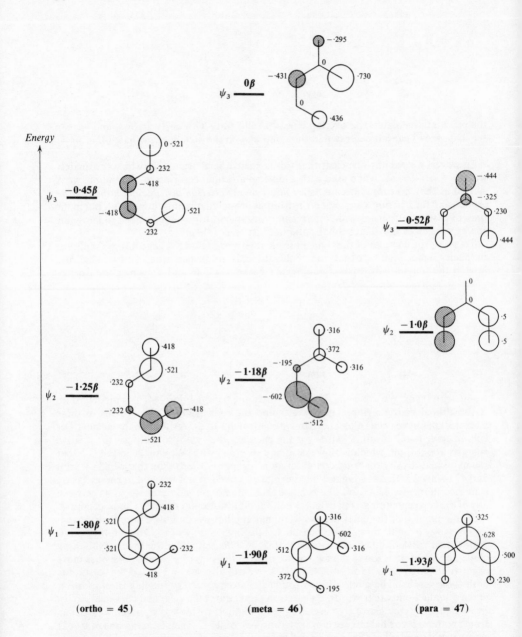

Fig. 3-5 Orbitals and Energies of the intermediates in the electrophilic substitution of the benzyl anion at the *ortho*, *meta*, and *para* positions

In a similar way, we can use the benzyl cation (**48**) as a model for a benzene ring having a Z-substituent. Again we have three possible intermediates (**49, 50** and **51**). The π-systems of these intermediates are the same as the ones we have just been looking at,

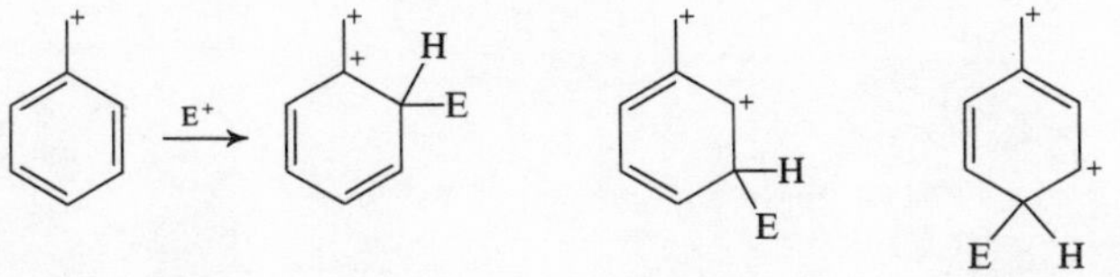

except that this time only ψ_1 and ψ_2 are filled in each case. We have, in fact, already observed that $\psi_1 + \psi_2$ is lowest for the intermediate (**50**) which is the result of attack in the *meta* position. This, then, is the product-development argument for *meta* substitution in Z-substituted benzenes.

Furthermore, we can explain the relatively slow rate of such substitutions and the relatively fast rate of the *ortho/para* substitutions in Ẍ-substituted benzenes, by using only the π-energies of the orbitals together with the argument based on the contribution of product-like character to the transition state. It is summarized in Fig. 3-6. The endothermicity of $1{\cdot}28\beta$ on the right is not much greater than that for benzene ($1{\cdot}27\beta$); however, the presence of *two* positive charges in the intermediate on the right has not been allowed for, and this will obviously raise the energy of this intermediate above that shown.

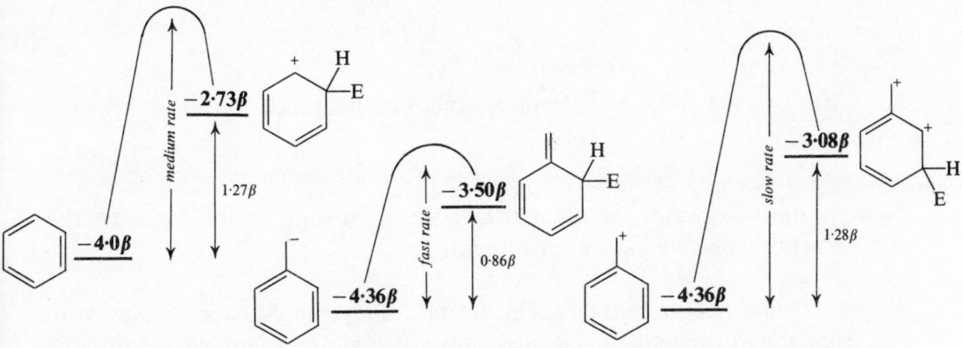

Fig. 3-6 Relative rates of aromatic substitutions based on product-like character in the transition state

It is now time to return to the frontier orbital theory and see how it copes with the other side of the reaction coordinate. To do this, obviously, we must know something about the orbitals of benzene rings.

3.2.2.3 The Orbitals of Benzene. The orbitals of benzene itself are shown in Fig. 3-7. There are two HOMOs of equal energy. Because electrophiles in general are characterized by having low-lying LUMOs, it is these HOMOs of benzene that are the important frontier orbitals in an electrophilic substitution.

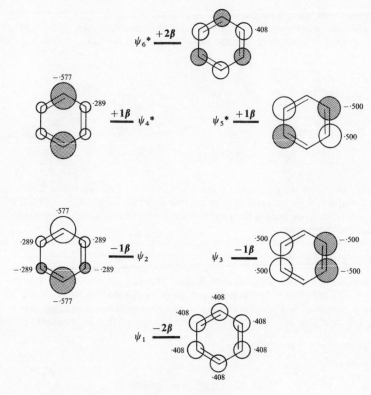

Fig. 3-7 π-Molecular orbitals on benzene[4]

Benzene itself, however, is not particularly interesting in this context: the interest arises when we have a substituent.

3.2.2.4 C-Substituted Benzenes. The orbitals of styrene (**52**), the simplest kind of C-substituted benzene (p. 47), are shown in Fig. 3-8, where we see that the three lowest-energy ones are very like those of benzene, except that ψ_1 and ψ_2 are lower in energy than the corresponding orbitals in benzene, because of the extra overlap between the orbital on the ring carbon atom and the orbital of the exocyclic carbon atom bonded to it. ψ_3 has a node through these atoms, and its energy is therefore unchanged from that of benzene. (The calculation used to get Fig. 3-8 used a *linear* styrene,[7] hence the node through these atoms and the carbon at the end of the side chain. The fact that styrene is not linear makes no difference to the conclusions.) The HOMO, ψ_4, is a new one, for which there is no counterpart in benzene, and this orbital is the important one for the third term of equation 2-7. We can see that it shows high c-values at the *ortho*, *para*, and *β*-positions. These are, of course, the sites at which C-substituted benzenes react with electrophiles:[42,43]

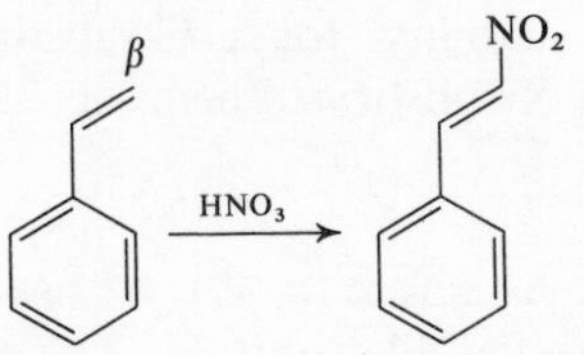

(52)

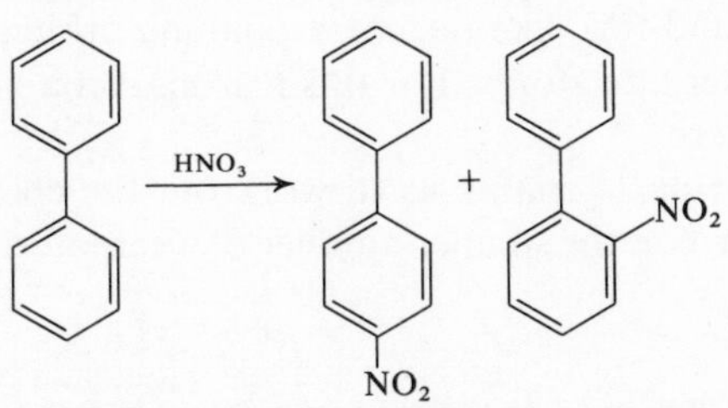

Furthermore, the HOMO energy of styrene is higher than the HOMO energy of benzene, and this will make the electrophilic substitution of styrene faster than that of benzene. The contribution from the second term of equation 2-7 (the Coulombic term) uses the total electron population on each atom: because styrene is a hydrocarbon, and because we have used a very simple Hückel theory, inevitably the sum of the squares of the coefficients of the filled orbitals must be equal on each atom. Thus the second term of equation 2-7 makes little contribution to the site-selectivity in this compound. A more elaborate theory, no doubt, would lead to a small unevenness of charge distribution, but the contribution of the Coulombic term will still be much less than that of the frontier orbital term.

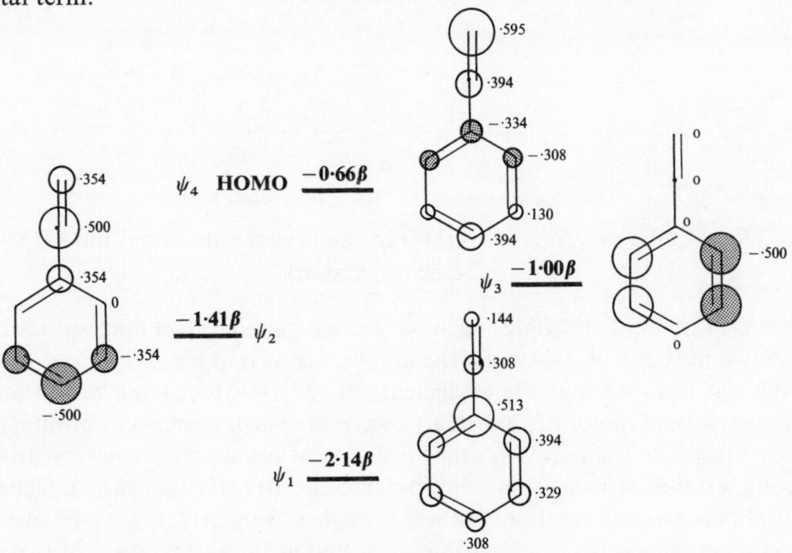

Fig. 3-8   The occupied π-molecular orbitals of styrene

The argument from product stability, for a C-substituted benzene, is a simpler one, very like that for an $\ddot{X}$-substituted benzene, given earlier. It is left as an exercise for the reader.

3.2.2.5 $\ddot{X}$-Substituted Benzenes. As before (p. 49), we use the benzyl anion as a model for $\ddot{X}$-substituted benzenes like anisole. The three lowest-energy orbitals (Fig. 3-10) are, like those of styrene, very similar to those of benzene. The HOMO is ψ_4, and this, like the corresponding orbital in the allyl system, has nodes on the alternate atoms. For this reason, it is a non-bonding orbital, and its π-energy is zero.

Two very simple rules[44] enable us to work out the coefficients in such orbitals. (1) Place a zero on the smaller number of alternate atoms; this identifies the nodes:

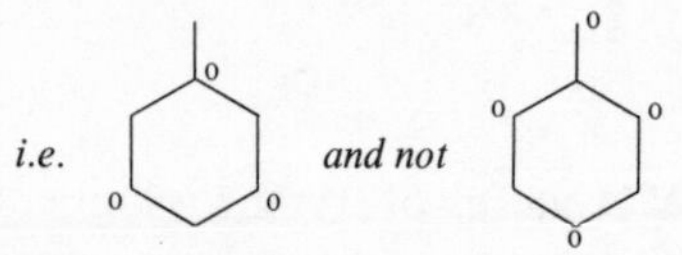

(2) The sum of the coefficients on all the unmarked atoms joined to any one of the marked atoms must be zero. Thus we can start at the *para* position and call the coefficient there a. The coefficients on the *ortho* positions must both be $-a$, in order that the second rule may be obeyed when applied to the (marked) *meta* positions. Now we look at the ring carbon which has the exocyclic carbon atom joined to it. It has a total of three unmarked atoms next to it, two of which, we have deduced, have coefficients of $-a$. The third atom, the exocyclic one, must therefore have a coefficient of $2a$, in order that the second rule may be obeyed. Thus the coefficients are those shown in Fig. 3-9a.

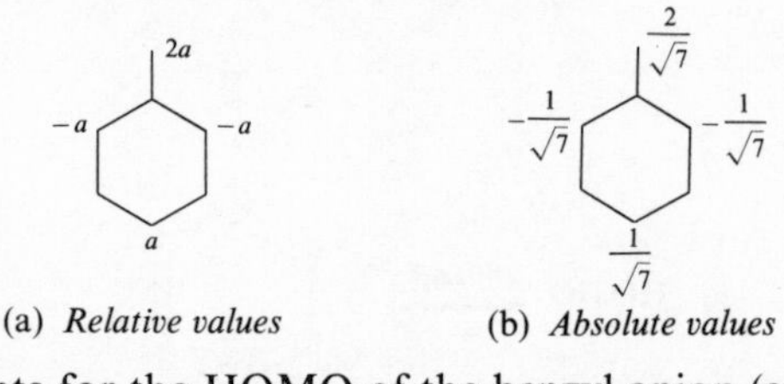

(a) *Relative values*   (b) *Absolute values*

Fig. 3-9 Coefficients for the HOMO of the benzyl anion (and the LUMO of the benzyl cation)

Since the sum of their squares must be one, we can give exact numbers to them, as shown in Fig. 3-9b. These are the numbers shown in Fig. 3-10.

We can now see that the coefficients of the HOMO of the benzyl anion, and therefore of molecules like it, are high on the *ortho* and *para* positions, and zero on the *meta* position. Thus the third term of equation 2-7 will very strongly favour reaction at these sites, and, because the HOMO energy is higher than that of benzene, the reaction rate will be higher. Similarly, the total π-electron population, which is shown on Fig. 3-11, makes the second term of equation 2-7 favour reaction at the same sites.

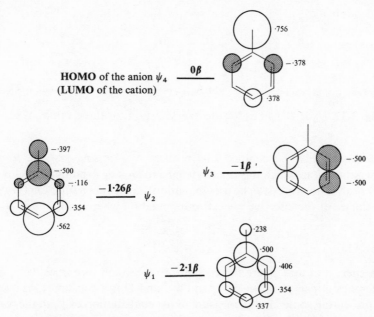

Fig. 3-10 The occupied π-orbitals of the benzyl anion[7]

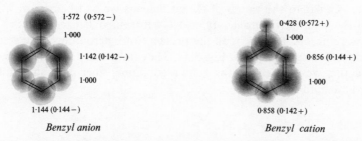

Fig. 3-11 Total π-electron population (and excess-charge distribution) in the benzyl anion and the benzyl cation

3.2.2.6 Z-Substituted Benzenes. If the substituent in the benzene ring is conjugated and electron-withdrawing, our model, as before, is the benzyl cation.
We now find that the HOMO is ψ_3 in Fig. 3-10, and we would, if we considered
this orbital alone, expect to get *ortho* and *meta* substitution equally. However,
the orbital ψ_2 is not far below ψ_3, and we are not safe in ignoring it. We should
instead assess the contribution of all the filled orbitals to the third term of
equation 2-7. We can do this by setting the energy of the LUMO of the electrophile arbitrarily at various levels, and then working out the value of $\sum c^2/E_r - E_s$
for each of the positions in the benzene ring. The results for three values of
E_{LUMO} are shown in Fig. 3-12. Clearly, when we take all the orbitals into
account in this way, the *meta* position is the most reactive, and the *ortho* and
para positions are much alike.

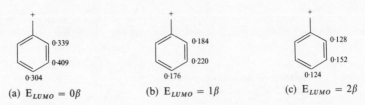

(a) $E_{LUMO} = 0\beta$ (b) $E_{LUMO} = 1\beta$ (c) $E_{LUMO} = 2\beta$

Fig. 3-12 $\sum c^2/E_r - E_s$ for the benzyl cation for three values of E_r

Fukui has explicitly examined this problem,[45] stressing just the *frontier* electron population, which this time is defined to include orbitals just below the HOMO itself. The expression he uses (equation 3-2) defines a function (f) which is an estimate of the effective π-electron population when both ψ_2 and ψ_3 are

$$f = 2\frac{c_3{}^2 + c_2{}^2\, e^{-D\Delta\lambda}}{1 - e^{-D\Delta\lambda}} \qquad 3\text{-}2$$

close in energy. c_3 and c_2 are the coefficients in ψ_3 and ψ_2 respectively, $\Delta\lambda$ is the difference in energy between ψ_3 and ψ_2, and D is a constant (3 is used in fact) representing some kind of measure of the contribution of ψ_2 to the overall effect. As it ought, this expression gives the higher-energy orbital, ψ_3, slightly greater weight. Using this expression, he gets values of f for styrene, which are, naturally, similar to the values of the coefficients we had before (p. 53). For benzonitrile and nitrobenzene, typical Z-substituted benzenes, more parameters were needed to cope with the presence of heteroatoms. The results of such calculations are shown in Fig. 3-13. They show that the analogy with the benzyl cation we used above was quite a good one, since *meta* substitution is expected for both of them.

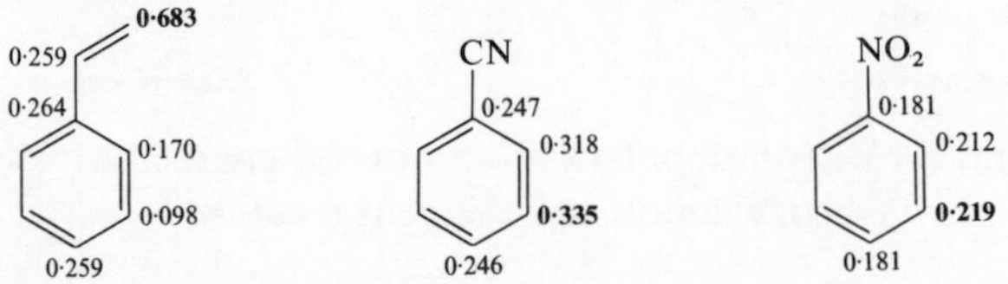

Fig. 3-13 Frontier electron population (f) for C- and Z-substituted benzenes

The total π-electron population for the benzyl cation is also shown on Fig. 3-11, where we see again that the Coulombic term favours reaction at the *meta* position. Finally, with ψ_2 and ψ_3 as the frontier orbitals, we can see why Z-substituted benzenes are less reactive than benzene: ψ_2 is lower in energy than ψ_2 in benzene. Furthermore, the Coulombic term must include allowance for the repulsion between the positive charge of the electrophile and any positive charge in the benzene ring. Most of this charge will be in the *ortho* and *para* positions, but it will still repel an electrophile even from the *meta* position.

We can now summarize our conclusions. Benzene rings with electron-donating substituents will be substituted by electrophiles in the *ortho* and *para* positions at a greater rate than that at which benzene itself will be substituted. Benzene rings with electron-withdrawing substituents will be substituted by electrophiles in the *meta* position at a slower rate than benzene itself. These conclusions follow from considering both the nature of the intermediates and the Coulombic and frontier orbital interactions of the starting materials. It hardly needs saying that the facts of aromatic electrophilic substitution are exactly those presented in these conclusions.

3.2.2.7 Frontier Orbitals of Other Aromatic Molecules. We can extend our analysis of reactivity to a very wide range of aromatic compounds. Thus, the preferred site of electrophilic attack in all the aromatic molecules shown in Fig. 3-14 is known: an arrow is shown pointing from it (or them, in some closely balanced cases). As before, the site of attack can generally be deduced quite satisfactorily by estimating the energy of the intermediate which would be produced and comparing it with the energy of alternative intermediates. Thus in pyrrole (53), for example, the intermediate (54) produced from attack at the 2-position will be lower in energy than the intermediate (55) from attack in the

3-position. Thus we have no need of further explanation. It is, however, extraordinarily gratifying to find that the frontier orbital approach also explains the site of electrophilic attack for each of these familiar compounds. There are even more examples on p. 171, where they are introduced in connection with another problem. The numbers on the structures drawn in Fig. 3-14 are either (for compounds 56 to 61) the coefficients of the HOMO of the molecule,[7,46] or (for compounds 62 to 66), the closely related frontier electron population (f) calculated by Fukui,[45] using equation 3-2, modified to take account of the heteroatoms. The experimentally observed site of nitration is represented by the arrow; in each case, except for the slightly anomalous one of pyrrocoline (63), the arrow does indeed come from the largest (or larger) number. Thus in all these cases we have the situation described on p. 24, where the lower-energy product and the lower-energy approach to the transition state are connected smoothly by what is evidently the lower-energy transition state. We can feel

58

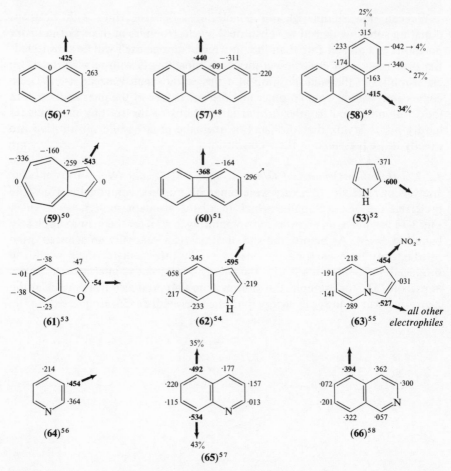

Fig. 3-14 Frontier electron populations of some aromatic compounds and the sites (arrowed) of nitration; the references are to the experimental work

confident, in a situation like this, that we have a fairly good qualitative picture of the influences which bear on the transition state.

3.2.2.8 Quantitative Support for the Molecular Orbital Theory of Reactivity. Some correlations with rate constants show that our understanding is also partly quantitative. Fukui[59] defined a term 'superdelocalisability', s, using an equation of the form 3-3.

$$s_r = \sum_j^{occ.} \frac{c_j^2}{E_j}$$

3-3

c_j is the coefficient at the atom r in the filled orbital j, and E_j is the energy of that orbital. This expression bears an obvious relation to the third term of equation

2-7. Using a single electrophile—the nitronium ion—a plot of the rate constant for nitration at particular sites in a very large range of aromatic hydrocarbons against s_r gives a good correlation over several powers of ten in rate constant.[37]

Similarly, many people have used the concept of 'localization energy' to account for the rates, and sites, of electrophilic substitution. The localization energy is a calculated value of the endothermicity in a reaction and is therefore part of an argument based on product development control. The plot of localization energy against rate constant is also a good straight line.[60] This is no place to try to estimate the *relative* success of these two approaches: they are obviously related in some deep-seated property of molecular orbitals.[61]

We can also add some experimental support for the soundness of some of the theoretical arguments. Photoelectron spectroscopy gives the ionization potentials (which are roughly the energies of the HOMOs) for the monosubstituted benzenes shown in Table 3-5. The C- and $\ddot{X}$-substituted benzenes do have higher-energy HOMOs than benzene, and the Z-substituted benzenes do have lower HOMOs than benzene. Furthermore, the HOMO energies do run in parallel with reactivity towards electrophilic substitution, with the exception of the halogenobenzenes, which are discussed separately later.

Table 3-5 Ionization potentials for some mono-substituted benzene compounds[8, 62]

R	IP (eV) (E_{HOMO})	
Me_2N $\left.\begin{array}{c} \\ \\ \\ \end{array}\right\}$ $\ddot{X}$- MeO Me	$-7{\cdot}51$ $-8{\cdot}54$ $(-8{\cdot}9)$	
Ph $\left.\begin{array}{c} \\ \\ \end{array}\right\}$ C- $C{=}C^a$	$-7{\cdot}42$ $-8{\cdot}13$	
H	$-9{\cdot}40$	
CHO $\left.\begin{array}{c} \\ \\ \\ \\ \end{array}\right\}$ Z- CF_3 CN NO_2	$-9{\cdot}8$ $-9{\cdot}9$ $-10{\cdot}02$ $-10{\cdot}26$	
I	$-8{\cdot}78$	$-9{\cdot}75^b$
Br	$-9{\cdot}25$	$-9{\cdot}78^b$
Cl	$-9{\cdot}31$	$-9{\cdot}71^b$
F	$(-9{\cdot}5)$	$-9{\cdot}86^b$

[a] The value is that for indene.
[b] Ionization potential for the next highest occupied molecular orbital (NHOMO).

Secondly, ESR measurements on the benzyl radical (**67**) clearly show that the coefficients of the singly occupied orbital are indeed high on the *ortho* and *para* positions; this orbital is the same as the HOMO of the benzyl anion, which is the orbital we have used as a model for an $\ddot{\text{X}}$-substituted benzene ring.

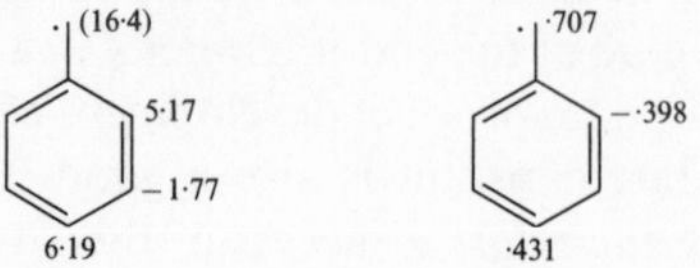

Fig. 3-15 ESR of the benzyl radical; the numbers on **67** are the hyperfine couplings to the corresponding ^{1}H nucleus in gauss,[9] and the numbers on **68** are the coefficients of ψ_4 derived from these numbers using the McConnell equation (p. 22) (ignoring the negative spin density)

3.2.2.9 Halogenobenzenes. It is well known that the halogenobenzenes are unusual in showing a mixture of the properties of the Z- and $\ddot{\text{X}}$-substituted benzenes: like Z-substituted benzenes, they undergo electrophilic substitution more slowly than benzene, but, like $\ddot{\text{X}}$-substituted benzenes, they are *ortho/para* directing. We shall see that there is no paradox: the factors affecting the overall rate are not all the same as those affecting orientation.

On the 'product' side of the reaction coordinate, we are on weak ground. The intermediates (**69**, **70** and **71**) should all be lower in energy than the corresponding intermediate in benzene, though not very much lower. Because the halogen is an $\ddot{\text{X}}$-substituent, there should be some covalent bonding gained by overlap from the lone pairs, but because a halogen is so electronegative, we can expect that the gain will be small. It is the same situation as that shown in Fig. 2-9; the energy gained by having overlap is dependent on the similarity of the electronegativities of the two elements. In our earlier

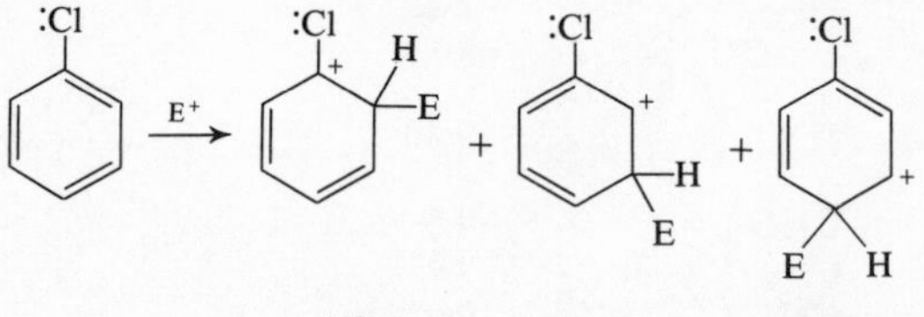

discussion of $\ddot{\text{X}}$-substituted benzenes, we rather ignored this point; we made the assumption—because it answered all our questions when we did so—that the electronegativity was not so large that the analogy with the benzyl anion would break down. With the halogenobenzenes the electronegativity is now large enough for the analogy to break down to some extent. Nevertheless, it does not do so completely: the overlap of the lone pairs on the halogen atoms is not absent, and it does lower the energies of the orbitals of the intermediates (**69** and **71**) more than those of the intermediate (**70**). However, this does not explain the lower rate of reaction. To do that we shall have to look at the Coulombic and frontier orbital effects.

On the starting-material side of the reaction coordinate, we are also in difficulty, because the non-mathematical description of the orbitals we have been using is inadequate. Even proper Hückel theory is not particularly good when there are strongly electronegative atoms present. However, we have no difficulty in seeing that the perturbation treatment leading to equation 2-7 makes it *possible* to have *ortho/para* substitution at a reduced rate: the *o/p* orientation could be mainly dependent on the coefficients and charges at each of the atoms, and the reduced rate could be largely determined by the energies of the higher occupied orbitals. But we *are* in difficulty in showing that this possibility is indeed true, unless we are prepared to do fairly elaborate calculations.

In the halogenobenzenes, the PES results (Table 3-5) show that there are two high-energy HOMOs, from which we are led to realise that the benzyl-anion analogy is grossly misleading. Presumably the very electronegative halogen atom holds so tightly, as it were, to its electrons that we can expect a highly polarized electron distribution. It is here that the qualitative arguments break down; and quantitative calculations have hardly begun. We shall use the results of a calculation[63] carried out on anisole (Fig. 3-16) as a model for a halogenobenzene. The benzyl-anion analogy worked quite well

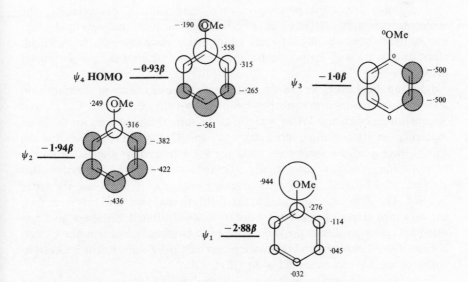

Fig. 3-16 Filled and LU π-molecular orbitals calculated for anisole

with anisole, but by using anisole as an example here, we shall see how having an electronegative element in place of carbon causes the orbitals to deviate from those of the benzyl anion. A halogenobenzene can then be expected to deviate in the same direction, but more so. The quantitative aspects of calculations like this are not to be relied upon, but the *directions* in which changes in energy and polarization occur, and the ordering of energy levels, are usually trustworthy.

In order to see the kind of polarization that has occurred, Fig. 3-16 should be compared with Fig. 3-10, which shows the orbitals of the benzyl anion. The coefficients shown on Fig. 3-16 agree with our early proposition that the benzyl anion was a good analogy to work from: thus, they show the high electron population in the HOMO on the *ortho* and *para* positions, and the high total electron population there as well. They also show that the HOMO is higher in energy than ψ_2 and ψ_3 of benzene (-1β), but, not surprisingly, that it is a good deal lower than the HOMO of the benzyl anion (0β). It will not be surprising, therefore, if the halogens, which are, like oxygen, more electronegative than carbon, also reduce the energy of some of the higher occupied orbitals to positions below those of the corresponding orbitals in benzene. Indeed the PE spectra of the halogenobenzenes (Table 3-5) show that, although the HOMO is slightly higher than ψ_2 and ψ_3 of benzene, the next orbital down is some way below them. When *both* orbitals are taken into consideration, in the manner used on p. 56, we can expect that the overall rate will be lower. But we can still have an electron distribution similar to that which causes any Ẍ-substituted benzene to undergo *ortho* and *para* substitution. For the Coulombic term, incidentally, a number of calculations[64] agree that the total electron population is in the order halogen > *ortho* > *para* > *meta*, but the details of the individual orbitals are not recorded. Thus it is entirely reasonable for halogenobenzenes to be *ortho/para*-directing at a reduced rate.

3.2.2.10 ortho/para *Ratios.*[20, 32] The proportion of *ortho* to *para* substitution ought to be susceptible to molecular orbital treatment, but we should not be surprised to find that such treatment has had only a little success as yet. The changes in *ortho/para* ratios are relatively small, and the differences in activation energies are, correspondingly, very small indeed. Furthermore, the molecular orbital treatment we have been using is far from complete in identifying all the factors which contribute to transition state energies. In this field, steric effects are well known to be important in reducing the proportion of *ortho* product.

Ignoring this factor, we can see that product-like character in the transition state clearly favours *ortho* substitution over *para* substitution for C-, Z-, and Ẍ-substituted benzenes. When we look at the sum of the energies for the filled molecular orbitals of the intermediates (**45** and **47**), we see (Fig. 3-5) that the total π-energy of the former $(-3\cdot50\beta$ for the intermediate derived from the benzyl anion) is more negative than that of the latter $(-3\cdot45\beta)$. Similarly with a Z-substituted benzene, the former gives a π-energy of $-3\cdot05\beta$ and the latter $-2\cdot93\beta$. The difference is greater in the Z-substituted case, and this is, in fact, the observed trend (Table 3-6): insofar as Z-substituted benzenes give any *ortho* and *para* products, the *ortho/para* ratio is greater than it is for C- and Ẍ-substituted benzenes, and the more powerfully the Z-substituent is electronwithdrawing, the more marked is the effect.

Now we should turn to equation 2-7 to see how the starting materials affect the transition state. The total electron population for a C-substituted benzene

Table 3-6 *o/2p* Ratios in aromatic nitration as a function of the substituent[65]

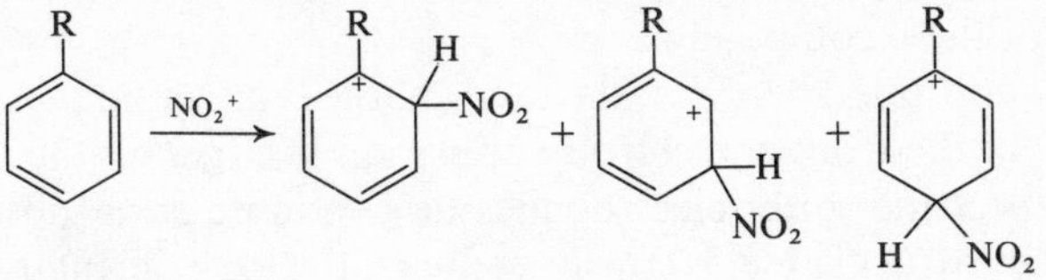

Type of substituent	R	%*o*	%*m*	%*p*	*o/2p*[a]
$\overset{..}{X}$-	MeO	17	—	83	0·10
C-	Ph	53	—	47	0·56
Z-	CO$_2$Et	28	68	3	4·3
Z-	NO$_2$	6	93	0·25	12·8

[a] There are two *ortho* positions and only one *para*, so a statistical correction is applied.

ring is the same (by simple Hückel theory, at any rate) on the *ortho* and the *para* position, but the frontier electron density is higher in the *para* position (pp. 53 and 56). We should therefore expect that the softer electrophiles will give more substitution in the *para* position. This is very much what is observed with biphenyl (Table 3-7). Nitration involves a fairly hard electrophile (NO$_2$$^+$), and so does protonation; the bromonium ion will be harder than neutral chlorine and neutral bromine, and mercuration involves a very soft electrophile (Table 3-1). The *o/2p* ratios fall in this order.

For a Z-substituted benzene ring, the total electron population is usually calculated to be higher on the *ortho* than on the *para* position (and much higher on the *meta* position, of course). The frontier electron population is probably

Table 3-7 *o/2p* Ratios for aromatic substitution as a function of the electrophile[66]

Electrophilic substitution	*o/2p* for toluene (Ph-$\overset{..}{X}$)	*o/2p* for biphenyl (Ph-C)
Hydroxylation	2·00	—
Chlorination with Cl$^+$	1·63	—
Bromination with Br$^+$	1·29	0·69
Proton exchange	1·06–0·3[a,b]	1·0–0·19[a,b]
Protodesilylation	0·84	2·14
Nitration	0·72	1·68
Chlorination with Cl$_2$	0·97–0·25[a]	0·32
Friedel-Crafts ethylation	0·47	0·41
Sulphonation	0·25	—
Mercuration	0·25	0·01
Bromination with Br$_2$	0·25–0·11[a]	0·03
Friedel-Crafts acetylation	0·0006	very small

[a] Dependent upon conditions.
[b] See also Table 3-8.

also higher on the *ortho* position (p. 56), so again all three molecular orbital contributions are in the same direction, which explains the observation of high $o/2p$ ratios for these compounds.

For an $\ddot{X}$-substituted benzene, the total charge is larger on the *ortho* position (Fig. 3-16), but the frontier electron population is larger on the *para* position. We again expect the softer electrophiles to give more *para* substitution. This fits moderately well (Table 3-7) with some of the experimental observations. Nitration and bromination with Br^+ give higher $o/2p$ ratios with toluene than the softer electrophiles involved in halogenation with molecular halogen and in mercuration. Furthermore, the halogens, whether as X^+ or X_2, are in the right order: chlorine is harder than bromine and gives higher $o/2p$ ratios. Friedel-Crafts acylation probably involves a species $RC{\equiv}O^+$; it seems likely that this species, although formally charged, will have a very low-energy LUMO and hence a high frontier orbital contribution.

But we cannot take this argument very far. Steric effects are well known to be important, and there are some striking anomalies. Thus hydroxylation (trifluoroperacetic acid) has a very soft electrophile, but gives a very high $o/2p$ ratio, and sulphonation has a relatively hard electrophile (usually solvated SO_3), but appears to have a rather low $o/2p$ ratio. This latter observation is particularly likely to be a steric effect, because sulphonation, not unexpectedly, is well known to be unusually sensitive to steric effects, and it is also known to be unusual in being reversible.

The proportion of *ortho* attack in any of these reactions is quite dependent upon the reaction conditions (thus the numbers in Table 3-6 are not the same as those in Table 3-7, the data coming from different sources). But none is more sensitive than proton exchange. There is a steady decrease in the proportion of *ortho* attack as the acid strength is reduced (Table 3-8). The nearer the electrophile is to being a free proton, the harder it is, and the more *ortho* substitution

Table 3-8 $o/2p$ Ratios for proton exchange as a function of acid strength[66]

Conditions	$o/2p$ for toluene (Ph-X)	$o/2p$ for biphenyl (Ph-C)
75% H_2SO_4	1·06	—
71% H_2SO_4	1·00	—
65% H_2SO_4	0·98	—
CF_3CO_2H/H_2O	0·6	0·63
CF_3CO_2H	0·49	0·60
liquid HI	0·50	0·25
liquid HBr	0·28	0·19

there is. The changes in the $o/2p$ ratio are unlikely to be all steric in origin, because the trend with toluene is also observed with biphenyl and with t-butylbenzene. Thus the frontier orbital theory is moderately successful in a field notoriously beset with confusing data and multifarious influences on transition-state energies.

3.2.2.11 Pyridine N-*oxide*. A special case of aromatic electrophilic substitution is provided by the ambient reactivity of pyridine *N*-oxide (**72**). Klopman[14] has used equation 2-7 to calculate the relative reactivity (ΔE values) for electrophilic attack at the 2-, 3- and 4-positions as it is influenced by the energy of the LUMO of the electrophile. He obtained a graph (Fig. 3-17) which shows that each position in turn can be the most nucleophilic. At very high values of $E_r - E_s$ (hard electrophiles), attack should take place at C-3; at lower $E_r - E_s$ values, it should take place at C-4; and, with the softest electrophiles, it should take place at C-2. Attack at each of these sites is certainly known: the hardest electrophile SO_3 (Table 3-1) does attack the 3-position,[67] the next hardest

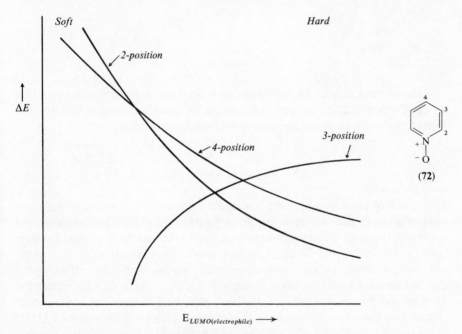

Fig. 3-17 Electrophilic substitution of pyridine *N*-oxide (**72**)

(NO_2^+) the 4-position,[68] and the softest ($HgOAc^+$) the 2-position.[69] This time, without the complicating steric effect, sulphur trioxide is showing the expected behaviour.

However, this reaction is really more complicated. The sulphonation, for example, almost certainly takes place on the *O*-protonated oxide rather than on the free *N*-oxide, and this must affect the relative reactivity of the 2-, 3- and 4-positions. The value of the exercise is not so much in the detail of this particular example as in the way in which it shows how a single nucleophile, such as pyridine *N*-oxide, can, in principle, be attacked at different sites, depending upon the energy of the LUMO of the electrophile.

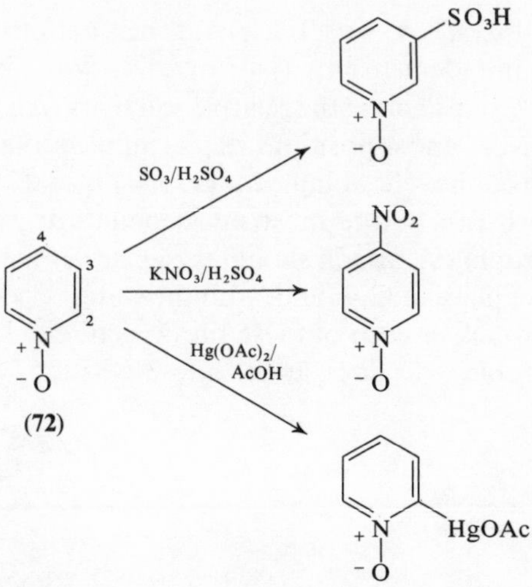

With this very simple, if possibly not quite accurate, explanation of a puzzling series of observations, we leave the subject of ambident nucleophilicity, and turn to the much smaller subject of ambident electrophilicity.

3.3 Ambident Electrophiles

The attack of a nucleophile on a conjugated system is susceptible to the same kind of analysis that we have just given to the attack of an electrophile on a conjugated system. This time, we shall use the LUMO of the conjugated system (and the HOMO of the nucleophile, of course) as the important frontier orbitals. In most cases, all the molecular orbital factors, both those affecting the product stability and those in the starting materials, point in the same direction. Thus each of the systems in Fig. 3-18 shows electrophilic reactivity at the site (or sites) where the arrow points; each of them has a high coefficient of the LUMO at the site of attack; each of them also has a high total electron deficiency at this site; and the product obtained from such attack is lower in energy than attack at the alternative sites.

We shall now concentrate our attention on those interesting cases where there is a choice of sites within the molecule.

3.3.1 Aromatic Electrophiles

3.3.1.1 The Pyridinium Cation. The pyridinium cation (**83**) is readily attacked by nucleophiles at C-2 and C-4. The *total* electron deficiency[37] at C-2 of $+0.241$ and at C-4 of $+0.165$ indicates that charge control (in other words with hard

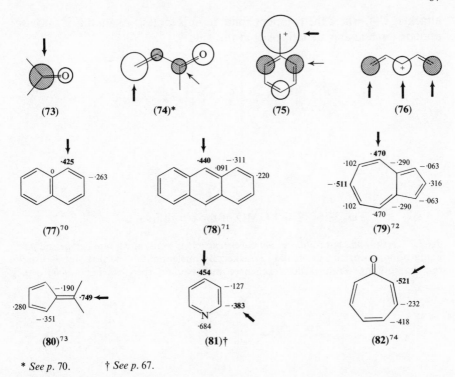

* See p. 70.   † See p. 67.

Fig. 3-18 LUMOs of some electrophilic compounds and the sites (arrowed) of nucleophilic attack upon them; the references are to the experimental work, mostly with organometallic nucleophiles

nucleophiles) will lead to reaction at C-2. This is the case with such relatively hard nucleophiles as hydroxide ion, amide ion, borohydride ion and Grignard reagents.[75]

However, if we look at the LUMO, we find that it has the form shown in Fig. 3-19, namely that of ψ_4^* of benzene, but polarized by the nitrogen atom. This polarization has reduced the coefficient at C-3, and the coefficient at C-4 is larger than that at C-2; for example, a simple Hückel calculation[7] for pyridine itself gives values of 0·454 and −0·383 respectively, and an energy of 0·56β (compare benzene with 1β for this orbital). Thus, soft nucleophiles should attack at C-4, where the frontier orbital term is largest. Again this is the case; cyanide ion and enolate ions react at this site.[76]

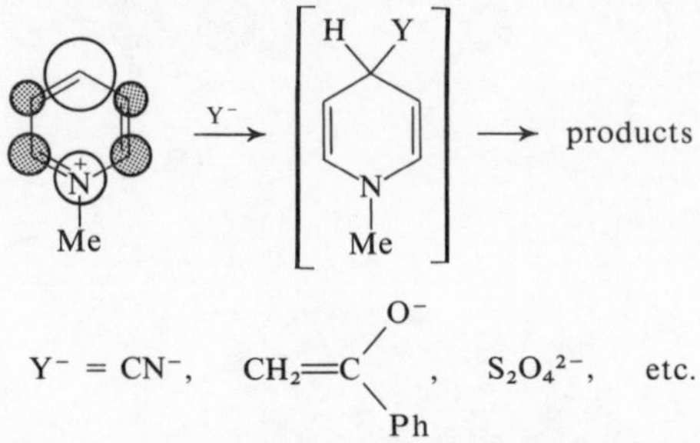

$$Y^- = CN^-, \quad CH_2{=}C{\underset{Ph}{\overset{O^-}{\diagup}}}, \quad S_2O_4{}^{2-}, \quad etc.$$

Fig. 3-19 The LUMO of the pyridinium cation[37]

3.3.1.2 ortho- and para-Halogenonitrobenzenes. It is well-known that *ortho-* and *para*-halogenonitrobenzenes are readily attacked by nucleophiles. The first step is usually rate-determining. Product development control should therefore have *ortho* attack

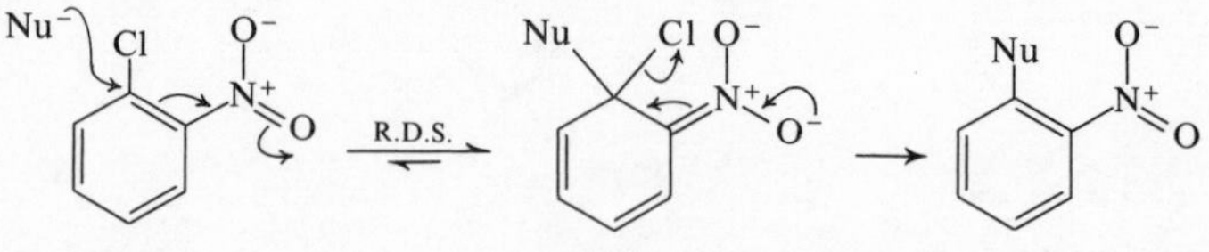

(84)

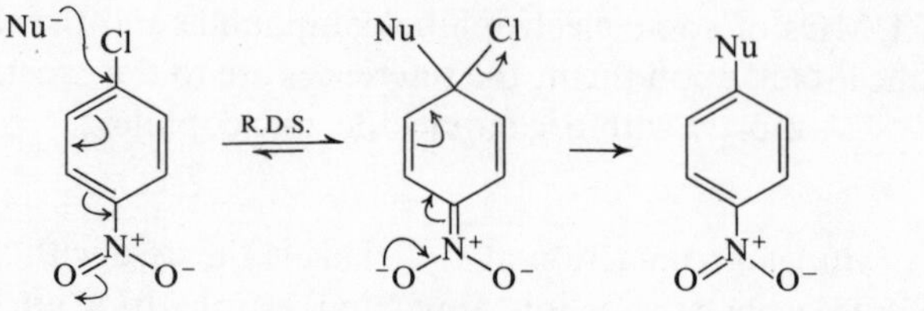

(85)

faster than *para* attack, because the intermediate (**84**) with the linear conjugated system will be lower in energy than the intermediate (**85**), other things being equal. The Coulombic term will also lead to faster reaction at the *ortho* than at the *para* position. The frontier orbital term, however, should favour attack at the *para* position. Thus the ESR spectrum of the benzyl radical (p. 60), which has the odd electron in an orbital which ought to be a model for the LUMO of a Z-substituted benzene, shows that there is a larger coefficient in the *para* position than in the *ortho*. There is some evidence[77] which supports this analysis. (a) With a *charged* activating group, such as the diazonium cation in **86** and **87**, attack at the *ortho* position is faster than attack at the *para* position, because of the large Coulombic contribution. With the *uncharged* activating groups in the compounds (**88** and **89**), the order is the other way round. (b) In the latter reaction (Z=NO$_2$), with the neutral (and hence softer) nucleophile (**90**), the preference for *para* attack is enhanced. (c) The ratio of the rates at which PhS$^-$ and MeO$^-$ react with 2,4-dinitrophenylhalogenobenzenes (**91**) is highest for the iodide and lowest for the fluoride. The former will make the Coulombic term least important, and the latter will make it most important.[78] (d) When the rate of the second step is *not* rate-determining, the aryl fluoride is much more readily attacked than the corresponding aryl chloride, bromide and iodide. This seems to be because of a large Coulombic contribu-

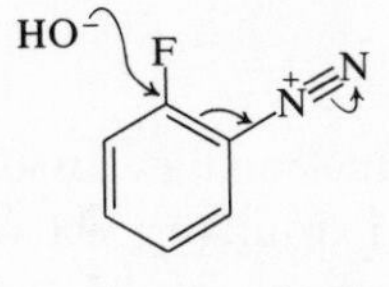

(86)   *10 times faster than*   (87)

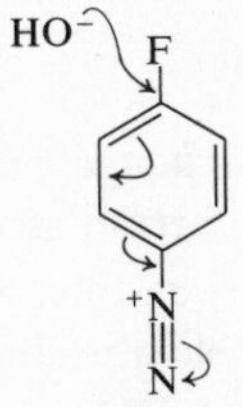

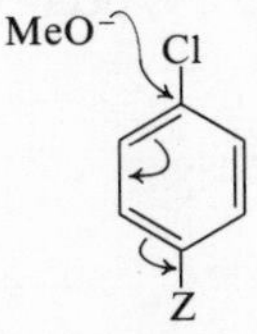

(88)   *between 1 and 4 times faster than*   (89)   *for Z = NO$_2$, CN,*
SO$_2$Me and COMe

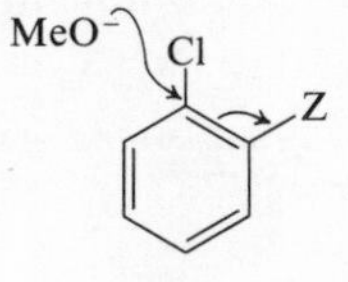

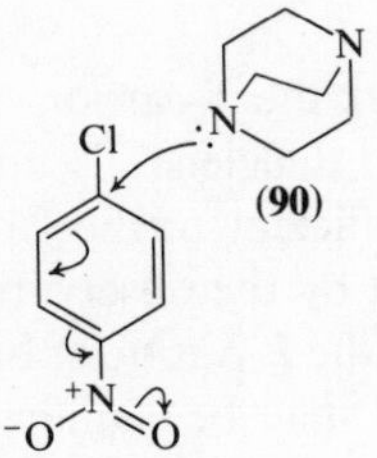

(90)   (90)   *250 times faster than*

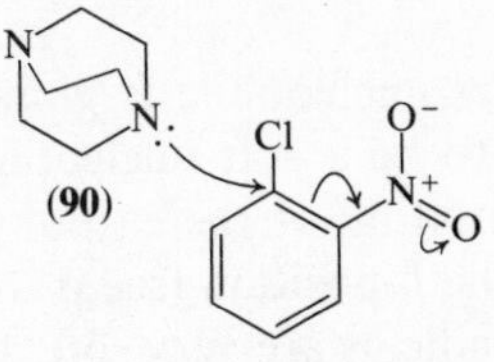

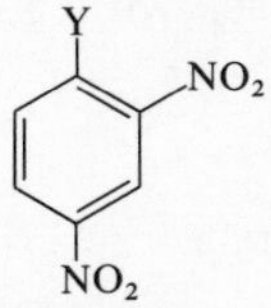

(91)
Y = *halogen*

+ PhS$^-$ *or*   MeO$^-$ $\dfrac{k_{PhS^-}}{k_{MeO^-}}$ *increases*   F < Cl < Br < I

tion: the rate-enhancement is, on the whole (Table 3-9), much less for neutral and soft nucleophiles; but the story is very complicated, because of the fact that the second step of the reaction, the loss of the fluoride ion, does become rate determining with some weak nucleophiles.

Table 3-9   Effect of the nucleophile on the relative rates of attack on the fluoro and chlorodinitrobenzenes (91)

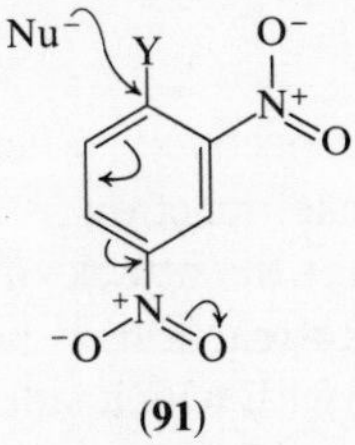

(91)

Y = F or Cl

Nucleophile	k_F/k_{Cl}	
$(H_2N)_2C{=}S:$	0·11	
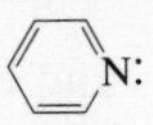 pyridine N:	3·26	
PhS$^-$·	33	soft
H$_3$N:	460	
MeO$^-$	890	hard
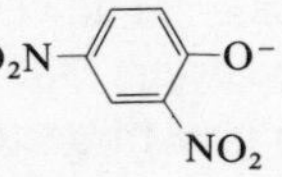 O_2N—C$_6$H$_3$(NO$_2$)—O$^-$	3160	

3.3.2 Aliphatic Electrophiles

3.3.2.1 αβ-Unsaturated Carbonyl Compounds. Most nucleophiles attack αβ-unsaturated ketones at the carbon atom of the carbonyl group (e.g. **93 → 94**); such attack is sometimes reversible. Attack at the β-carbon (e.g. **93 → 92**) is commonly the result of a slower, but thermodynamically more favourable, reaction. This is well known to be the case with cyanide ion, for example. If we look at the balance of Coulombic and frontier orbital terms, we can expect[79]

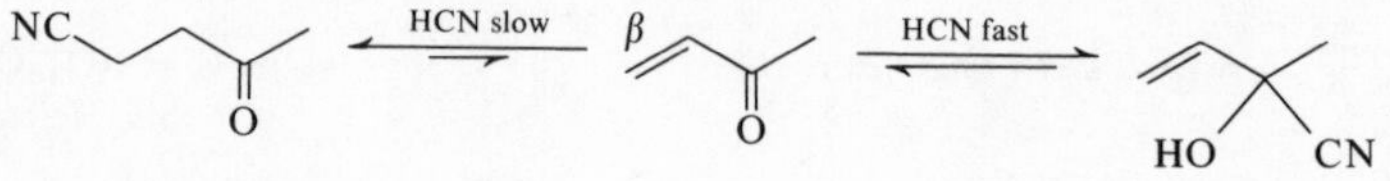

(92)   (93)   (94)

that, if any nucleophile is going to attack directly at the β-carbon atom, it will have to be a soft nucleophile: the total electron deficiency is greater at the carbon atom of the carbonyl group, but the coefficient of the LUMO is larger at the β-position (see p. 163). This is borne out by the observation that radicals (radicals are very soft, see Chapter 5) add at the β-position; but there are few clear cut pieces of evidence on this point. It may be significant that hydroxide ion[80] and alkoxide ion[81] (hard nucleophiles) react with ethyl

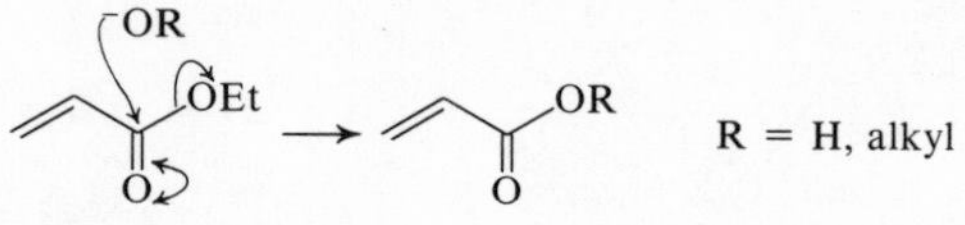

R = H, alkyl

(95)

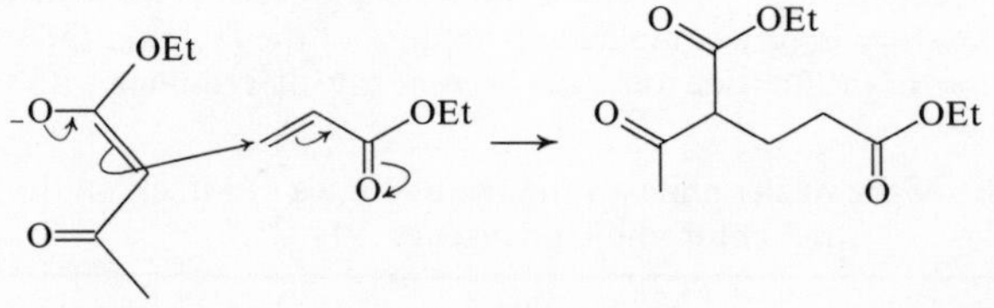

(96)   (95)

acrylate (**95**) to give ester hydrolysis and ester exchange respectively, whereas the enolate ion (**96**) (a soft nucleophile) undergoes a Michael reaction.[82] Unfortunately there is no certainty, in this latter reaction, that the attack of the enolate anion (**96**) on the carbonyl group, in a Claisen-like reaction, is not a more rapid (and reversible) process.[83] The very great ease with which sulphur nucleophiles add to αβ-unsaturated esters (**98 → 97**) is also ambiguous: thiolate anions do not react with esters to give thioesters (**99**), because the equilibrium lies in the other direction; so we cannot tell what are the relative rates of *attack* at the two sites of an $\alpha\beta$-unsaturated ester.

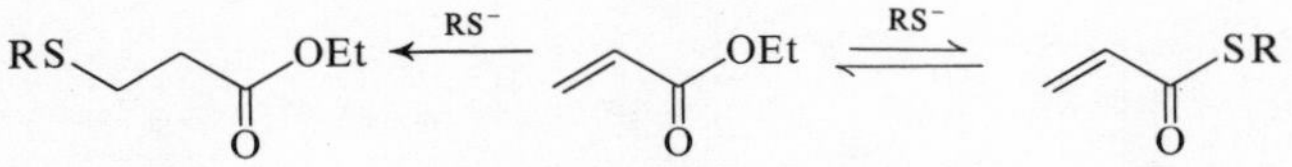

(97)　　　　　　　(98)　　　　　　　(99)

One case, however, seems clear. Ammonia and amines do react with ordinary esters to give amides, and it is known[84] that the attack at the carbonyl group is rate-determining and effectively irreversible above pH 7. Ammonia and amines (neutral and therefore relatively soft nucleophiles) in methanol react with acrylic ester (**100**) at the β-position.[85]

A calculation (see p. 163) indicates that although acrolein has a larger co-efficient in the LUMO at the β-position than at the carbonyl carbon atom, protonated acrolein has the larger coefficient at the carbonyl carbon atom. Thus, the more electrophilic the carbonyl group, the more likely it is that all nucleophiles will attack at the carbonyl carbon atom. In contrast to its behaviour with acrylic ester, ammonia does react[86] with acryloyl chloride (**101**), which has a very electrophilic carbonyl group, at the carbonyl carbon atom.

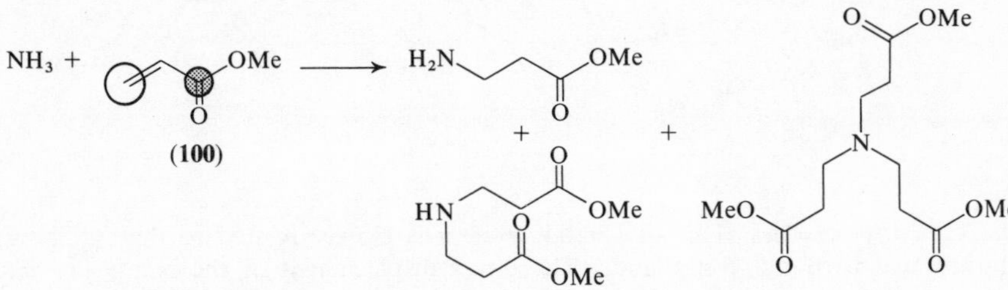

(100)

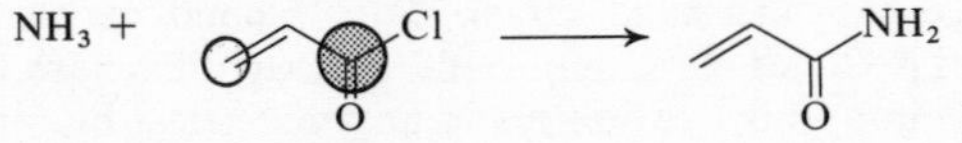

(101)

The reduction of unsaturated carbonyl compounds by metal hydrides, and the reaction of organometallic nucleophiles with them, is a complicated story.[87] It is more common than not, in each case, to get direct attack at the carbonyl group, but reaction in the conjugate position is well known. Conjugate reduction of $\alpha\beta$-unsaturated ketones by metal hydrides increases[88] in the sequences: Bu_2^iAlH < $LiAlH_4$ < $LiAlH(OMe)_3$ < $LiAlH(OBu^t)_3$ and

$LiAlH_4$ < $NaBH(OMe)_3$ < $NaBH_4$ < $LiAlH_4$ (in pyridine). (The active species in the last of these reagents is of the type:

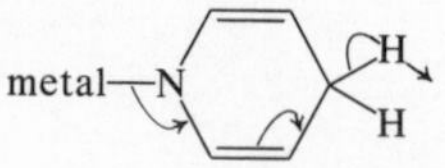

The hydride ion is delivered, in other words, from carbon, and not from a metal atom.) These trends appear to agree with our frontier orbital analysis. In particular, we may note the striking tendency for the delivery of hydride from carbon to give conjugate reduction, whereas delivery from a metal atom usually gives direct attack at the carbonyl group. The metal hydrogen bond will be much more polarized, and the hydride should therefore be harder when delivered from a metal than from carbon. Similarly, the delivery of hydride from boron will make it softer than when it is delivered from the more electropositive metal, aluminium. It also seems that, amongst $\alpha\beta$-unsaturated carbonyl compounds, the amount of conjugate reduction increases in the sequence: ketones < esters < acids < amides; but there are far too few examples to be sure.

With two activating substituents, as in the ester (**102**), conjugate reduction is fast,[89] but product stability, and Coulombic and frontier orbital factors, can all readily explain this observation.

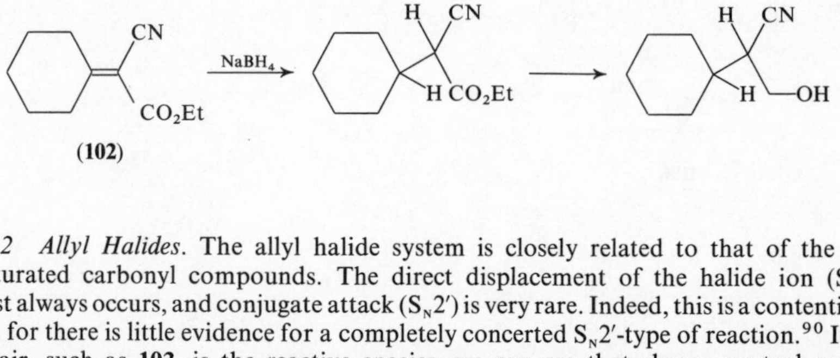

(**102**)

3.3.2.2 *Allyl Halides.* The allyl halide system is closely related to that of the $\alpha\beta$-unsaturated carbonyl compounds. The direct displacement of the halide ion (S_N2) almost always occurs, and conjugate attack (S_N2') is very rare. Indeed, this is a contentious issue, for there is little evidence for a completely concerted S_N2'-type of reaction.[90] If an ion-pair, such as **103**, is the reactive species, we can see that charge control would strongly favour direct displacement. On the other hand, attack at the 'π-bond' may well be preferred if frontier orbital control becomes more important, although it is hard to specify what exactly the orbitals are, in an allyl cation made unsymmetrical by ion-pairing at one end of the conjugated system. It is perhaps significant that the few examples

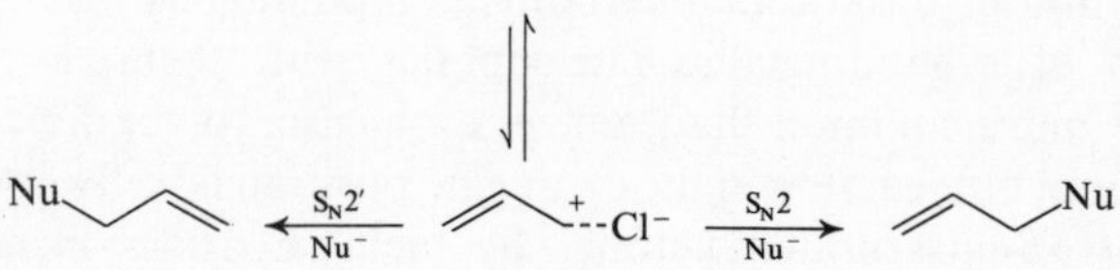

(**103**)

of conjugate reaction which have been observed are with very soft nucleophiles such as phenylthioxide ion,[91] cyanide ion, azide ion and secondary amines,[92] all in non-polar solvents.

3.3.2.3 Arynes. Nucleophiles readily attack the reactive intermediate 2,3-pyridyne (**104**) entirely at C-2.[93] A calculation [94] shows that the coefficient in the LUMO of the pyridyne

(**104**)

LUMO   *Total charge-distribution*

(**105**)   (**106**)

(**105**) is larger at C-2 than at C-3. Also, the total charge distribution (**106**) is such that C-2 bears a partial positive charge. We can rationalize the polarization of the LUMO by comparing the lobes of the p-orbitals in the plane of the ring (**105**) with the π-system of the allyl anion. The large coefficient in the LUMO of the allyl anion is on the central atom, just as it is here. The net result is that nucleophiles attack at C-2 because both Coulombic and frontier orbital forces favour attack at that site. They also react at the 2-position because the anion formed, a 3-pyridyl anion, is more stable than the alternative anion, a 2-pyridyl anion.[94]

2,3-pyridyne is very difficult to trap with a diene; nucleophilic attack takes place much faster, even with relatively poor nucleophiles like acetic acid.[95] For the reaction with a diene, Coulombic forces are small, and large coefficients on *both* C-2 and C-3 would help. Since, 2,3-pyridyne is very polarized, the ionic reaction is made much easier than the cycloaddition.

3,4-pyridyne (**109**) is much less polarized, because the p orbital on nitrogen is too far from those on C-3 and C-4 to exert much influence through space, and therefore it does so primarily through the bonds.[94] Though this takes us beyond the scope of this book, we should note the result, which is that C-4 has a slightly larger coefficient in the LUMO than C-3 (**107**), and the total charge-distribution (**108**) also makes C-4 the more

LUMO   *Total charge-distribution*

(**107**)   (**108**)   (**109**)

electrophilic site. The polarization however is quite a bit smaller than that of 2,3-pyridyne. Nucleophiles do attack C-4 faster than C-3, but both types of product (**110** and **111**) are formed,[96] in agreement with this analysis. In addition, the more even polarization in 3,4-pyridyne reduces the forces favouring ionic reactions and increases the frontier orbital forces favouring a cycloaddition: 3,4-pyridyne is quite easily trapped by Diels-Alder reaction with dienes.[95]

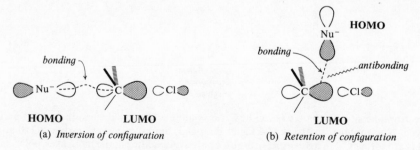

$$(109) \qquad (110) \quad (111)$$

2 parts to 1 part

In spite of their very high total energy, arynes in general are quite selective towards different nucleophiles; thus benzyne (112) easily captures the anion (113) of acetonitrile,

(112)

even in the presence of an excess of amide ion.[97] The reactivity of various nucleophiles is roughly $R_3C^- \approx RS^- > R_2N^- > RO^-$ and $I^- > Br^- > Cl^-$ (where R_3C^- represents reagents like butyl lithium and phenylacetylide ion). Clearly this is an order of softness, and arynes must be soft electrophiles. The splitting of the HOMO and LUMO of an aryne will be very small, because of the poor overlap of the p orbitals in the plane of the ring (calculation[98] suggests a gap between these orbitals of about 1·5 eV). The result is that the LUMO of an aryne is very low in energy, so much so that its interaction with the HOMO of a nucleophile will often be a first-order perturbation. This makes the aryne both very electrophilic and very responsive to the energy of the HOMO of the nucleophile. Since it is also uncharged, it will necessarily be a very soft electrophile.

3.4 Substitution at a Saturated Carbon Atom

3.4.1 Stereochemistry

It is well known that bimolecular nucleophilic substitution (the S_N2 reaction) takes place with inversion of configuration. We can easily explain this by looking at the frontier orbitals,[99] which will be the HOMO of the nucleophile and the LUMO of the electrophile. The overlap is bonding when the nucleophile approaches the electrophile from the rear (Fig. 3-20a), but the approach from

Fig. 3-20 Frontier orbitals for the S_N2 reaction

the front (Fig. 3-20b) is both bonding and antibonding. The former is clearly preferred.

In electrophilic substitution, the frontier orbitals will be the HOMO of the nucleophile (the C-metal bond) and the LUMO of the electrophile. In this case,[99] the overlap (Fig. 3-21a and b) is bonding for attack on *either* side of the carbon atom. In agreement with this, electrophilic substitutions with retention

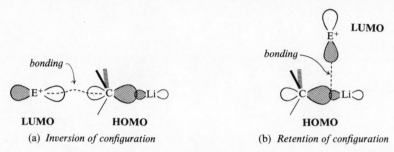

(a) *Inversion of configuration*　　　(b) *Retention of configuration*

Fig. 3-21　Frontier orbitals for the $S_E 2$ reaction

(e.g. **115 → 114**) and with inversion of configuration (**115 → 116**) have been observed.[100]

$$
\textbf{(114)} \quad \textbf{(115)} \quad \textbf{(116)}
$$

3.4.2　The Orbitals of a Methyl Halide

In the last section we saw how readily frontier orbitals explain some striking features of substitution reactions. Earlier (pp. 38–39) we learned that alkyl halides are soft electrophiles and that they attack ambident nucleophiles at the softer site. Nevertheless, it remains seemingly anomalous that an alkyl halide undergoing nucleophilic displacement should be a soft electrophile, when a carbonyl group undergoing nucleophilic addition is a hard electrophile. The former involves breaking a σ-bond (which has a relatively high-energy LUMO), and the latter involves breaking a π-bond (which has a lower-energy LUMO). So far we have explained this (p. 47) first by recognizing that the actual charge carried by the carbon atom is greater in the case of the carbonyl group, and hence that the Coulombic term is made larger in this case, and, secondly, by relying on Hudson's rule (p. 46) that when both Coulombic and frontier orbital terms are small, the latter assumes the greater importance.

This anomaly is made more disturbing if we are wedded to the idea of hybridization (p. 8). An examination of the molecular orbitals of a methyl halide

(Fig. 3-22a) may make the anomaly less disturbing. Were we to use hybridized orbitals (Fig. 3-22b), we should have an antibonding orbital (sp^{3*}_{C-X}) which was at least as much antibonding as the bonding orbital (sp^{3}_{C-X}) was bonding. It is;

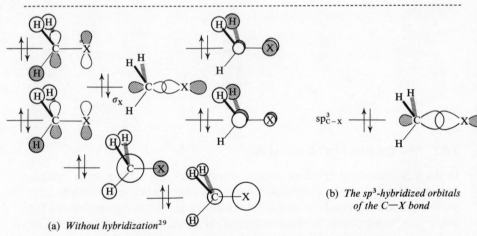

Fig. 3-22 The molecular orbitals of a methyl halide

but this antibonding orbital is no more the LUMO than σ_x in Fig. 3-22a is the HOMO. The latter is only one of the bonding orbitals, and a proper measure of the total C—X bond strength would come much lower in energy than σ_x, and would correspond to that of the bonding hybrid orbital (Fig. 3-22b). There will be a complementary situation among the antibonding orbitals. This inbalance, in which the true LUMO is lower in energy than the antibonding hybrid orbital, is not found in the corresponding π-bond of a carbonyl group, because that orbital, localized on the carbon and oxygen atoms, is not made up of hybridized orbitals. Thus, in a comparison of alkyl halide chemistry with carbonyl chemistry, the use of hybridization appears to exaggerate how high is the energy of the LUMO of the carbon-halogen bond. The LUMO of an alkyl halide is usually higher in energy than that of a carbonyl group, but it is not, perhaps, quite as high as one might think. Furthermore, the LUMO (σ_x^*) does have quite a large lobe pointing in the direction from which the nucleophile approaches. It actually has a larger lobe in the opposite direction, but any attack from that direction is inhibited by the unavoidable antibonding overlap with the adjacent lobe of the leaving group (Fig. 3-20b). When these orbitals are drawn in textbooks, the size of the lobe being attacked by the nucleophile is often made much too small. This is because they have usually been constructed from sp^3 hybrids.

3.5 The α-Effect[37]

The solvated proton is a hard electrophile, little affected by frontier orbital interactions. For this reason, the pKa of the conjugate acid of a nucleophile is a good measure of the rate at which that nucleophile will attack other hard electrophiles like the carbonyl group. We have already seen (p. 47) that carbonyl groups are somewhat responsive to frontier orbital effects, more so, anyway, than solvated protons. Thus a thioxide ion, RS$^-$, is more nucleophilic towards a carbonyl group than one would expect from its pKa: a plot of the log of the rate constant for nucleophilic attack on a carbonyl group against the pKa of the nucleophile is a good straight line only when the nucleophilic atom is the same. In other words, there is a series of straight lines, one for oxygen nucleophiles, one for sulphur nucleophiles, and yet another for nitrogen nucleophiles.

But some nucleophiles, HO$_2^-$, ClO$^-$, HO$\ddot{\text{N}}$H$_2$, $\dot{\text{N}}_2$H$_4$, and R$_2$S$_2$, stand out because they do not fit on their respective lines: they are much more nucleophilic towards such electrophiles as carbonyl groups than one would expect from their pKa values. These nucleophiles all have a nucleophilic site which is flanked by a heteroatom bearing a lone pair of electrons (an $\ddot{\text{X}}$-substituent, in other words). If the orbital containing the electrons on the nucleophilic atom overlaps with the orbital of the lone pair of the $\ddot{\text{X}}$-substituent, then

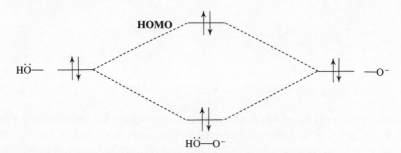

Fig. 3-23 The filled orbitals of a nucleophile having an $\ddot{\text{X}}$-substituent

a splitting occurs (Fig. 3-23). The result is that the highest occupied orbital is raised in energy relative to its position in the unsubstituted nucleophile. Consequently, the denominator of the third term of equation 2-7 is reduced, and the importance of this term is increased. The result is an increase in nucleophilicity towards electrophiles with any soft character at all. The effect is quite dramatic (Table 3-10). What is more, the order of the effect is right: the LUMO of the triple bond of the nitrile will be lower than that of the double bond of the carbonyl group, which will be lower than that of the σ-bond of the bromide. Hence the frontier orbital term is most enhanced in the case of benzonitrile, and least enhanced for benzyl bromide.

Table 3-10 Relative reactivity of HOO^- and HO^-

Electrophile	k_{HOO^-}/k_{HO^-}
$PhC\equiv N$	10^5
$p\text{-}O_2NC_6H_4CO_2Me$	10^3
$PhCH_2Br$	50

	K_{HOO^-}/K_{HO^-}
H_3O^+	10^{-4}

This observation is related to the problem mentioned earlier (on p. 47), of what happens when charge and orbital terms are either both low or both high. In this case, we have a thousandfold increase in nucleophilicity towards a carbonyl group, accompanied by a tenthousandfold decrease in basicity. It serves to remind us that carbonyl groups, although very responsive to basicity (charge control), are not unresponsive to frontier orbital effects.

The increased nucleophilicity found among these nucleophiles with α-lone-pairs is called the α-effect. It is observed not only as a kinetic effect; there is a thermodynamic α-effect as well.[101] For example, we may compare an ester (**117**) with a perester (**118**). The overlap of a lone pair with the π^* orbital of a

(117)   (118)

carbonyl group (**117** arrows) is an important part of the reason why esters are 'stabilized' relative to ketones. The effect of another lone pair is to raise the energy of the first lone pair (Fig. 3-23) and hence to make the overlap with the π^* orbital more energy-lowering.

3.6 Some Applications of Perturbation Theory to the Structures of Organic Molecules

By mentioning the *thermodynamic* α-effect in the last section, we have again strayed from the main concern of this book—chemical reactions—into an area beyond its scope, namely the static properties of a molecule. Nevertheless, it is a large and growing area of study, and since it is in fact closely related to the general subject of frontier orbital theory, further digression on the subject will not be inappropriate. The interactions of orbitals *within a molecule* account for many features of chemical structure, much as the interactions of frontier orbitals account for many features of chemical reactivity. Just as frontier orbital theory is especially successful when it is used to compare the relative reactivity of two closely related systems, so its application to structural problems is most successful when the energies of two closely related molecules are to be compared. Here are two examples.

3.6.1 The Anomeric Effect

In cyclohexanol (**119**), the hydroxyl group is mainly equatorial. However, in 2-hydroxytetrahydropyran (**120**) it is mainly axial. The best-known illustration

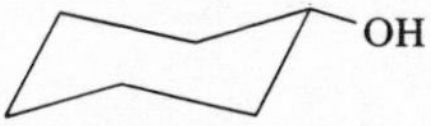

(**119**) 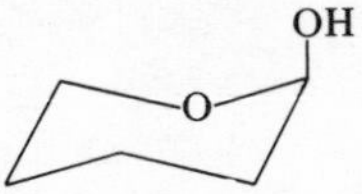   (**120**)

of this *anomeric effect,* as it is called,[102] is in the equilibrium of the methyl glucosides (**121** and **122**).[103] One explanation[104] for this is the stabilization pro-

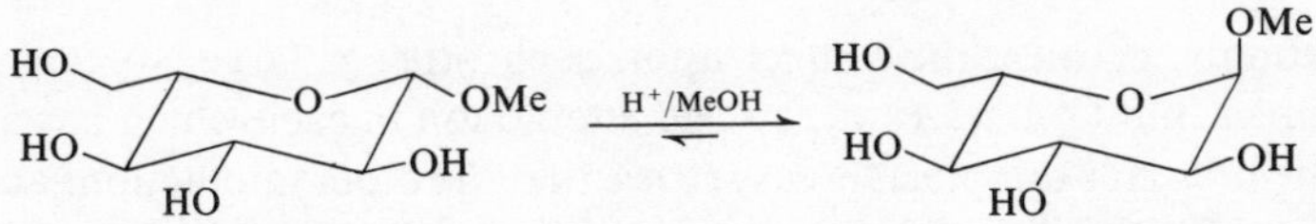

(**121**)   (**122**)

vided by overlap of a lone pair on the ring oxygen atom with the σ* orbitals of the exocyclic C—O bond. When the C—O bond is axial, its orbitals overlap well with the p-type lone pair (Fig. 3-24a); but when it is equatorial, it is the orbitals of the C—H bond which overlap well with the p-type lone pair (Fig. 3-24b). Since the σ* levels for a C—O bond will on average be lower than the

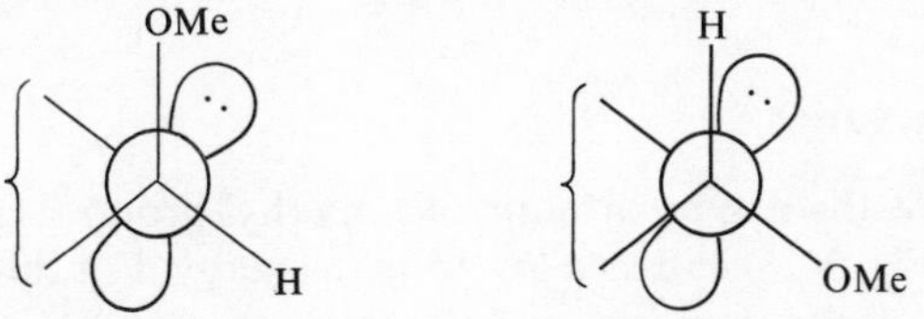

(a) *Axial* OMe   (b) *Equatorial* OMe

Fig. 3-24 Conformations of 2-methoxytetrahydropyrans

$\sigma*$ levels for a C—H bond, the overlap of the lone pair is more effective with the C—O $\sigma*$ orbitals than with the C—H $\sigma*$ orbitals (Fig. 3-25). Thus there is a greater gain in bonding from having the electronegative atom axial. Strictly speaking, to assess the relative energies of the two conformations (Fig. 3-24a and b) and of the two esters (**117** and **118**), we should have looked at all the orbitals. But just as we can, for many reactions, limit ourselves to looking at

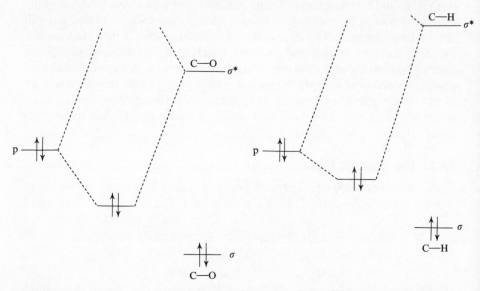

Fig. 3-25 Overlap of a lone pair with $\sigma*$ orbitals of a C—O and a C—H bond

frontier orbitals, because their effect upon each other is large, so can we, in these examples, limit ourselves to the one interaction in each which has a large effect. In the present example, it is easy to see that the other interactions actually enhance the effect. Thus, the bonding orbitals of the C—O bond, like the corresponding antibonding orbitals, are on average lower in energy than those of the C—H bond; hence the net repulsive interaction of the filled p orbital on oxygen with the C—O bonding orbitals is less than the corresponding interaction with the C—H orbitals. This again makes the conformation with an axial methoxy group the preferred one. Evidently the sum of these effects overrides the usual *steric* preference for an equatorial position.

3.6.2 Hyperconjugation[39]

The interaction of the π-type orbitals of a methyl group[29] (see p. 9) with an *empty* p orbital (Fig. 3-26) causes a lowering in energy. For this reason a methyl group stabilizes a carbonium ion, and the effect is known as *hyperconjugation*. (Because the two π-type orbitals, π_z and π_y, are of the same energy, the interactions in the two conformations shown in Fig. 3-26 are, to a first approxima-

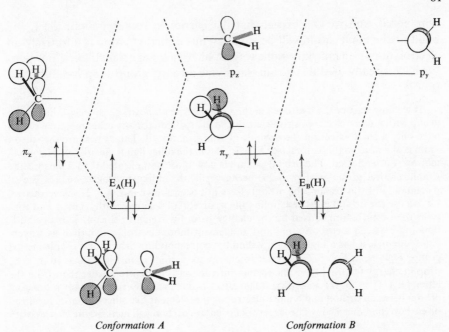

Fig. 3-26 Orbital interactions stabilizing two conformations of the ethyl cation

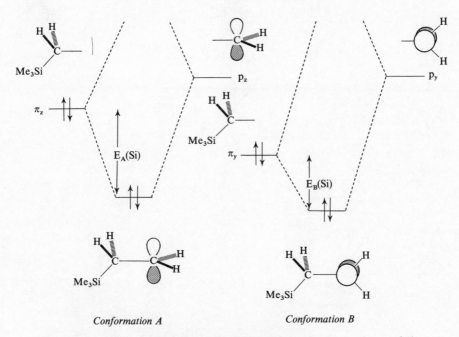

Fig. 3-27 Orbital interactions stabilizing two conformations of the
β-trimethylsilylethyl cation

tion, equal, and we can expect that the barrier to rotation about the C—C bond of the ethyl cation will be small.) Thus a methyl group is effectively an electron donor, in much the same way as, but to a lesser extent than, a lone pair. We have already used this fact in classifying an alkyl group as an $\ddot{X}$-substituent (p. 47).

If we now replace the hydrogen in the xz plane (the plane of the paper) of the methyl group with a more electropositive element, such as silicon (or any other metal), we shall have only a small effect on the orbitals of conformation B. This is because the silicon atom is at a node in this orbital (see Fig. 3-27). On the other hand, the orbitals of conformation A are affected. The orbital picture is now more complicated than it was for the simple methyl group of Fig. 3-26, because silicon, unlike hydrogen, makes use of p orbitals. For this reason, the orbitals have not been drawn in Fig. 3-27. Nevertheless, we can use the orbital most resembling the π_z orbital of a simple methyl group and consider the perturbation caused by the change from hydrogen to silicon. This orbital is bonding between a carbon atom and an element more electropositive than hydrogen. Consequently, it has a higher energy than the corresponding orbital of a simple methyl group. Thus π_z in Fig. 3-27 is closer in energy to p_z than π_z in Fig. 3-26 is to p_z. The drop in energy [$E_A(Si)$] from the former interaction is therefore greater than from the latter [$E_A(H) = E_B(H) \approx E_B(Si)$]. (The effect is enhanced by the fact that a bonding orbital between carbon and a more electropositive element like silicon is more localized on carbon than on silicon. The larger coefficient on carbon will increase the overlap with

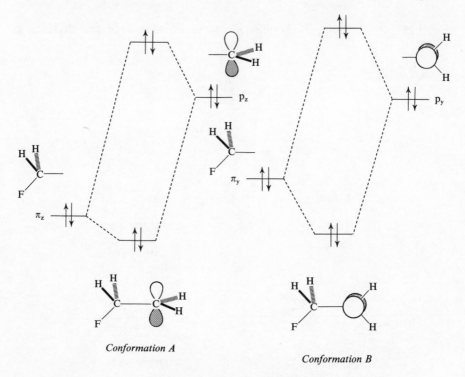

Conformation A

Conformation B

Fig. 3-28 Orbital interactions between a filled p-orbital and the filled π-orbitals of a fluoromethyl group

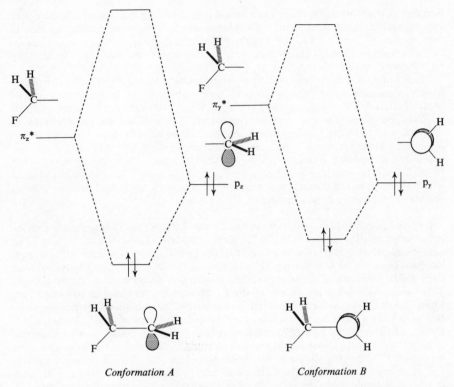

Conformation A   Conformation B

Fig. 3-29 Orbital interactions between a filled p-orbital and the
π*-orbitals of a fluoromethyl group

the p_z orbital.) There are two consequences: one is that conformation A will be favoured, the other that the cation in Fig. 3-27 will be thermodynamically more 'stable' than the ethyl cation. There is evidence to support both the conformational point[105] and the ability of a metal-carbon bond to stabilize a cation.[106]

We shall now change the silicon atom to an element, such as fluorine, which is more *electronegative* than hydrogen. Again the overlap of conformation B is unaffected. The $π_z$-type orbital of a fluoromethyl group will be lowered in energy relative to that of a methyl group, and the coefficient on carbon in the bonding molecular orbital will be small. Hence, the overlap of this orbital with the empty p orbital will be poor, and conformation A will be destabilized. There is no evidence on the conformational point,[107] but a group like the trifluoromethyl group is well-known to destabilize a cation.

On the other hand, a trifluoromethyl group does stabilize an anion. To explain this, we must compare the overlap of a filled p orbital (instead of an empty one) with the π-type orbitals of the methyl and fluoromethyl groups. In fact, the anions will be tetrahedral at the carbon atom bearing the negative charge, but for simplicity we shall ignore this, since it makes little difference to the argument. With a filled p orbital, we must look at the interaction not only with the bonding π-type orbitals (Fig. 3-28), but also with the corresponding antibonding orbitals (Fig. 3-29). The occupied π orbital of the fluoromethyl group of conformation A is lower in energy than the π orbital of conformation B, as we saw above. Thus, the interactions in conformation A are smaller than those in conformation B. Since the interactions will be antibonding in total, it is likely that those in conformation A are somewhat less antibonding than those in B. To a first approximation, the orbital picture in conformation B is the same whether fluorine is there or not. It therefore follows that the fluoromethyl group does not destabilize an anion as much as a methyl group does. Also, we can expect conformation A, in a rather negative way, to be favoured.

The interaction of the corresponding π^* orbital with the filled p orbital is, like a frontier orbital interaction, more straightforward. The π^* orbital of the fluoromethyl group in conformation A is lower in energy than the undisturbed π^*-orbital of conformation B. Its lower energy, and the large coefficient on carbon in this orbital, make its overlap with the filled p orbital more bonding than when the fluorine was not there. The net result is again twofold: the presence of fluorine makes conformation A the preferred one in the anion, and the anion is stabilized relative to the ethyl anion. The stabilization provided by the fluorine is sometimes called negative hyperconjugation. The well-known electron-withdrawing power of the trifluoromethyl group is at least partly, and perhaps wholly, explained by negative hyperconjugation.[108]

Hyperconjugation also plays a direct part in reactivity. An interesting situation arises in the series of alkyl halides Bu^tBr, Pr^iBr, EtBr and MeBr. For an S_N2 reaction of these compounds, the LUMO is the important frontier orbital, and we have already seen how localized this is on the C-halogen bond in methyl halides (Fig. 3-22). For t-butyl bromide there will be overlap with the empty (antibonding) p orbitals on the three methyl groups. This overlap will *lower* the energy of the LUMO, and would lead us to expect that t-butyl bromide would be more reactive than methyl bromide towards nucleophiles. It is well-known that this is not the case: the order of reactivity in S_N2 reactions is MeBr > EtBr > Pr^iBr > Bu^tBr. This is usually explained, of course, as a consequence of steric hindrance to attack on the more substituted carbon atoms, but it has also been explained[109] by making allowance for the change of the *coefficient* on the carbon atom. The same hyperconjugation which lowers the energy of the LUMO of the C—Br bond in t-butyl bromide more than that in methyl bromide also reduces the *coefficient* on carbon (because the new orbital is now delocalized over more atoms). This effect on the coefficients may contribute to the lower reactivity (in S_N2 reactions) of t-butyl bromide relative to methyl bromide.

The same hyperconjugation that lowers the coefficient on the carbon of the C—Br bond also lowers the coefficient on the hydrogen atom of the C—H bond. But the more C—H bonds there are involved in such hyperconjugation, the less the coefficient on each one is lowered. Hence the coefficients on hydrogen in the LUMO of the C—H bonds will fall in the order Bu^t > Pr^i > Et. This may explain the well-known order for the ease of *elimination*.

In general, a C—H bond will have a higher energy LUMO than a C-halogen bond. Soft nucleophiles, therefore, are more likely to do an S_N2 reaction than to initiate elimination. On the other hand, in an elimination reaction, the proton which is breaking free of the C—H bond will, because it is so small, have a relatively concentrated partial positive charge. This will make hard nucleophiles attack it rather than the carbon atom of the C—Br bond. This is the usual observation: the harder the base, the more elimination there is relative to substitution.

Soft nucleophiles

Hard nucleophiles
(strong bases)

Hyperconjugation has had a chequered history. The valence-bond representation of it has misled many people, including the inventors of the idea. They offered it in the 1930s[110] as an explanation for the Baker–Nathan order (Me > Et > Pr^i > Bu^t) of *apparent* electron-releasing ability of alkyl groups. Today, the Baker–Nathan order is almost always best explained by steric hindrance to solvation: t-butyl compounds are not as well solvated as methyl, and the device of placing the alkyl group *para* to the site of reaction does not, as it was supposed to, remove it from solvation sites. For this reason, hyperconjugation was quite widely discredited in the 1950s.[111]

Today, it enjoys a more soundly based popularity. We can see that, because carbon has about the same electronegativity as hydrogen, there should not be much change in hyperconjugating ability as we replace C—H bonds by C—C bonds. If anything, carbon is more electropositive than hydrogen, and we should expect, by analogy with the argument used above for the trimethylsilyl-ethyl cation, that the true hyperconjugation order should be Bu^t > Pr^i > Et > Me. Formulated in molecular orbital terms, and used to explain the electron-donating and electron-withdrawing effects of alkyl and substituted alkyl groups, hyperconjugation is widely accepted.

CHAPTER 4

Thermal Pericyclic Reactions

Note. Several people have contributed to this field, but in the account that follows, their names have not always been placed in the section corresponding to the work they did. The version of each topic presented here is not always that of any one of them—nor is it proper to link their names with some of the over-simple arguments used. In roughly chronological order, the principal contributors are R. B. Woodward and R. Hoffmann,[1] and K. Fukui,[3] for the frontier orbital theory of the Woodward-Hoffmann rules, and W. C. Herndon,[112] R. Sustmann,[113] N. T. Anh,[114, 115] K. N. Houk[40, 116, 117] and N. D. Epiotis,[118] for the various aspects of selectivity in cycloaddition reactions.

4.1   The Woodward-Hoffmann Rules

The Woodward-Hoffmann rules[1] for pericyclic reactions can be explained by frontier orbital theory, as Fukui[3] has demonstrated. If you already know anything about frontier orbital theory, it is quite likely that you know it as one of the ways in which the Woodward-Hoffmann rules are accounted for. If this is the case, you should leave out the next 23 pages, and turn to page 109.

One of the questions posed in Chapter 1 sets the scene: why does maleic anhydride (**124**) react easily with butadiene (**123**), but not at all easily with ethylene (**125**)? In the former reaction, two new σ-bonds are made, and it is believed

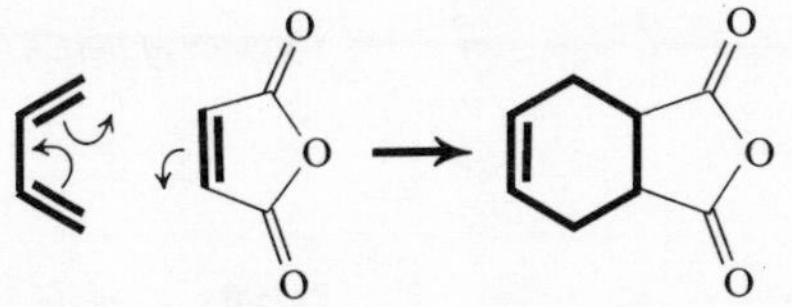

(**123**)   (**124**)

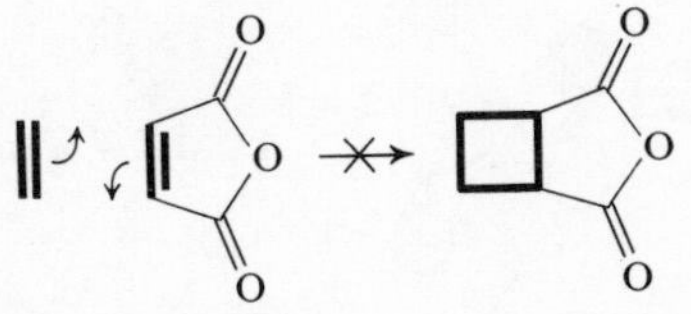

(**125**)   (**124**)

that they are made *at the same time*; furthermore, the electrons mobilized in the reaction complete a circuit. Concerted and cyclic reactions like this are called *pericyclic*, and this particular kind of pericyclic reaction is called a *cycloaddition*.

Most pericyclic reactions, though of course not all, are little influenced by Coulombic forces: for example, it is well known that the polarity of the solvent has little effect on the rate of Diels-Alder reactions. We can therefore expect that a major factor influencing reactivity will be the size of the frontier orbital interaction represented by the third term of equation 2-7, p. 27. This is why this chapter is much the largest in this book: the most dramatic successes of frontier orbital theory have been the explanations it has given to an amazingly large number of observations in pericyclic chemistry.

4.1.1 Cycloadditions

In order to answer the question first posed in Chapter 1 and repeated above, we begin by ignoring the substituents and counting only those parts of the conjugated system directly involved in the reaction. (We shall return to the crucial role of the substituents later in the chapter.) Thus the Diels-Alder reaction is simplified to that of butadiene reacting with ethylene; the former component has four π-electrons and the latter two, and these are the only electrons directly involved, as we can see from the curly arrows. Such a reaction is called a $[4 + 2]$ cycloaddition. We now examine the signs of the coefficients of the frontier orbitals on the atoms which are to become bonded (Fig. 4-1). We are not yet concerned with the magnitude of the coefficients of the frontier orbitals, and therefore in this section all orbitals are drawn the same size, so as not to

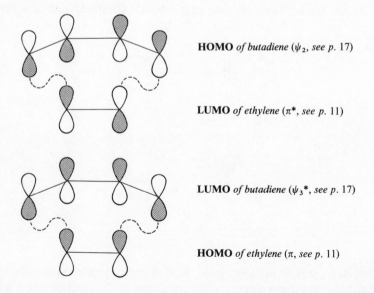

HOMO *of butadiene* (ψ_2, *see p. 17*)

LUMO *of ethylene* (π^*, *see p. 11*)

LUMO *of butadiene* (ψ_3^*, *see p. 17*)

HOMO *of ethylene* (π, *see p. 11*)

Fig. 4-1 Overlap of the frontier orbitals of a Diels-Alder reaction. The dashed lines identify the bonding overlap which can develop as the reaction proceeds

detract from our main concern, the *sign* of the coefficients. If the interacting lobes of the frontier orbitals are of like sign, as they are in this reaction, there is no impediment to the continuous development of overlap. Thus we see in Fig. 4-1 that in the Diels-Alder reaction the developing overlap is *bonding* at *both* sites where new bonds are being formed. For this purpose, it does not matter which pair of frontier orbitals we take, so long as we take the HOMO of one component and the LUMO of the other.

It is quite different when we examine the cycloaddition of ethylene and maleic anhydride. Once again we ignore the substituents and look only at the simplest components. The reaction, if it were to take place, would be a [2 + 2] cycloaddition, but this time the frontier orbitals (Fig. 4-2) have an *antibonding*

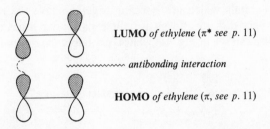

Fig. 4-2 Overlap of the frontier orbitals for a [2 + 2] cycloaddition. The dashed line represents the only bonding overlap which could easily develop

interaction where one of the new bonds is to be made. Thus a *concerted* reaction of this type is not favoured. (A *stepwise* reaction, in which only the bonding interaction, on the left of Fig. 4-2, develops, meets no such barrier, and many of the known [2 + 2] cycloadditions may be of this type.)

However, we do not need to have precisely the reaction shown in Fig. 4-2. Instead we could examine the possibility of overlap developing in the manner shown in Fig. 4-3. When two new bonds are formed on *opposite* sides of a

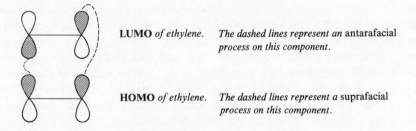

Fig. 4-3 Overlap of the frontier orbitals for a [π2a + π2s] cycloaddition. The dashed lines identify the bonding overlap which could develop as the reaction proceeded

π-bond (or conjugated system), as in the upper component of Fig. 4-3, we call the process taking place on that component an *antarafacial* one. When new bonds are formed on the same side of a π-bond (or conjugated system), as in the lower component of Fig. 4-3 and in both components of Fig. 4-1, we call the process a *suprafacial* one. Thus the reaction shown in Fig. 4-3 is classified as a [π2a + π2s] cycloaddition, whereas the reaction shown in Fig. 4-2, would, if it ever took place, be classified as a [π2s + π2s] cycloaddition. Similarly, the Diels-Alder reaction of Fig. 4-1 is classified as a [π4s + π2s] cycloaddition.

We can see from Fig. 4-3 that the [π2a + π2s] cycloaddition is one in which smooth bonding overlap could, in principle, develop between the frontier orbitals as the reaction proceeded; nevertheless, the reaction is still not observed (except, just possibly, in some very special cases[119]). The reason is that, although the orbitals which are bonding together have the right signs, it is geometrically very difficult indeed for these particular lobes to overlap with one another. We can see in Fig. 4-4 what is probably the optimum arrangement

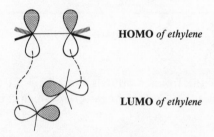

HOMO *of ethylene*

LUMO *of ethylene*

Fig. 4-4   Preferred geometry for an early stage in a [π2a + π2s] cycloaddition

for achieving this kind of overlap, and we can also see there how the substituents sticking up from the lower component would prevent the lobes from getting near to each other. Thus the geometrically easy process (Fig. 4-2) has no gain in bonding from the interaction of the frontier orbitals, and the process which does have a good frontier orbital interaction (Fig. 4-3) cannot take advantage of it. A nomenclature has grown up in this field, by which we call a reaction like the one shown in Fig. 4-2 a *symmetry-forbidden* one, and reactions like those shown in Figs. 4-1 and 4-3 *symmetry-allowed* ones; rather often, these words are shortened simply to forbidden and allowed. However, we have to be careful here: a reaction which is 'forbidden' may still take place if nothing easier is available. What is meant by 'forbidden' is that the interaction of the orbitals presents an energy barrier which the 'allowed' reactions do not have. It has turned out that, in most situations, the barrier is quite substantial—examples of allowed reactions are abundant, but forbidden reactions are few and far between. Because of this, the terms 'allowed' and 'forbidden' have come to have more force than they ought. If we remember that we should read 'symmetry-allowed' whenever we come across the word 'allowed' in connection with pericyclic reactions, we shall not go far wrong.

So far we have dealt only with the best known of all pericyclic reactions, the

Diels-Alder reaction, and we have contrasted it with the rarely observed [2 + 2] cycloaddition. There are many more pericyclic reactions than these. Thus thermal [π4s + π4s] cycloadditions are unknown, but several [π6s + π4s] and [π8s + π2s] are known. Here is an example of each:[120, 121]

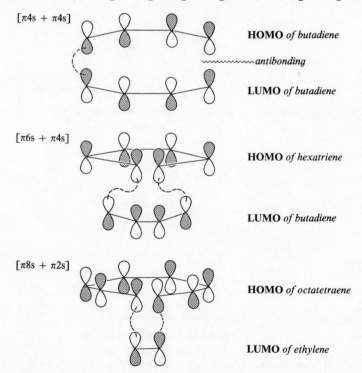

In Fig. 4-5, we can see how this pattern of reactivity is explained, using frontier orbitals. The same answer would have been obtained by taking the other pairs of frontier orbitals (in the [6 + 4] and [8 + 2] cases; in the [4 + 4] case there

Fig. 4-5 Frontier orbitals for [π4s + π4s], [π6s + π4s] and [π8s + π2s] cycloadditions

(a) $[\pi 2s + \pi 2s]$

Frontier orbitals:

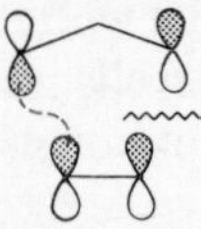

LUMO *of an allyl cation*

~~~~~*antibonding*

HOMO *of an olefin*

There are no known examples of a pericyclic reaction of an allyl cation with an olefin.

(b) $[\pi 4s + \pi 2s]$

Frontier orbitals:

HOMO *of an allyl anion*

LUMO *of an olefin*

An example:[122]

(c) $[\pi 4s + \pi 2s]$

Frontier orbitals:

LUMO *of an allyl cation*

HOMO *of a diene*

An example:[123]

(d) $[\pi 4s + \pi 2s]$

Frontier orbitals:

LUMO *of a pentadienyl cation*

HOMO *of an olefin*

An example:[124]

(e) $[\pi 4s + \pi 4s]$

Frontier orbitals:

LUMO *of a pentadienyl cation*

~~~~~*antibonding*

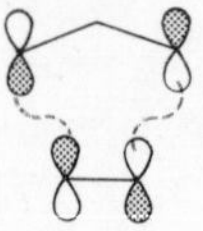

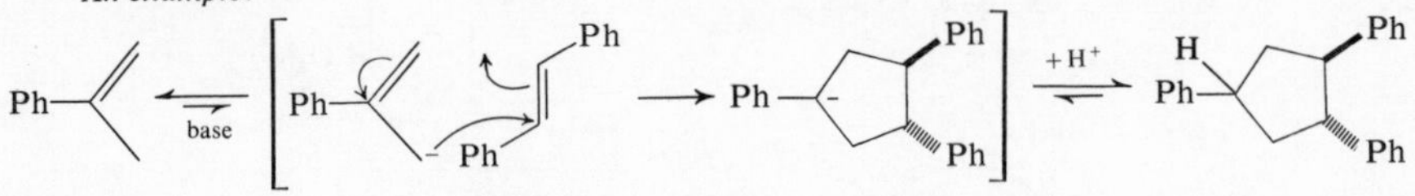

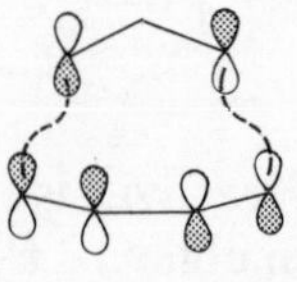

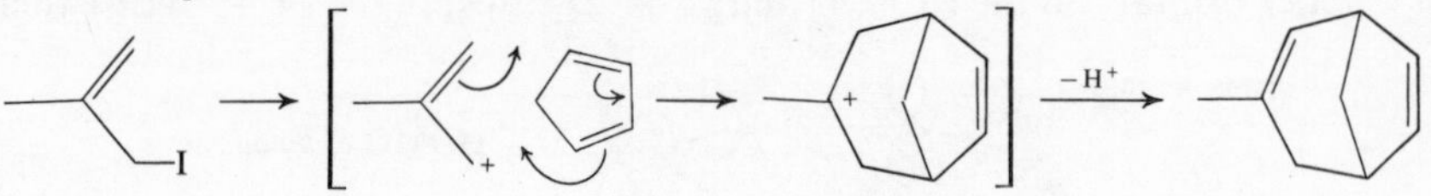

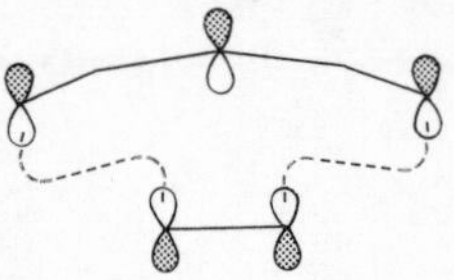

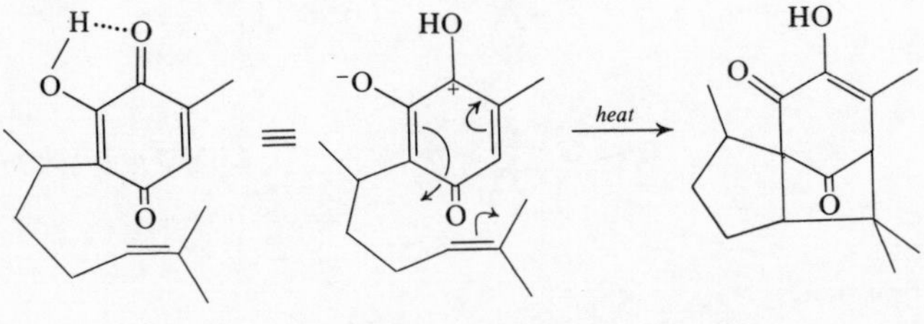

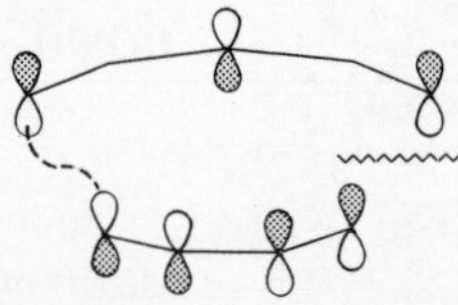

HOMO *of a diene*

There are no known examples of a pericyclic reaction of a pentadienyl cation with a diene.

Fig. 4-6 Ionic cycloadditions

is no other pair to take), as you can easily check for yourself. This is so generally true, that only one pair of frontier orbitals will be drawn from now on.

We are not restricted to the reactions of neutral molecules. Allyl anions can react with olefins, allyl cations with dienes, and pentadienyl cations with olefins, as we can see from the examples in Fig. 4-6, in which we can also see why the reactions of allyl cations with olefins, of allyl anions with dienes, and of pentadienyl cations with dienes are not observed.

The reaction of ozone (126) with olefins, and the reaction of diazomethane (127) with methyl acrylate (128), are examples of a very large class of reactions,

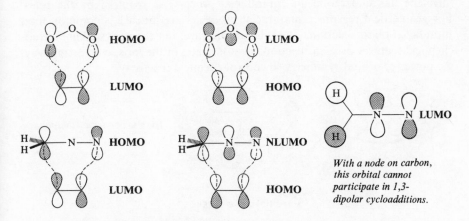

known as 1,3-dipolar cycloadditions.[125] The ozone and the diazomethane are the 1,3-dipoles, and their frontier orbitals resemble those of the allyl anion. Again, we can quickly see from Fig. 4-7 that these reactions are allowed. The only complication here is that the LUMOs of linear dipoles like diazomethane have a node through one of the terminal atoms. This orbital is shown on the right of Fig. 4-7; we have to ignore it and use the next lowest unoccupied orbital (NLUMO) instead.

With a node on carbon, this orbital cannot participate in 1,3-dipolar cycloadditions.

Fig. 4-7 Frontier orbitals for 1,3-dipolar cycloadditions

All the reactions we have looked at so far involve only the geometrically easy, and hence commonly observed, suprafacial processes. If we restrict ourselves to all-suprafacial processes, we may note that the allowed reactions involve an aromatic (4n + 2) number of electrons, and the forbidden reactions an anti-aromatic (4n) number of electrons. (The number of electrons involved is easily counted; it is twice the number of curly arrows.) This is the simplest version of the Woodward-Hoffmann rules; it was first pointed out by Evans[126] in 1939 and recalled many years later by Dewar.[127]

Moving on to reactions in which antarafacial processes occur, we find that if we incorporate one antarafacial process, and keep the other suprafacial, then those reactions in which a total of 4n electrons are involved become allowed, and those in which a total of (4n + 2) electrons are involved become forbidden. To take just one of the relatively small number of reactions in this class, we can consider the remarkable reaction of heptafulvalene (**129**) with tetracyano-ethylene (**130**).[128] This reaction is a [14 + 2] cycloaddition; in other words,

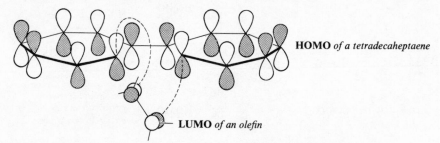

it involves a total of sixteen electrons, and sixteen is a 4n number. To be allowed, a [14 + 2] cycloaddition (Fig. 4-8) must have one of the components undergo reaction in an antarafacial manner. In the product (**131**), one of the new bonds (shown on the left and numbered 1 on the drawing) has been made to the upper surface of the conjugated system of the heptafulvalene, and the other new bond (on the right, numbered 2) to the lower surface. Thus we see that the hepta-fulvalene has undergone an antarafacial process, as required by the rules. For geometrical reasons, antarafacial processes are much less common than suprafacial ones; reactions like this one are rare, but the very striking stereo-chemical features of such reactions demonstrate, in the most convincing way, the power of orbital symmetry to control chemical reactivity.

HOMO *of a tetradecaheptaene*

LUMO *of an olefin*

Fig. 4-8   Frontier orbitals for a [π14a + π2s] cycloaddition

Formulated in the most general way, the Woodward-Hoffmann rule for thermal pericyclic reactions states:

A ground-state pericyclic change is symmetry-allowed when the total number of $(4q + 2)_s$ and $(4r)_a$ components is odd.

All the cycloadditions we have examined obey this rule. Thus, to take a single example, the last reaction is a $[\pi 14a + \pi 2s]$ process. It has one $(4q + 2)_s$ process (the only suprafacial process is a 2-electron one, and 2 is a $4q + 2$ number) and it has no $(4r)_a$ processes (the only antarafacial process involves 14 electrons, and 14 is not a $4r$ number). The total of $(4q + 2)_s$ and $(4r)_a$ processes is thus one, and one is an odd number.

This rule is almost always obeyed, and we shall find that the signs of the coefficients of the frontier orbitals regularly account for it. Frontier orbital theory is not the only way to explain the patterns of reactivity covered by the Woodward-Hoffmann rules, but it is one of the easiest.

4.1.2   Other Pericyclic Reactions

So far, cycloadditions have been our only examples of pericyclic reactions. There are several other classes of pericyclic reactions, of which the most notable are *cheletropic* reactions, *sigmatropic* rearrangements and *electrocyclic* reactions. In essence, frontier orbital theory treats each of them as a cycloaddition reaction.

4.1.2.1   Cheletropic Reactions. Cheletropic reactions are really a sub-class of cycloadditions. The only difference is that, on one of the components, both new bonds are being made to the same atom. The best-known reaction of this type is the general reaction of a carbene (**133**) with an olefin. At first sight, this

(**132**)   (**133**)

reaction appears to be disobedient to the Woodward-Hoffmann rules. If the carbene approaches the olefin in a straight line, the HOMO/LUMO interactions (Fig. 4-9a) are clearly antibonding, whichever way round we take them. Nevertheless, the reaction is common. The explanation for this anomaly is that the carbene probably does not approach in a linear manner, but that overlap begins to develop at a stage when the carbene is sideways-on to the olefin. If the carbene approaches this way, it is clear (Fig. 4-9b) that the initial interactions are bonding, and all that is required is that the electrons reorganize themselves into the new bonds as the carbene moves into place. As it does so, the plane containing the bonds to the two substituents moves from being parallel to the C—C bond of the olefin to being at right angles to it.

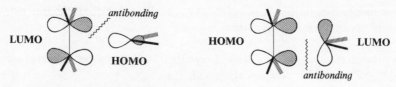

(a) *The linear approach of a carbene to an olefin*

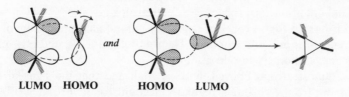

(b) *The non-linear approach of a carbene to an olefin*

Fig. 4-9   Two possible approaches of a carbene to an olefin

Another cheletropic reaction is the addition of sulphur dioxide under pressure to the hexa-2,4-dienes (**134** and **136**) to give the dihydrothiophen dioxides (**135** and **137** respectively).[129] These reactions are easily reversible simply by heating

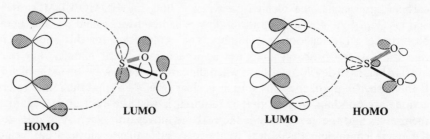

the dioxides (**135** and **137**). The frontier orbital interactions for the forward reaction (Fig. 4-10) show that the simple linear approach is allowed, and we do not need any special explanation. Because of the presence of the methyl groups, we can actually see that the process taking place on the dienes is supra-facial, as it ought to be; we cannot, of course, prove that the sulphur dioxide adopted the linear approach. We should note here that, since the forward

Fig. 4-10   Frontier orbitals for the reaction of sulphur dioxide with a diene

reaction and the back reaction take the same path, the explanation using frontier orbitals as in Fig. 4-10 could have been used to show that the back reaction is allowed. This is easier than trying to deal with the back reaction itself.

We can apply this technique to another cheletropic reaction—the extrusion of sulphur dioxide from the sulphones (**138** and **140**).[130] In this case, the reaction

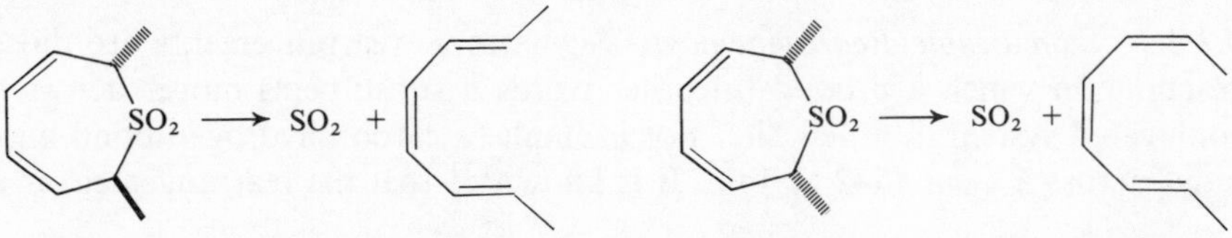

(138)   **(139)**   **(140)**   **(141)**

of sulphur dioxide with the trienes (**139** and **141**) does not take the reverse path, but finds something else to do instead. Nevertheless, we can use the reverse pathway to look at the frontier orbitals, and our conclusions will apply to the extrusion reaction. Two pathways for the reverse reaction are favourable, as we can see from the frontier orbitals in Fig. 4-11. One (a) is suprafacial on the

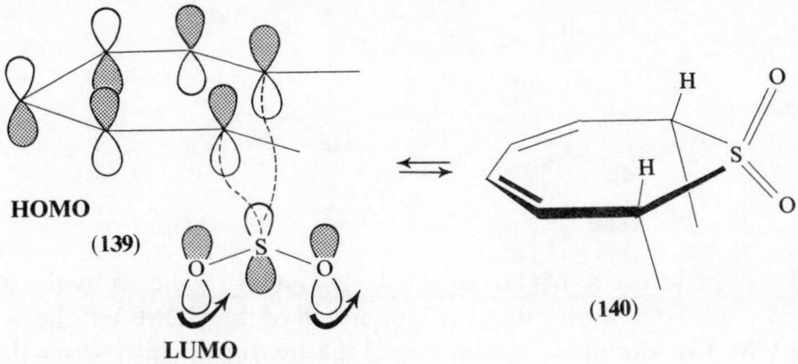

(a) *Non-linear and suprafacial on the triene*

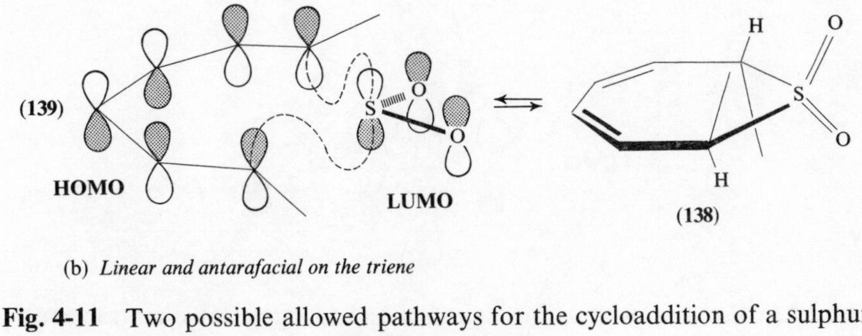

(b) *Linear and antarafacial on the triene*

Fig. 4-11 Two possible allowed pathways for the cycloaddition of a sulphur dioxide to a triene

98

triene and would require non-linear approach of the sulphur dioxide; the other (b) is antarafacial on the triene and requires a linear approach of the sulphur dioxide. The stereochemistry of the products (**139** and **141**) show that the latter is favoured, though once again we cannot prove that the sulphur dioxide is actually adopting the linear approach.

4.1.2.2 Sigmatropic Rearrangements. Sigmatropic rearrangements are those reactions in which a σ-bond (in other words a substituent) moves across a conjugated system to a new site. For example, a carbon-hydrogen bond may move across a diene (**142** ⇌ **143**). It is known[131] that the rearrangement is a

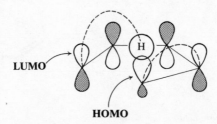

$$(142) \qquad\qquad (143)$$

suprafacial one: the σ-bond is made and broken on the same side of the conjugated system (**144** ⇌ **145**). This reaction can be treated as a cycloaddition of

the C—H σ-bond to the π-orbitals of the diene system, as shown by the dotted lines in Fig. 4-12. Here we see that the interaction of the HOMO of the σ-bond and the LUMO of the diene is bonding if the hydrogen shifts from the top surface to the top surface, as it is known to do. This kind of shift is called suprafacial, by analogy with the suprafacial processes we have seen earlier, and the reaction is called a [1,5] sigmatropic rearrangement.

Fig. 4-12 Frontier orbitals for suprafacial [1,5] sigmatropic rearrangement of hydrogen

One of the best-known examples of this type of reaction is the rearrangement[132] of monosubstituted cyclopentadienes (**146** ⇌ **147** ⇌ **148**); this rearrangement

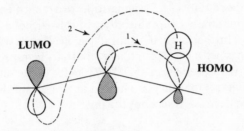

is so fast that it takes place at an appreciable rate at room temperature, presumably because the hydrogen atom is moving to a carbon atom held permanently close to it.

If we replace the diene component by a simple π-bond, we can see in Fig. 4-13 that the frontier orbitals will favour a rearrangement in which the hydrogen

Fig. 4-13 Frontier orbitals for antarafacial [1,3] sigmatropic rearrangement of hydrogen. The reaction is not observed

atom leaves the upper surface and moves to the lower surface. Such a shift would be called antarafacial, and the reaction a [1,3] sigmatropic rearrangement. Such rearrangements are virtually unknown[133] in thermal reactions; although they are allowed, there is obviously little hope of maintaining the overlap marked 1 on Fig. 4-13 at the same time as developing the overlap marked 2. To increase one would diminish the other. As with [π2a + π2s] cycloadditions, the geometrical requirement prevents the allowed reaction, and the orbital symmetry frustrates the geometrically easy [1,3] suprafacial shift.

If we have a longer conjugated system, as in the triene (**149**), then a [1,7] shift takes place on heating.[134] In Fig. 4-14 we can see that this is favoured only if the hydrogen leaves the top surface and arrives on the lower surface, in other words, in an antarafacial manner. Unlike the case of the [1,3] shift, there is a helical shape, shown in the figure, in which continuous overlap can be maintained. The known examples of this kind of reaction take place in open-chain trienes, in which helical geometry, like that shown, is possible. There is no proof that the shift is in fact antarafacial, but it is known that in cyclic trienes, like the

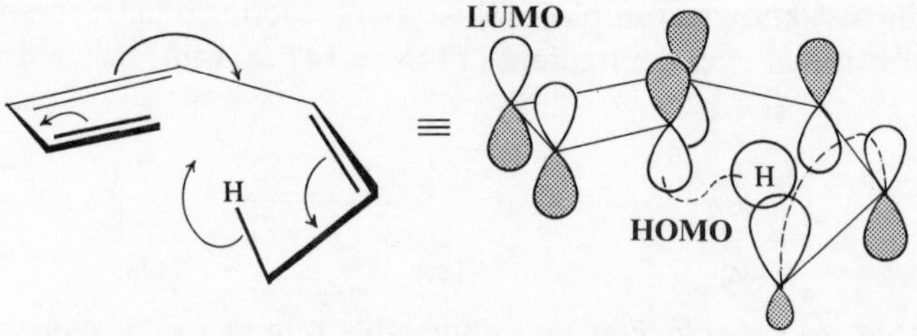

Fig. 4-14 Frontier orbitals for the [1,7] antarafacial shift of hydrogen in a triene

cycloheptatriene (**150**), a [1,7] sigmatropic rearrangement does not take place (at temperatures above 150° a [1,5] sigmatropic rearrangement of the hydrogen atom occurs,[135] presumably suprafacially). In this triene, a helical transition state is impossible; furthermore, if a suprafacial [1,7] shift were to be allowed, it ought, by analogy with cyclopentadiene (**146**), to be especially easy in this compound. Thus the observation that [1,7] shifts occur only in open-chain trienes strongly suggests that they are indeed antarafacial. The frontier orbital interactions, as we have seen, account for this.

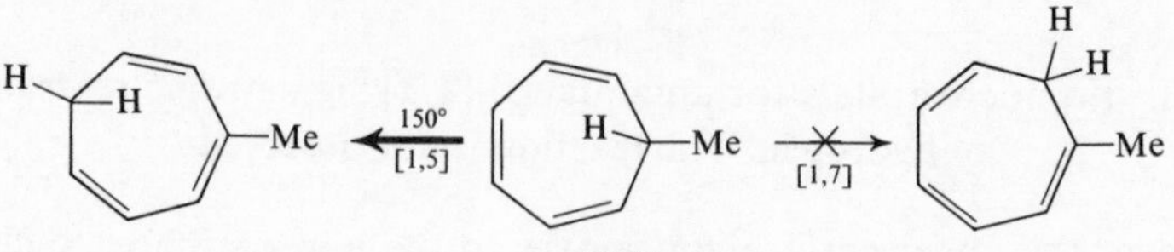

(**150**)

Hydrogen is not the only group which can migrate, and there are therefore many other kinds of sigmatropic rearrangement. As usual, those involving a total of (4n + 2) electrons are allowed in the all-suprafacial mode, and these are the common reactions. Here are some examples.[136–142]

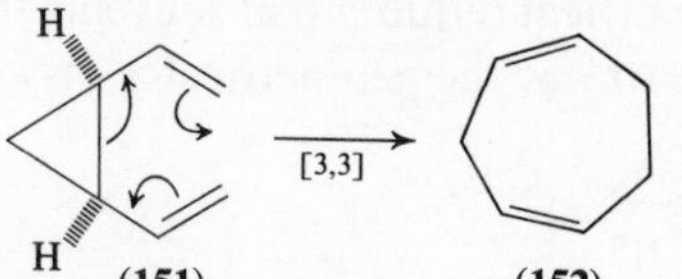

(**151**)   (**152**)

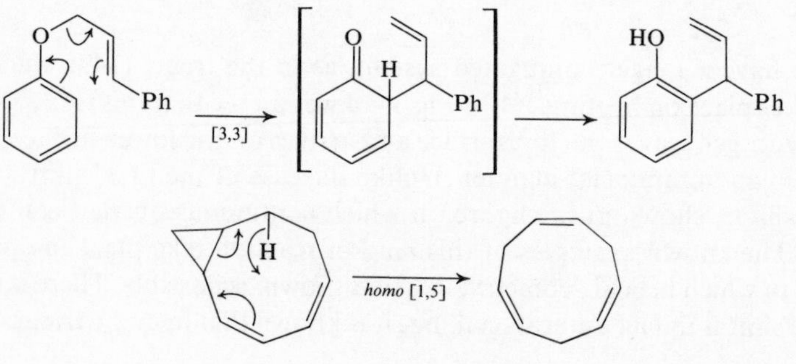

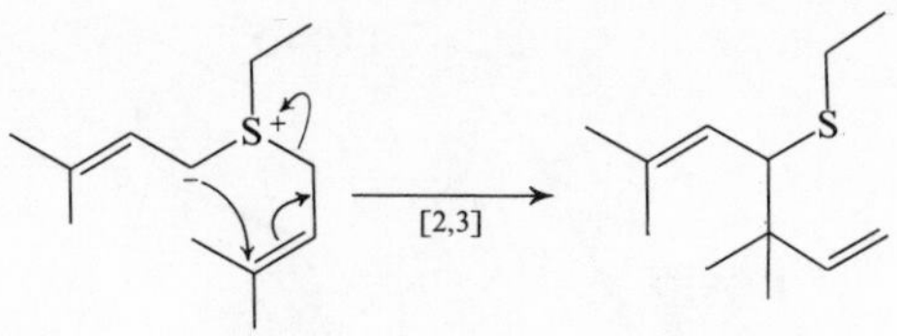

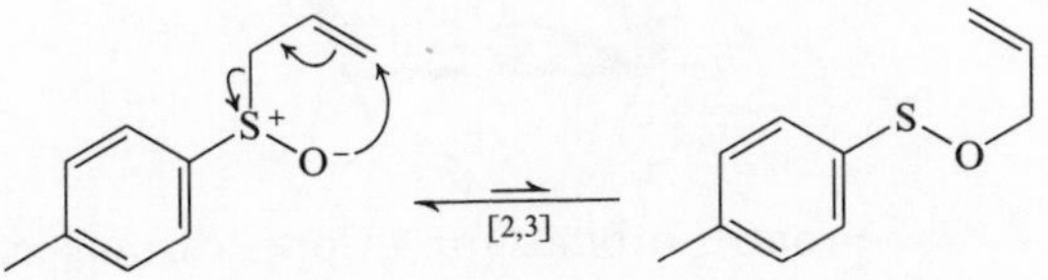

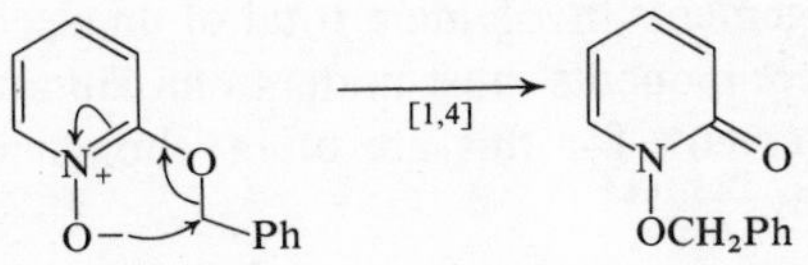

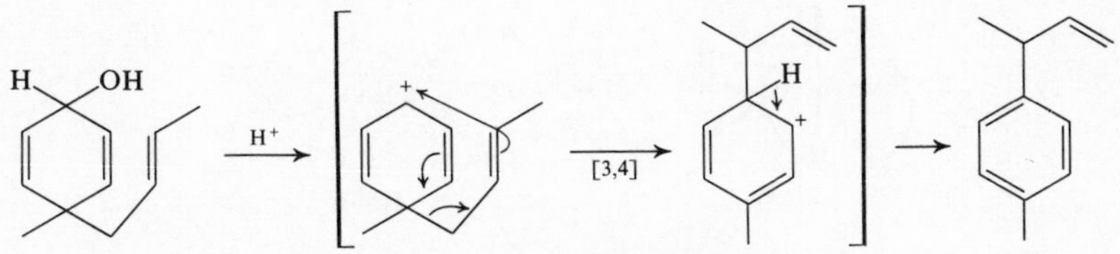

In all these reactions, frontier orbital considerations explain the observations. Let us take one of them, the [3,3] sigmatropic rearrangement (**151** → **152**) known as the Cope rearrangement. As usual we simplify the reaction to the minimum (**153** → **154**), and the problem is to identify the 'components' with

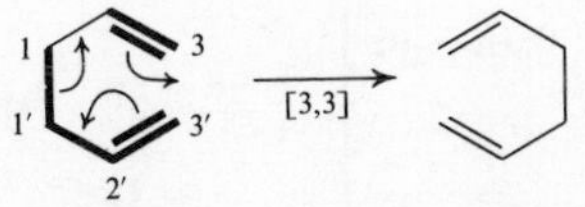

which to set up the frontier orbitals. This example serves to illustrate the wide latitude we can allow ourselves. One way is to take the π-bond between C-2 and C-3 as the 2-electron component, and the σ-bond and the other π-bond as a 'conjugated' 4-electron component. Looked at in this way, the reaction is a [4 + 2] cycloaddition. Although the 4-electron component is made up of one σ-bond and one π-bond, the nodal properties of its orbitals, which is all we are concerned with at the moment, will closely resemble those of butadiene. Thus the HOMO of this conjugated system will, like ψ_2 of butadiene, have one node. This is the orbital shown on Fig. 4-15. We can now see that the new bonds are formed where bonding interactions exist between this orbital and the complementary frontier orbital, the LUMO of the 2-electron component.

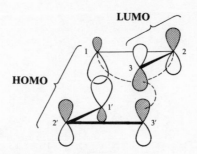

Frontier orbitals for a Cope rearrangement

Sigmatropic rearrangements involving a total of 4n electrons are much less common—one of the components must undergo an antarafacial process, and the geometrical requirements for this are often, though not always, severe. Here are some examples.[143-145]

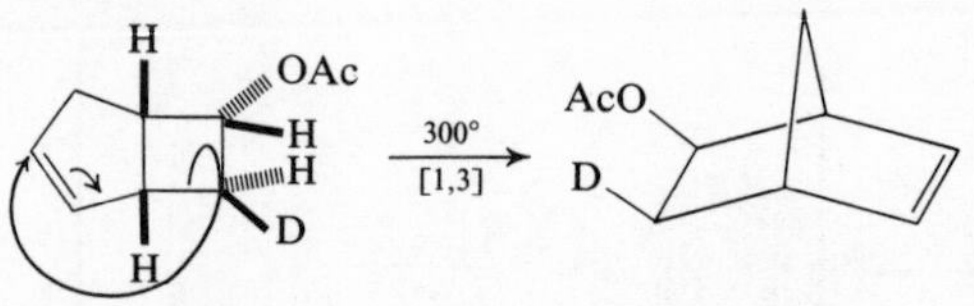

(155) $\xrightarrow[{[1,3]}]{300^\circ}$ (156)

$\xrightarrow[{[1,3]}]{120^\circ}$

$\xrightarrow{[1,4]}$

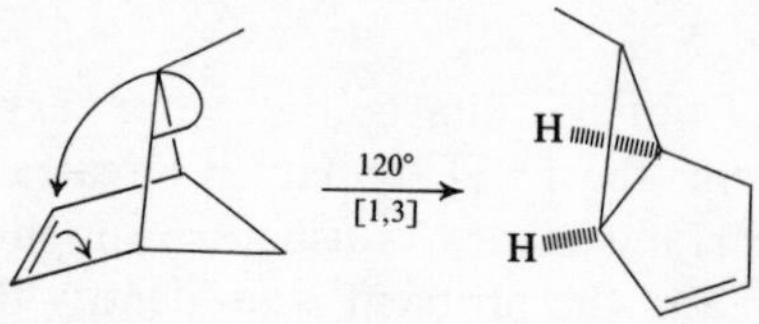

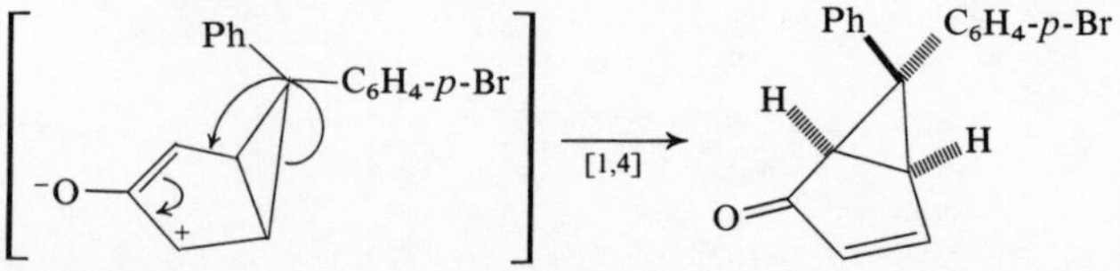

In all these reactions too, frontier orbitals explain the stereochemistry observed. The remarkable stereospecific [1,3] rearrangement (**155 → 156**), discovered by Berson,[143] is the example we shall examine. As in the case of hydrogen migration (Fig. 4-13), the alkyl group is *formally* allowed to migrate antarafacially across the π-bond (Fig. 4-16a), but, as usual, this is geometrically impossible. The alternative allowed pathway is for the carbon atom (marked * in Fig. 4-16b) to undergo *inversion of configuration* as it migrates suprafacially across the π-bond. When we were dealing with [1, 3] shifts of hydrogen (p. 99), this possibility did not arise: the p orbitals of hydrogen are much too high in energy to be involved, and p orbitals are needed for 'inversion of configuration' to take place at a migrating centre. As we can see from the substituents on the

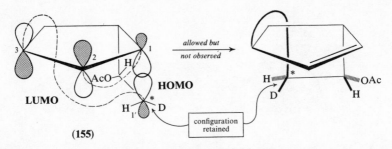

Fig. 4-16a Frontier orbitals for a [1,3] antarafacial shift with retention of configuration

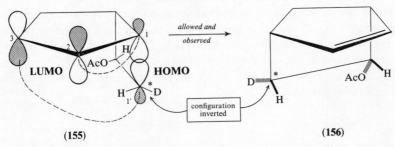

Fig. 4-16b Frontier orbitals for a [1,3] suprafacial shift with inversion of configuration

migrating carbon, inversion of configuration has taken place, and the reaction is evidently concerted, in spite of the improbable-looking overlap which must develop between the lobes on C-1' and C-3.

4.1.2.3 Electrocyclic Reactions. Electrocyclic reactions are those pericyclic reactions in which a ring is formed (or opened). Thus, cyclobutene (**157**), on heating, gives butadiene (**158**), and hexatriene (**159**) gives cyclohexadiene (**160**).

Once again, we can most easily treat these reactions as though they were cyclo-additions. It is generally most convenient to look at them in the direction of the ring-opening reaction, no matter in which direction the reaction actually goes (see p. 97).

Thus the opening of cyclobutene to butadiene is, in a sense, the cycloaddition of the σ-bond to the π-bond (Fig. 4-17). The HOMO and the LUMO are then smoothly connected, with a suprafacial component on the π-bond and an antarafacial component on the σ-bond. It is therefore a $[\pi 2s + \sigma 2a]$ reaction

104

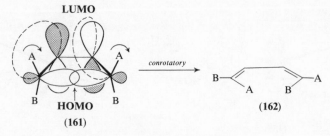

Fig. 4-17 The electrocyclic ring-opening of cyclobutene seen as a cycloaddition of a σ-bond to a π-bond

and, as such, fits the Woodward-Hoffmann rule (p. 95). The developing overlap shown by the dashed lines implies that the substituents A on the cyclobutene (**161**) should move in the same direction as each other, as the reaction proceeds, and that the product should be the diene (**162**). This kind of movement is called *conrotatory*, and it is exactly what is observed in the opening of substituted cyclobutenes: the dimethyl cyclobutenes (**163** and **165**) give the butadienes (**164** and **166** respectively).[146]

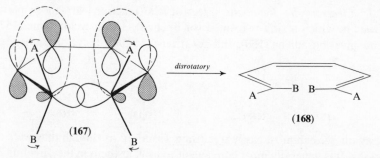

If we treat the cyclohexadiene ⇌ hexatriene reaction (**159** ⇌ **160**) in the same way (Fig. 4-18), we discover that the stereochemical outcome is different. In order to fit the rule on p. 95, both components must undergo suprafacial

Fig. 4-18 The electrocyclic ring-opening of cyclohexa-1,3-diene

processes, and the frontier orbitals show that such a reaction is favourable. As the overlap shown by the dashed lines develops, the substituents A on the cyclohexadiene (**167**) should move in opposite directions, away from each other, to give the triene (**168**); this kind of movement is called *disrotatory*. Except that the reaction takes place in the other direction, this is exactly what is observed: the trienes (**169** and **171**) give the cyclohexadienes (**170** and **172** respectively).[147]

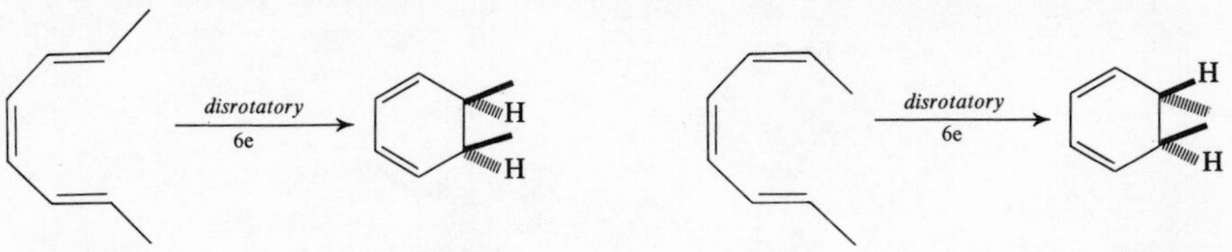

| (169) | (170) | (171) | (172) |

Here are several more electrocyclic reactions, each of which can be similarly explained:[148-152]

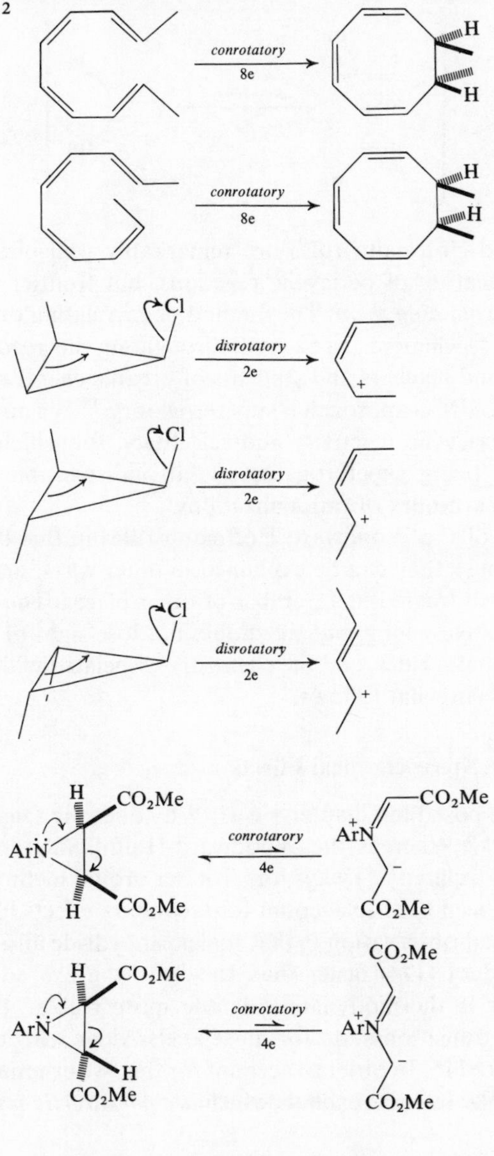

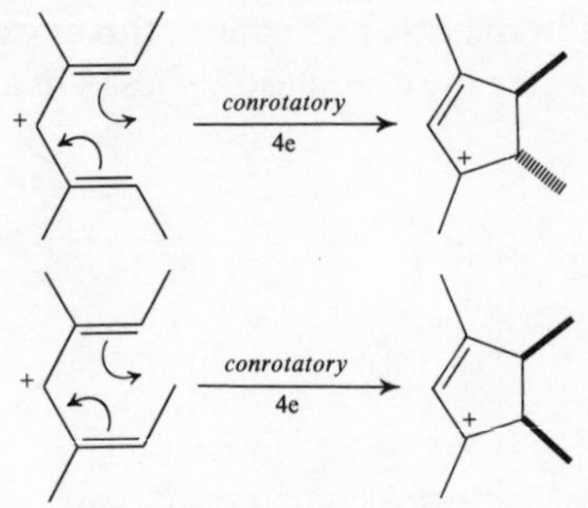

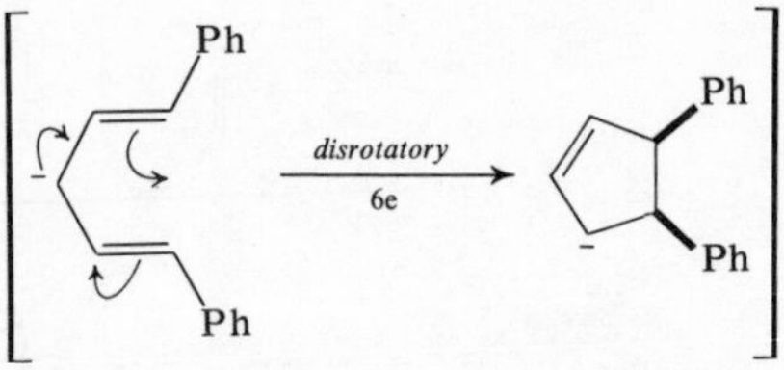

The Woodward-Hoffmann rules are remarkably well obeyed. This is the most important feature of pericyclic reactions, but frontier orbital theory *is only one way of explaining them*. The method of correlating orbitals in relation to those symmetry elements preserved throughout the reaction[153] is much more satisfying, and teachers and students of organic chemistry are apt to feel that the HOMO/LUMO approach is supererogatory.[11] We now move on to the finer points of pericyclic reactivity and selectivity, for which frontier orbital theory, far from being supererogatory, is an indispensible addition to the organic chemist's armoury of rationalizations.

The treatment of the Woodward-Hoffmann rules in this book is relatively short, partly because they can be explained in other ways, and partly because they have been well treated in a number of other places. The treatment of the finer points is relatively long, but we should not lose sight of the fact that the Woodward-Hoffmann rules, and the chemistry associated with them, are much more important than what follows.

4.1.3 Secondary, Stereochemical Effects

Another question posed in Chapter 1 was: Why does the Diels-Alder reaction give *endo* adducts? Whereas the Woodward-Hoffmann rules have been explained in several (related[154]) ways, the frontier orbital method is virtually the only one to have been used to account for secondary effects like this.[155]

The experimental observation is that maleic anhydride and cyclopentadiene give the *endo* adduct (174) faster than they give the *exo* adduct (176), even though the latter is thermodynamically the more stable. This is a general observation; the transition state for most Diels-Alder reactions must be like 173 rather than like 175. In order to account for this, we examine the interaction of those parts of the frontier orbitals which *are not directly involved in forming*

(173)   **(174)**

(175)   **(176)**

new bonds. If we change the ethylene of Fig. 4-1 to maleic anhydride and the butadiene of Fig. 4-1 to cyclopentadiene, we get Fig. 4-19. The dashed lines are the primary interactions we saw before—they represent the sites of the new bonds. We can now see that the dotted lines identify further bonding interactions. Although this overlap does not lead directly to new bonds, it does lower the energy of the transition state relative to that of the *exo* transition state (**175**), where these interactions are absent; hence the *endo* adduct is the one obtained under kinetically controlled conditions.

An interesting stereochemical feature of the Cope rearrangement (**177** → **178**) can be explained similarly. It is known[156] that a chair-like transition state (**177**) is preferred to a boat-like one (**179**). If we look at the frontier orbitals of Fig. 4-20b, we see that there is an antibonding interaction between the lobes

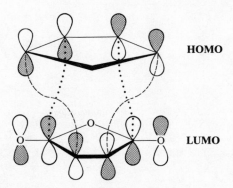

HOMO

LUMO

Fig. 4-19 Secondary overlap of the frontier orbitals of Diels-Alder reactions. The dotted lines show the bonding overlap which stabilizes the *endo* transition state

108

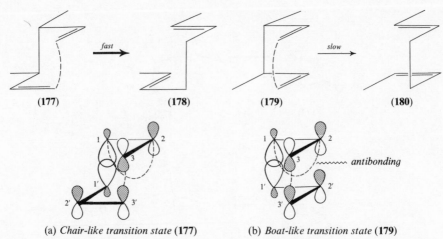

(177)　　　　　　**(178)**　　　　　　**(179)**　　　　　　**(180)**

(a) *Chair-like transition state* (**177**)　　　(b) *Boat-like transition state* (**179**)

Fig. 4-20　Frontier orbitals and secondary interactions in the Cope rearrangement

on C-2 and C-2′. In the chair-like shape (Fig. 4-20a), these orbitals are too far apart to interact, and this shape is consequently the preferred one.

The cycloaddition of cyclopentadiene and tropone (p. 91) gives the *exo*-adduct (**181**) rather than the *endo*-adduct (**182**), because the secondary interactions (Fig. 4-21b, wavy line) of the frontier orbitals are antibonding.

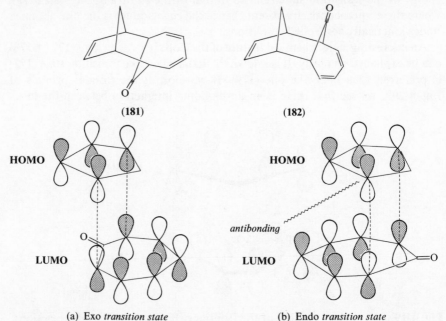

(181)　　　　　　　　　**(182)**

(a) Exo *transition state*　　　(b) Endo *transition state*

Fig. 4-21　Frontier orbitals and secondary interactions for the cycloaddition of tropone to cyclopentadiene

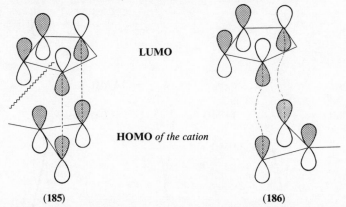

(183)   (184)

There is some evidence[157] that the cycloaddition of the allyl cation (183) to cyclopentadiene (184) takes place with a transition state like 186 rather than 185. This is clearly in agreement with a frontier orbital analysis, provided that we look at the HOMO of the allyl cation and the LUMO of the diene.

LUMO

HOMO *of the cation*

(185)   (186)

Finally the stereochemistry of a 1,3-dipolar cycloaddition (187 + 188 → 189) has been explained[158] by the favourable overlap of the frontier orbitals in the transition state (Fig. 4-22) leading to it.

(187)   (188)   (189)

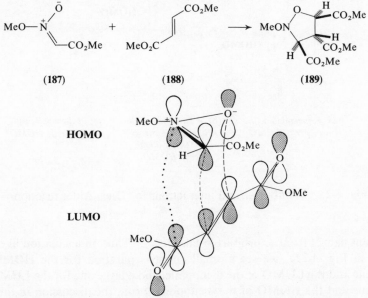

HOMO

LUMO

Fig. 4-22   Secondary interaction in a 1,3-dipolar cycloaddition

110

4.2   The Rates of Cycloaddition Reactions

In discussing the Woodward-Hoffmann rules, we were indifferent as to which of the two components of a cycloaddition would provide the HOMO and which the LUMO. The two pairs of frontier orbitals bear a complementary relationship to each other; they both invariably give the same answer, and we could safely make an arbitrary choice. To explain the effects of substituents on the rates of Diels-Alder reactions, however, we need to know which is the more

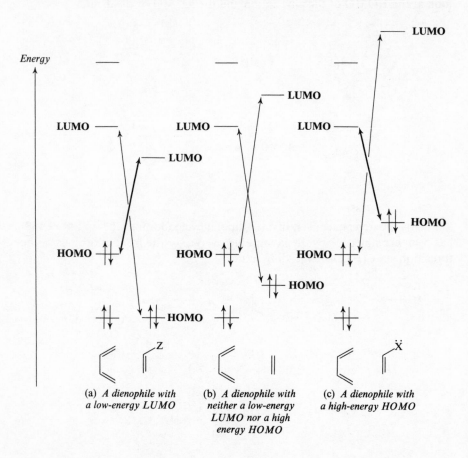

(a) *A dienophile with a low-energy LUMO*

(b) *A dienophile with neither a low-energy LUMO nor a high energy HOMO*

(c) *A dienophile with a high-energy HOMO*

Fig. 4-23   Frontier orbital interactions for Diels-Alder reactions

important pair of frontier orbitals to take.[112,113] Thus, in a situation like that shown in Fig. 4-23a, we see a small energy-separation for the HOMO of butadiene and the LUMO of the dienophile, and a large one for the LUMO of butadiene and the HOMO of the dienophile. From the discussion in the last chapter, it is obvious that the former interaction is the one to concentrate on.

The smaller the energy-gap is in any particular case, the faster the reaction ought to be, because two medium-sized interactions (Fig. 4-23b) are not as effective at lowering the transition-state energy as a strong and a weak one. (This may not be immediately obvious, but it follows from the fact that the energy-separation, $E_r - E_s$, is in the *denominator* of equation 2-7.) Therefore, the rate of a Diels-Alder reaction like that shown in Fig. 4-23b will be slower than the rate of a Diels-Alder reaction like that shown in Fig. 4-23a.

This explains why maleic anhydride (191a) reacts with butadiene 190) so much faster than ethylene (191b) reacts with butadiene: the LUMO of maleic anhydride is much lower in energy than that of ethylene, and we are in the situation shown in Fig. 4-23a. We shall see later in this chapter why an electron-withdrawing substituent (Z-) lowers the energy of the LUMO of the dienophile.

(190)   (191)

a R = —CO—O—CO—; 100% yield in 24h at 20° [159a]
b R = H;   78% yield in 17h at 165° and 900 atmospheres[159b]

In principle, it ought to be possible to increase the rate of a Diels-Alder reaction by adding electron-donating ($\ddot{X}$-) groups to the dienophile; these raise the energy of the HOMO, as we shall see later in the chapter, and we might expect to approach the situation in Fig. 4-23c. In practice, this does not often happen with butadiene itself; the separation in energy between the LUMO of butadiene and the HOMO of a dienophile with electron-donating substituents is never small enough for this frontier orbital interaction to be the major one. However, we can have a situation resembling that shown in Fig. 4-23c, by changing the diene from butadiene to one with a lower-energy LUMO. In Fig. 4-24, the energy levels of the frontier orbitals of the three dienophiles are the same as in Fig. 4-23, but the HOMO and the LUMO of the hypothetical diene have been set much lower in energy than those of butadiene. With such a diene, the interaction between the HOMO of the dienophiles and the LUMO of the diene will be greater than the other interaction for all three dienophiles. When this situation obtains, a dienophile with an electron-donating substituent will react faster than one with an electron-withdrawing substituent. This has been observed a few times and is usually described as a Diels-Alder reaction going with 'reverse electron demand'.[160]

For example, the 'diene' (192) reacts faster with ketene acetal (193a, R = R' = OEt), an electron-rich dienophile, than with acrylonitrile (193c, R = CN, R' = H), an electron-deficient dienophile.[161] Since allyl alcohol (193b, R = CH_2OH, R' = H) probably has HOMO and LUMO energies very close to those of ethylene itself, and since it reacts at an intermediate rate,

112

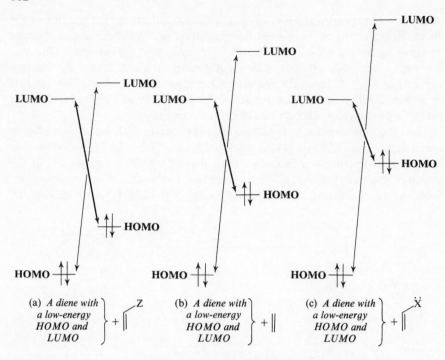

Fig. 4-24 Frontier orbitals for the Diels-Alder reactions of a diene with a lower-energy HOMO and LUMO

(192) + **(193)** → **(194)**

a R = R' = OEt; 75% reaction in 4 min at 25°
b R = CH₂OH, R' = H; 75% reaction in 275 min at 100°
c R = CN, R' = H; 75% reaction in 1080 min at 100°

this particular diene (**192**) probably has frontier orbitals with energies very like those for the hypothetical diene in Fig. 4-24.

We can also create many intermediate situations by adjusting the energies of the HOMO and LUMO of both the diene and the dienophile. For example, Konovalov[162] has found one diene (**195**) which in its reactions with substituted styrenes falls nicely into a pattern exactly like that of Fig. 4-23. The energies of the HOMO and LUMO of this diene are evidently so placed, with respect to those of the styrenes, that Fig. 4-23b describes the situation when R (in **196**) is H, Fig. 4-23a the situation when R is an electron-withdrawing substituent, and Fig. 4-23c the situation when R is an electron-donating substituent; for he

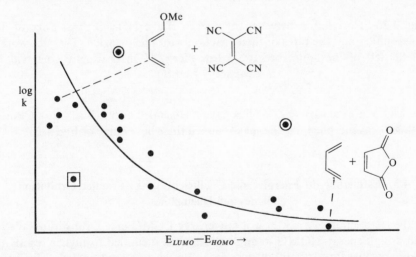

R:	p-NMe$_2$	p-OMe	H	p-Cl	m-NO$_2$	p-NO$_2$
k_2 ($\times$ 10^6 $mol^{-1}s^{-1}$):	338	102	73	78	79	88

$\xleftarrow{\text{electron-donating}}$ | $\xrightarrow{\text{electron-withdrawing}}$

observed that the reaction was slowest for the unsubstituted styrene, and faster when either Z- or Ẍ-substituents were present.

Sustmann[163] has collected data for a wide range of Diels-Alder reactions of normal electron demand. Using the electron affinities of the dienophiles and the ionization potential of the dienes, he estimated the separation in energy between the LUMO of the dienophile and the HOMO of the diene and plotted this against the log of the rate constant (Fig. 4-25). The highest rates were found

Fig. 4-25 Correlation between the energy separation of the frontier orbitals and the rates of Diels-Alder reactions

for the reactions having the smallest energy gap. The points fitting least well on the graph are for cyclopentadiene (circles, too fast) and cycloheptatriene (square, too slow).

Sustmann[164] has also collected data for 1,3-dipolar cycloadditions. Phenyl azide (197) reacts fast with electron-poor and with electron-rich olefins, but slowly with a simple olefin. A plot of the rate constant against the energy of the

114

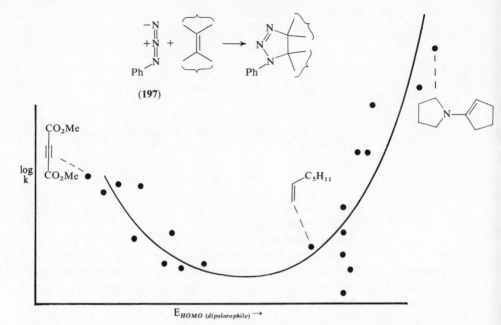

(197)

Fig. 4-26 Correlation between the energy of the HOMO of a range of dipolarophiles and the rates of their reaction with phenyl azide. The plot works on the left of the figure because a low-energy HOMO usually goes with a low-energy LUMO

HOMO of a wide variety of olefins gave a U-shaped curve (Fig. 4-26). Clearly a single reagent, phenyl azide, has spanned the whole range of Fig. 4-23.

4.3 Estimating the Energies and Coefficients of the Frontier Orbitals of Dienes and Dienophiles

In the last section, we saw how a low-energy LUMO in a Z-substituted olefin and a high-energy HOMO in an Ẍ-substituted olefin led to higher reactivity (with appropriate partners) in each case. We must now rationalize this claim. We want to work out how the energies of the HOMO and LUMO of ethylene are affected by adding electron-withdrawing and electron-donating substituents, and, for dienes, we want to work out how the same substituents affect the HOMO and LUMO of butadiene. This can, of course, be done properly, with the aid of a computer, some fairly simple mathematics and a good many assumptions and approximations. Alternatively, in some cases, the information is available from an experimental measurement, such as PES or ESR provides. But these aids are not always to hand. Furthermore, neither a computer programme nor the simple mathematics makes immediate *chemical* sense to everyone, and an experimental measurement still needs an explanation. The discussion in the following pages shows that we can work out the effects of substituents in an easy, non-mathematical way. Although the procedure used is legitimate (and works), it is perhaps worth bearing in mind that it does not find much favour with theoreticians, nor does it resemble the method used in proper calculations.

4.3.1 The Energies of the Frontier Orbitals

4.3.1.1 C-Substituted Olefins. First let us consider the effect of merely adding conjugation (C-), as we do in going from ethylene to butadiene. This is easy, because we know how the molecular orbitals of butadiene lie in relation to those of ethylene (Fig. 2-12): the HOMO is raised in energy and the LUMO lowered.

4.3.1.2 Z-Substituted Olefins. Now let us consider adding an electron-withdrawing group, such as a carbonyl group, to ethylene. If we ignore the fact that one of the atoms is an oxygen atom and not a carbon atom, we shall simply have the orbitals of butadiene. But obviously we cannot ignore the oxygen atom. One way to take it into consideration is to regard the carbonyl group as a kind of carbonium ion, highly stabilized by an oxyanion substituent (**199**). Normally we do not draw it this way, because such good stabilization is better expressed by drawing the molecule (as in **198**) with a full π-bond between

$$\text{(198)} \qquad \text{(199)}$$

the oxygen atom and the carbon atom. The truth is somewhere in between, and organic chemists have usually to make a mental reservation about the meaning of such a drawing as **198**. We make the mental reservation that the butadiene-like system, implied by the drawing of a localized structure (**198**), is only one extreme approximation of the true orbital picture for acrolein. The other extreme approximation is an allyl cation, substituted by a non-interacting oxy-anion, as implied by the localized drawing (**199**). The molecular orbitals for these two extremes are shown in Fig. 4-27, and the true orbital picture is shown in between. We can expect the true structure to be more like the butadiene system than the allyl cation system (for the same reason that we prefer to draw it as **198** rather than **199**). What we immediately learn from Fig. 4-27 is that the effect of mixing in some allyl cation like nature to the butadiene orbitals is to *lower* both the HOMO and the LUMO, relative to the HOMO and LUMO of butadiene. Also, because it is butadiene-like, the HOMO and the LUMO will be closer in energy than they are in ethylene. What we have done is to superimpose the orbitals of an allyl cation on those of butadiene, and, with suitable weighting, to add the two together.

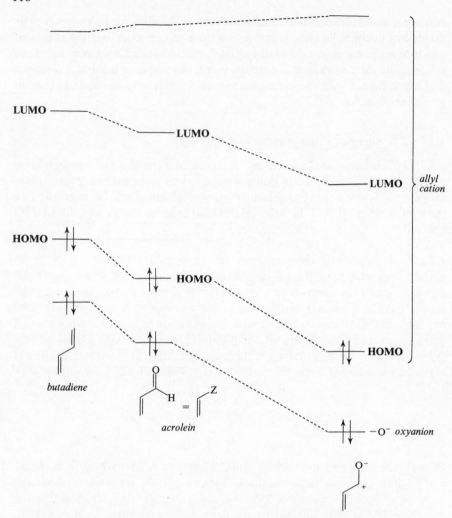

Fig. 4-27 The orbitals of acrolein seen as a weighted sum of the orbitals of butadiene and a hypothetical oxyanion-substituted allyl cation. For the energies of the orbitals of the allyl cation relative to those of butadiene, see Figs. 2-12 and 2-14 and the discussion in Chapter 2

4.3.1.3 Ẍ-Substituted Olefins. In an Ẍ-substituted olefin like methyl vinyl ether (**200**), we have a lone pair of electrons brought into conjugation with the double bond. Our model, therefore, will be the allyl anion (**201**), just as the benzyl anion was a model for an Ẍ-substituted benzene in Chapter 3. The molecular orbitals of the allyl anion are the same (Fig. 4-28) as those of the allyl cation, except that ψ_2 is now the HOMO and ψ_3^* is the LUMO. In other words, both the HOMO and the LUMO of an allyl anion, and, hence, of an Ẍ-substituted olefin, are raised in energy relative to the HOMO and LUMO of ethylene.

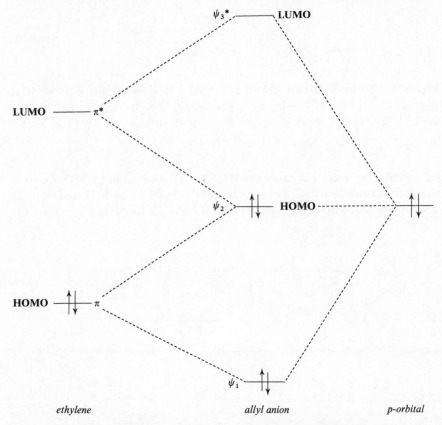

Fig. 4-28 Energies of the molecular orbitals of the allyl anion

We can summarize these trends in Fig. 4-29, which gives the relative energies of the three types of substituted ethylenes. The numbers given for the energies in Fig. 4-29 are experimentally derived; they were chosen by Houk[40] as representative of each kind of olefin. The HOMO-energies were measured by PES, and the LUMO-energies were estimated from electron affinities, charge transfer spectra, polarographic reduction potentials and π–π^* absorption spectra. The trends in these experimentally derived energies agree very well with our simple-minded deductions.

4.3.1.4 Dienes with Substituents at C-1. We can make the same qualitative approach to dienes as we have just made to olefins. The result (Fig. 4-30) is

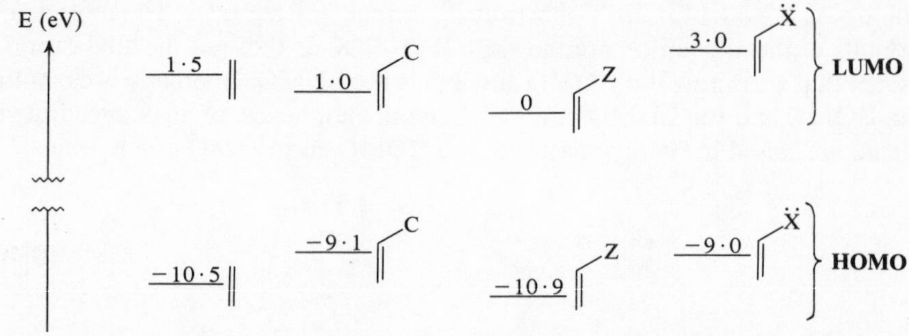

118

E (eV)

1·5 ‖   1·0 ⟨⟨ C   0 ⟨⟨ Z   3·0 ⟨⟨ Ẍ⟩ LUMO

−9·1 ⟨⟨ C   −9·0 ⟨⟨ Ẍ⟩ HOMO
−10·5 ‖   −10·9 ⟨⟨ Z

Fig. 4-29 Energies of the frontier orbitals of dienophiles. C = vinyl or phenyl; Z = CHO, CN, NO$_2$ etc.; X = MeO, Me$_2$N, Me, etc. The energies (in eV) are typical values for each class of dienophile. (1 eV = 23 kcal = 96·5 kJ)

very similar: conjugation raises the HOMO and lowers the LUMO, electron-withdrawing substituents lower both the HOMO and the LUMO, and electron-donating substituents raise both the HOMO and the LUMO.

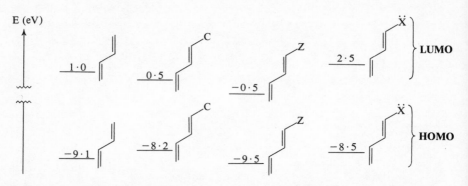

Fig. 4-30 Energies of the frontier orbitals of 1-substituted dienes. The energies are typical values for each class of diene

4.3.1.5 Dienes with Substituents at C-2. For dienes with substituents at C-2, the same arguments are used and similar results obtained (Fig. 4-31). The difference is that the C-, Z- and Ẍ-substituents are now attached to the carbon atom of the diene which has the smaller coefficient in the HOMO and the LUMO (see p. 17). The effect is that the energies are shifted in the same direction, but to a lesser extent. The HOMO and LUMO are brought closer in energy by extra conjugation, the HOMO and LUMO are lowered by a Z-substituent, and the HOMO and LUMO are raised by an Ẍ-substituent; but none of these effects is as large as it is in a 1-substituted diene.

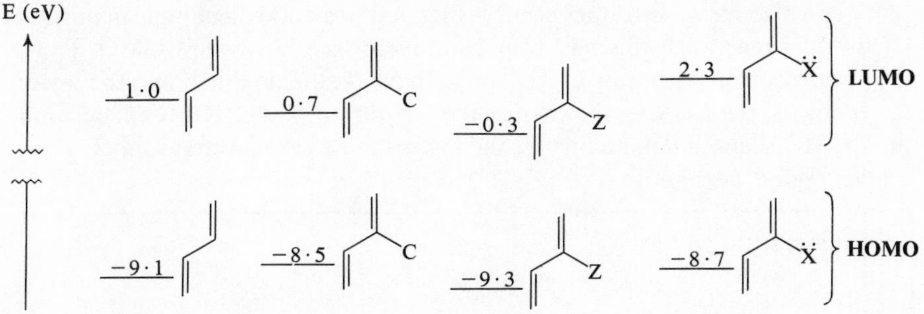

E (eV)

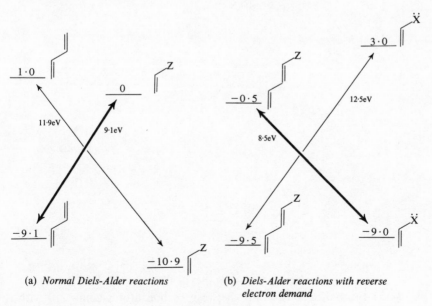

Fig. 4-31 Energies of the frontier orbitals of 2-substituted dienes.

We are now in a position to see (Fig. 4-32a) that a normal Diels-Alder reaction *is* dominated by the interaction of the HOMO of the diene (at −9·1 eV) and the LUMO of the dienophile (at 0 eV). The difference in energy is 9·1 eV, whereas the difference in energy of the LUMO of the diene (at 1·0 eV) and the HOMO of the dienophile (at −10·9 eV) is 11·9 eV. For reverse electron demand, a Z-substituted diene has its LUMO at −0·5 eV and an Ẍ-substituted dienophile has its HOMO at −9·0 eV, giving a separation of 8·5 eV. The other way round gives a separation of 12·5 eV; so the important interaction in this case (Fig. 4-32b) is indeed HOMO (dienophile)/LUMO (diene).

(a) *Normal Diels-Alder reactions*

(b) *Diels-Alder reactions with reverse electron demand*

Fig. 4-32

We can also, now, see rather clearly what makes a good dienophile in normal Diels-Alder reactions: the most important factor is a low-lying LUMO. Thus, the more electron-withdrawing groups we have on the double bond, the lower the energy of the LUMO, the smaller the separation of the HOMO(diene) and the LUMO(dienophile), and hence the faster the reaction. Tetracyanoethylene is a very good dienophile.

In summary:

C-
Extra conjugation   *raises the energy of the HOMO*
                                           lowers the energy of the LUMO

Z-
An electron-withdrawing group lowers the energy of the HOMO
                                           lowers the energy of the LUMO

Ẍ-
An electron-donating group   *raises the energy of the HOMO*
                                           raises the energy of the LUMO

Another way of producing a low-lying LUMO is to have an oxygen or nitrogen atom in the π-bond. Because p orbitals on these atoms lie so much lower in energy than those on carbon, the π molecular orbital that they make will inevitably have low-energy HOMOs and LUMOs, as we can see from Fig. 4-33. This is what happens with O=O and —N=N— double bonds,

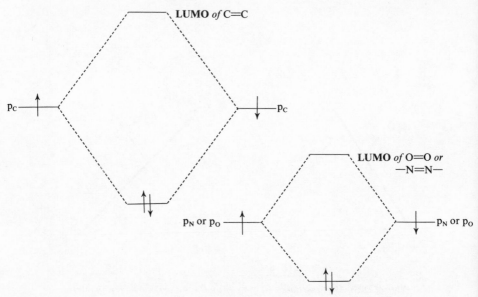

Fig. 4-33 Formation of π-bonds between carbon atoms compared with the formation of π-bonds between heteroatoms

which is one reason why singlet oxygen and azadienophiles like **202** are such good dienophiles.[165]

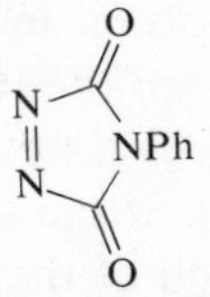

(202)

4.3.2 The Coefficients of the Frontier Orbitals

In discussing pericyclic reactions so far, we have only been looking at the denominator of the third term of equation 2-7. However, the coefficients of the atomic orbitals also play their part. They particularly influence the *regioselectivity*, the *site-selectivity* and the *periselectivity* of cycloaddition reactions. The former term refers to the orientation of a cycloaddition: for example, methoxybutadiene (**204**) gives[166] the 'ortho' adduct (**206**) rather than the 'meta' adduct (**203**) with acrolein (**205**). *Site-selectivity* and *periselectivity*,

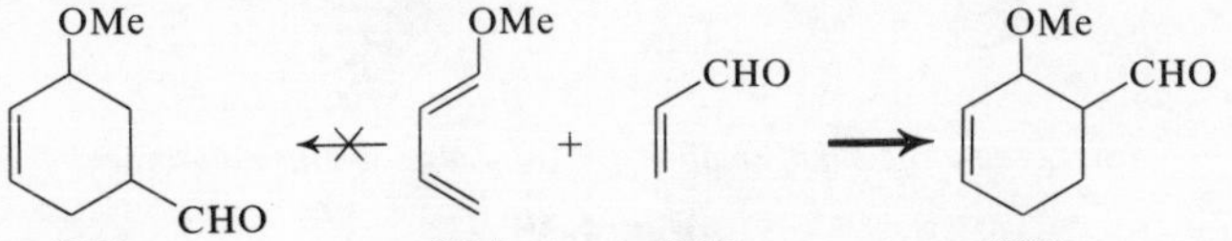

(203)   **(204)**   **(205)**   **(206)**

which are related terms, we shall define and deal with later (pp. 165 and 173). In summary, the explanation for regioselectivity is that if we look at the coefficients of the atomic orbitals of the mono-substituted diene and of the mono-substituted dienophile, we find that they are not equal at each end, as they necessarily were in the unsubstituted or symmetrically substituted cases. For methoxybutadiene and acrolein the frontier orbitals are polarized as shown in Fig. 4-34, where the size of the circle is roughly in proportion to the size of the coefficient. The circles represent the lobes of the p orbitals above the plane of the paper, and the shaded and unshaded ones are of opposite sign in the usual way. We should perhaps remind ourselves that the *sign* of the lobe that is overlapping with another lobe is a much more important factor in determining the energy change than is the second-order effect of its *size*. From now on, most of the

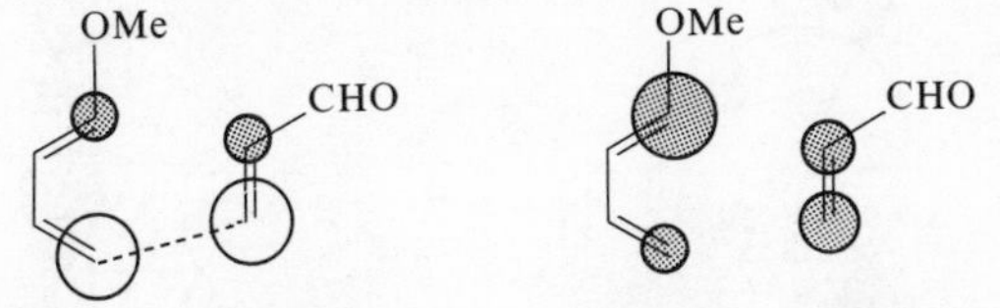

HOMO	LUMO	LUMO	HOMO
−8·5 eV	0 eV	2·5 eV	−10·9 eV

Fig. 4-34 Coefficients of the frontier orbitals of methoxybutadiene and acrolein

122

discussion is about the size of the orbital; we shall ignore the sign because we shall only be looking at allowed (and hence observed) reactions.

We already know that the important interaction in the methoxybutadiene/acrolein reaction will be that between the HOMO of the diene and the LUMO of the dienophile ($E_r - E_s = 8.5$ eV), not the other way round ($E_r - E_s = 13.4$ eV); so we need to look only at the left-hand combination in Fig. 4-34. There we see that, if the interaction of the two larger atomic orbitals is paramount (dotted line), then we shall get the right answer, for these two atoms are bonded to each other in the observed product (**206**). We now have to justify the polarization of the orbitals shown in Fig. 4-34, using just the same kind of arguments that we used earlier in estimating the energies of the frontier orbitals.

It is not self evident that the choice of the large–large interaction in Fig. 4-34 was the best one. Here is a simple theorem which proves that it was right. Consider two interacting molecules X and Y in Fig. 4-35: let the square of the terminal coefficients on X be x

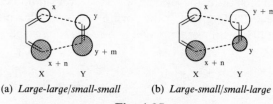

(a) *Large-large/small-small*   (b) *Large-small/small-large*

Fig. 4-35

and $x + n$, and let the square of the coefficients on Y be y and $y + m$. For the large–large/small–small interaction (Fig. 4-35a), the contribution to the numerator of the third term of equation 2-7 will be:

$$xy + (x + n)(y + m)$$

For the large–small/small–large case (Fig. 4-35b), the contribution will be:

$$x(y + m) + (x + n)y$$

Subtracting the latter from the former gives nm. In other words, the former interaction is greater so long as n and m are of the same sign; that is, $x + n$ and $y + m$ are either the two large (as shown) or the two small lobes. Pictorially, this conclusion can be even less rigorously demonstrated by Fig. 4-36.

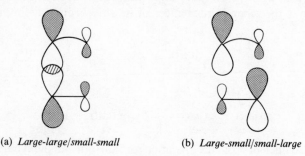

(a) *Large-large/small-small*   (b) *Large-small/small-large*

Fig. 4-36

4.3.2.1 C-Substituted Olefins. For the coefficients of a conjugated olefin we need only look at those of butadiene. We saw what the coefficients of butadiene were like on p. 17; clearly the HOMO and the LUMO are polarized as shown on Fig. 4-37. The coefficients of some simple conjugated systems are listed in Table 4-1, which we shall refer to several times.

	HOMO		LUMO

Fig. 4-37 Coefficients of the frontier orbitals of a C-substituted olefin

4.3.2.2 Z-Substituted Olefins. When we change from butadiene to acrolein, for the Z-substituted-olefin case, we have first the contribution from the conjugation, which is just like that in butadiene (Fig. 4-37), and we must add to it a

Table 4-1 The coefficients of the atomic orbitals in the molecular orbitals of conjugated systems[7]

Ethylene	ψ_2*	·707	−·707						LUMO
	ψ_1	·707	·707						HOMO
Allyl	ψ_3*	·500	−·707	·500					LUMO in anion
	ψ_2	·707	0	−·707					HOMO in anion, LUMO in cation
	ψ_1	·500	·707	·500					HOMO in cation
Butadiene	ψ_4*	·371	−·600	·600	−·371				
	ψ_3*	·600	−·371	−·371	·600				LUMO
	ψ_2	·600	·371	−·371	−·600				HOMO
	ψ_1	·371	·600	·600	·371				
Pentadienyl	ψ_5*	·288	−·500	·576	−·500	·288			
	ψ_4*	·500	−·500	0	·500	−·500			LUMO in anion
	ψ_3	·576	0	−·576	0	·576			HOMO in anion, LUMO in cation
	ψ_2	·500	·500	0	−·500	−·500			HOMO in cation
	ψ_1	·288	·500	·576	·500	·288			
Hexatriene	ψ_6*	·232	−·418	·521	−·521	·418	−·232		
	ψ_5*	·418	−·521	·232	·232	−·521	·418		
	ψ_4*	·521	−·232	−·418	·418	·232	−·521		LUMO
	ψ_3	·521	·232	−·418	−·418	·232	·521		HOMO
	ψ_2	·418	·521	·232	−·232	−·521	−·418		
	ψ_1	·232	·418	·521	·521	·418	·232		
Heptatrienyl	ψ_7*	·191	−·354	·462	−·500	·462	−·354	·191	
	ψ_6*	·354	−·500	·354	0	−·354	·500	−·354	
	ψ_5*	·462	−·354	−·191	·500	−·191	−·354	·462	LUMO in anion
	ψ_4	·500	0	−·500	0	·500	0	−·500	{ HOMO in anion, LUMO in cation
	ψ_3	·462	·354	−·191	−·500	−·191	·354	·462	HOMO in cation
	ψ_2	·354	·500	·354	0	−·354	−·500	−·354	
	ψ_1	·191	·354	·462	·500	·462	·354	·191	
Octatetraene	ψ_8*	·161	−·303	·408	−·464	·464	−·408	·303	−·161
	ψ_7*	·303	−·464	·408	−·161	−·161	·408	−·464	·303
	ψ_6*	·408	−·408	0	·408	−·408	0	·408	−·408
	ψ_5*	·464	−·161	−·408	·303	·303	−·408	−·161	·464 LUMO
	ψ_4	·464	·161	−·408	−·303	·303	·408	−·161	−·464 HOMO
	ψ_3	·408	·408	0	−·408	−·408	0	·408	·408
	ψ_2	·303	·464	·408	·161	−·161	−·408	−·464	−·303
	ψ_1	·161	·303	·408	·464	·464	·408	·303	·161

contribution from the allyl-cation-like nature of acrolein (**199**). The HOMO of the allyl cation (ψ_1) is the one which has the coefficient on the central atom larger than those at the other two (p. 20). The two effects therefore act in opposite directions—the conjugation causing a reduction in the coefficient on the carbon atom carrying the formyl group, and the allyl-cation-like contribution causing an increase in this coefficient. We have already seen (p. 115) that acrolein is better represented by the drawing (**198**) than by the drawing (**199**), from which we may guess that it is the butadiene-like character which makes the greater contribution to the HOMO. The result is that acrolein will have its HOMO coefficients polarized in the same way as those of butadiene, but to a lesser extent (Fig. 4-38). (Epiotis[118] actually came to the opposite conclusion for acrylonitrile i.e. Z = CN; he has evidently given greater weight to the allyl cation-like nature of the system. This shows that the situation is a very delicately balanced one. It may well be that some Z-substituents do give the opposite polarization in the HOMO to that shown in Fig. 4-38.) For the

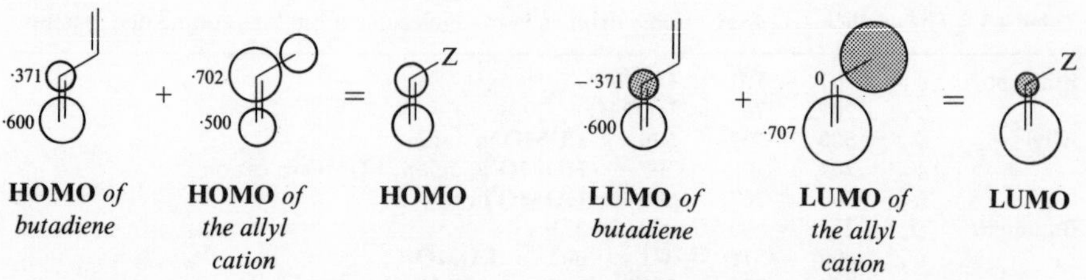

Fig. 4-38 Coefficients of the frontier orbitals of a Z-substituted olefin

LUMO of acrolein, both contributions are in the same direction: the carbon atom with the Z-substituent on it already has the lower coefficient from the butadiene-like contribution, and the allyl cation has a coefficient of zero on this atom. Putting these together makes the coefficients of the LUMO of acrolein polarized like those of butadiene, but this time to a greater extent (Fig. 4-38).

4.3.2.3 Ẍ-Substituted Olefins. For an Ẍ-substituted olefin, we look at the unperturbed olefin and add a bit of allyl anion-like character to it. Methyl vinyl ether is somewhere between the two in character and in reactivity. For the HOMO, the unperturbed olefin has (necessarily) equal coefficients on each

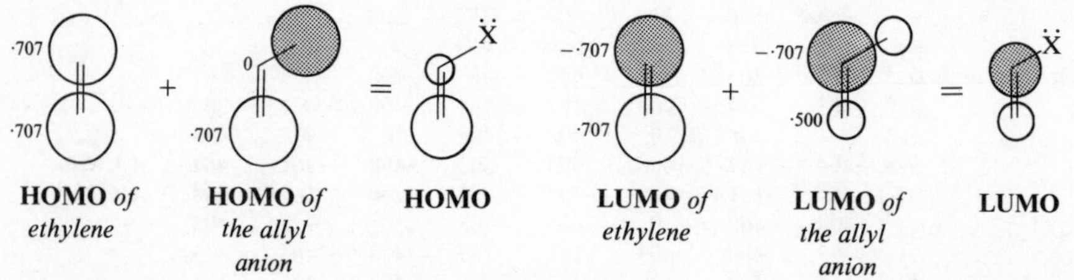

Fig. 4-39 Coefficients of the frontier orbitals of an Ẍ-substituted olefin

atom, and the allyl anion has a zero coefficient on the atom bearing the Ẍ-substituent. The result of mixing these two is a relatively strongly polarized orbital (Fig. 4-39). For the LUMO, the unperturbed olefin again has equal coefficients, but the allyl anion has a larger coefficient on the carbon atom carrying the Ẍ-substituent than on the other one. The result (Fig. 4-39) is a mildly polarized orbital.

4.3.2.4 1-C-Substituted Dienes. For the coefficients of conjugated trienes, we can simply look at Table 4-1 for hexatriene. Both the HOMO and the LUMO are polarized the same way, as shown in Fig. 4-40.

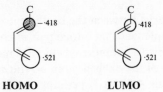

HOMO   **LUMO**

Fig. 4-40 Coefficients of the frontier orbitals of a 1-C-substituted diene

4.3.2.5 1-Z-Substituted Dienes. For dienes substituted with electron-withdrawing groups on C-1, we must add the polarization from Fig. 4-40, above, to that of the pentadienyl cation. For the HOMO, the dienyl cation has equal coefficients (of opposite sign), and so the polarization of the conjugated system is little changed. For the LUMO, the pentadienyl cation has a node at C-4, and the polarization is therefore in the same direction as that of the simple conjugated system, but stronger.

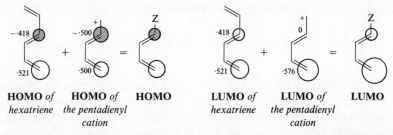

HOMO *of*   **HOMO** *of*   **HOMO**   **LUMO** *of*   **LUMO** *of*   **LUMO**
hexatriene   *the pentadienyl*   *hexatriene*   *the pentadienyl*
                cation   *cation*

Fig. 4-41 Coefficients of the frontier orbitals of a 1-Z-substituted diene

4.3.2.6 1-Ẍ-Substituted Dienes. For dienes with electron-donating substituents, the HOMO is straightforward: we look at ψ_3 of the pentadienyl anion, which has a node at C-4. The result is a strongly polarized orbital. But with the LUMO of these dienes we meet our first complete failure in the simple qualitative approach we have been using. Both the simple diene (necessarily) and the dienyl anion have equal coefficients on C-1 and C-4, and we cannot, by adding them, get any clue as to which actually has the larger coefficient.

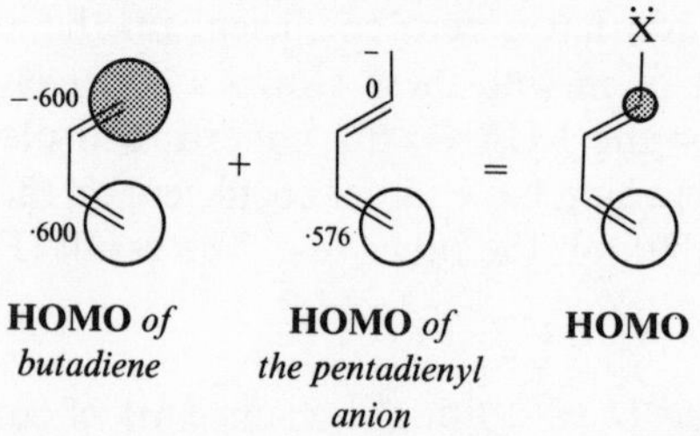

HOMO *of*
butadiene

HOMO *of*
the pentadienyl
anion

HOMO

Fig. 4-42 Coefficients of the HOMO of a 1-$\ddot{X}$-substituted diene

Such difference as does exist between the coefficients on C-1 and C-4 arises at least partly from the simple fact that a lone pair in conjugation with a diene is not accurately modelled by the pentadienyl anion. A more elaborate calculation, which includes the effect of the oxygen atom on the σ-framework, leads to the polarization shown in Fig. 4-43. It is perhaps comforting to observe that the carbon atom next to the $\ddot{X}$-group is the one with the larger coefficient, just as it is in the LUMO of olefins with an $\ddot{X}$-substituent (Fig. 4-39). Fortunately, the LUMO of a diene with an electron-donating substituent is very high in energy (Fig. 4-30), and it will therefore almost never be an important frontier orbital.

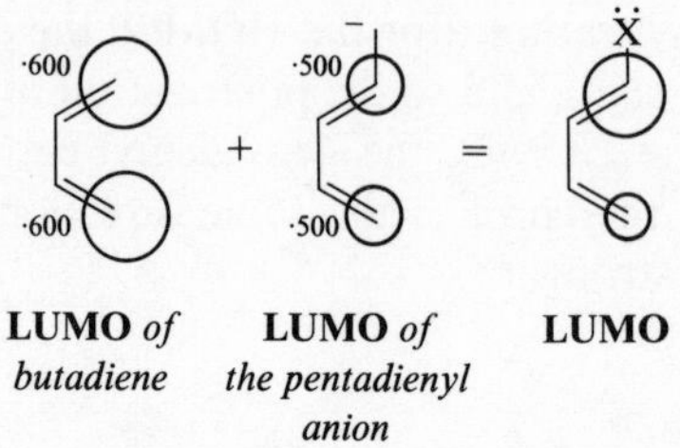

LUMO *of*
butadiene

LUMO *of*
the pentadienyl
anion

LUMO

Fig. 4-43 Coefficients of the LUMO of a 1-$\ddot{X}$-substituted diene

4.3.2.7 2-Substituted Dienes. The C-, Z-, and $\ddot{X}$-substituents on the 2-position can be thought of as affecting the π-bond to which they are attached more than they affect the other π-bond. Thus, to take just one case, the HOMO of the 2-$\ddot{X}$-substituted diene is made by mixing a butadiene and an allyl anion, as

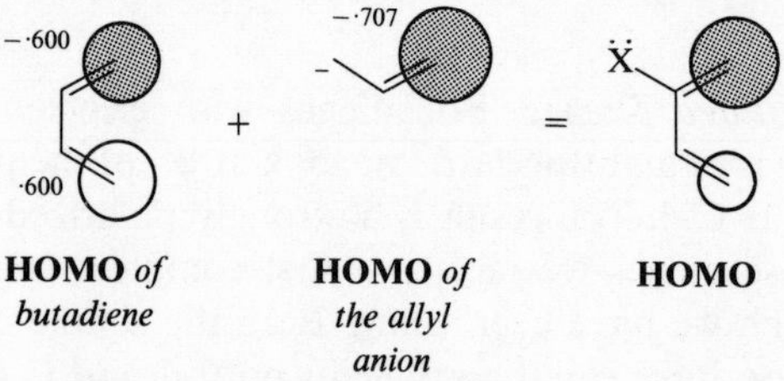

HOMO *of*
butadiene

HOMO *of*
the allyl
anion

HOMO

Fig. 4-44 Coefficients of the HOMO of a 2-$\ddot{X}$-substituted diene

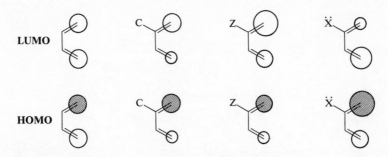

Fig. 4-45 Coefficients of the frontier orbitals of 2-substituted dienes

shown in Fig. 4-44. We can continue easily in this way for all the 2-substituted dienes, and the result is shown in Fig. 4-45.

Because we shall need to refer frequently both to the energies and to the polarizations of all the substituted olefins and dienes discussed above, the conclusions of the arguments we have just seen are collected together in a convenient form in Fig. 4-46. This picture is essentially due to Houk.[40]

We are now in a position to look back at our original example of a regioselective Diels-Alder reaction (Fig. 4-34), where we find that, by taking the HOMO(diene)/LUMO(dienophile) interaction, by taking the correct polarization of orbitals for a 1-Ẍ-substituted diene and a Z-substituted olefin, and by allowing the large-large/small-small interaction to predominate, we do indeed get the right answer. But we now see that each step of the argument is easily justifiable and easily capable of being worked out on the back of an envelope.

4.4   Examples of Regioselectivity

You may feel that we have laboured hard to justify an example of regioselectivity which any experienced organic chemist would have predicted would go this way round. He would have drawn curly arrows (or canonical structures) to express his feeling that C-4 of the diene is a nucleophilic carbon and C-3' of the acrolein an electrophilic carbon. He might still acknowledge that the cycloaddition is likely to be concerted, with both bonds forming at the same time, but, following Woodward and Katz,[167] he would say that the bonding between C-4 and C-3', represented by the curly arrows (**207**), was advanced over that which develops between C-1 and C-2'. This reasoning is fine, but it cannot be

(207)

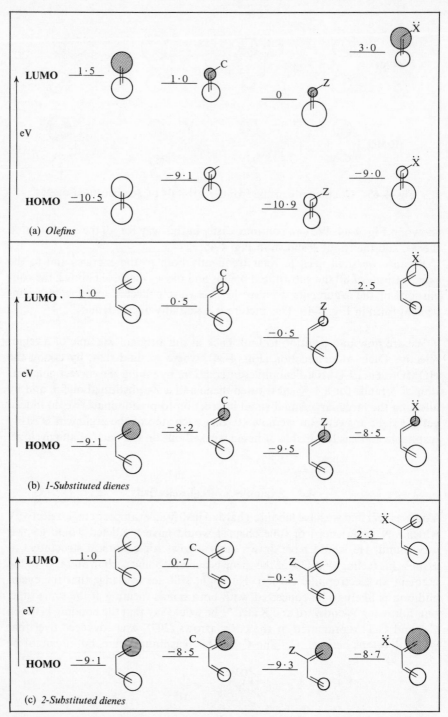

Fig. 4-46 Frontier orbital energies and coefficients of olefins and dienes. Energies are typical values for each class of olefin and diene.[40] (1 eV = 23 kcal = 96·5 kJ)

applied in all cases. For example, it would not work in a case mentioned earlier, the reaction of acrylonitrile (209) and the azoniaanthracene cation (208 = 192), giving a mixture of two stereoisomers (210). These *exo* and *endo* adducts were the only ones detected, even though this means that the two carbon atoms marked * have become bonded to each other. *Both* of these atoms are strongly *electrophilic* in ordinary ionic reactions.

We can use the pyridinium cation (211) as a model for the 'diene' (212 = 208): C-2 and C-5 of pyridine have the same character as C-10 and C-9 respec-

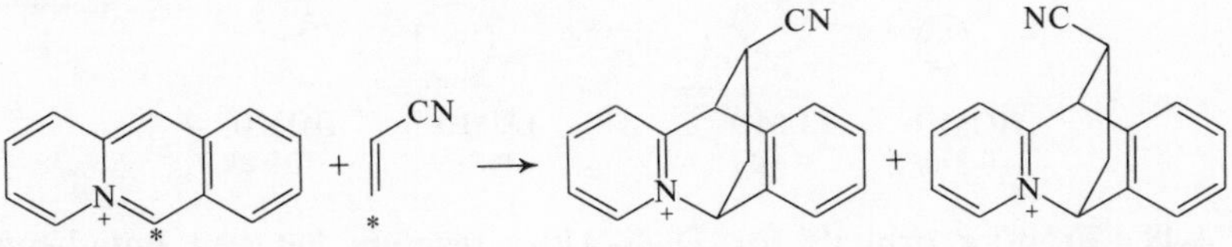

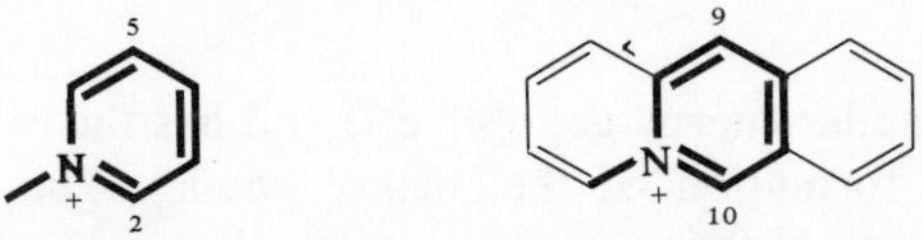

tively in the azoniaanthracene ion (212). We saw on p. 112 that the major frontier orbital interaction is almost certainly that between its LUMO and the HOMO of the dienophile, even with acrylonitrile. We now look to see what the coefficients are like on C-2 and C-5 of the pyridinium cation. We have in fact already met them on p. 68, where we saw that C-2 has the larger coefficient. Acrylonitrile is a simple Z-substituted olefin, and the frontier orbitals are therefore those shown in Fig. 4-47; this immediately explains the regioselectivity shown by this reaction.

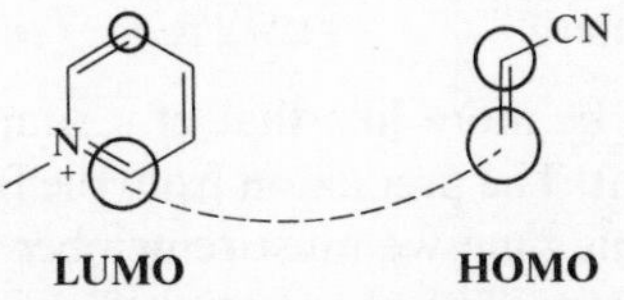

LUMO   **HOMO**

Fig. 4-47 Frontier orbitals of the pyridinium cation as a model for the Diels-Alder reaction of **208** with acrylonitrile

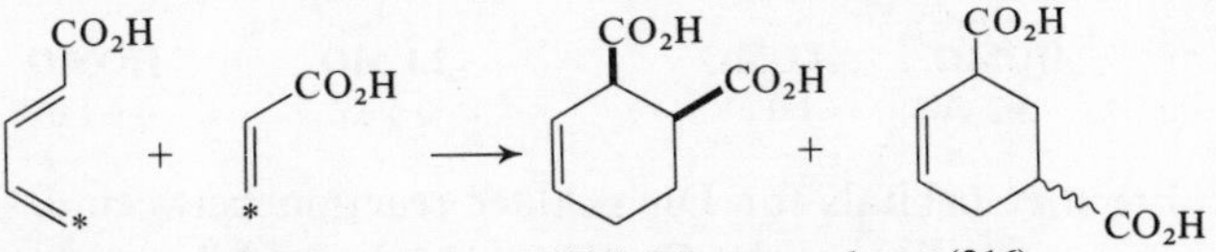

Another example, in which the simple 'curly arrow' argument would not have predicted the right answer, is the reaction of butadiene-1-carboxylic acid (213) and acrylic acid (214). Both adducts (215 and 216) are formed, but the

'ortho' adduct (**215**) is the major one,[168] even though, again, this means that the electrophilic carbon atoms (*) have become bonded to each other in this adduct.

We can use the data from Fig. 4-46 directly in this case; they are repeated in Fig. 4-48, where we see that the HOMO(diene)/LUMO(dienophile) combination

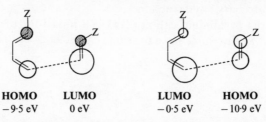

HOMO	LUMO	LUMO	HOMO
$-9 \cdot 5$ eV	0 eV	$-0 \cdot 5$ eV	$-10 \cdot 9$ eV

Fig. 4-48 Frontier orbitals for Diels-Alder reaction between butadiene-1-carboxylic acid and acrylic acid

on the left has the smaller energy-gap (9·5 eV) and has the orbitals polarized so as to favour the formation of the 'ortho' adduct. The LUMO(diene)/HOMO(dienophile) combination on the right has an energy-separation (10·4 eV) which is not so much larger that we can safely ignore it. As it happens, this combination also favours the formation of the 'ortho' adduct.

Even more striking in this series is the case of the anions of these acids (**217** and **218**). The contribution of a carboxylate ion group ($-CO_2{}^-$) to the

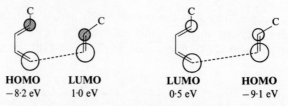

(**217**)   (**218**)   (**219**) *1 part to 1 part* (**220**)

frontier orbital will now be more like that of a simple C-substituent and less like that of a Z-substituent. The prediction from the frontier orbitals (Fig. 4-49) is again the same, but this time we must remember that the negative charges will strongly repel each other. The observation[169] of a 50:50 mixture of 'ortho' and 'meta' adducts (**219** and **220**) shows how powerful a directing effect the frontier orbital contribution must be.

HOMO	LUMO	LUMO	HOMO
$-8 \cdot 2$ eV	$1 \cdot 0$ eV	$0 \cdot 5$ eV	$-9 \cdot 1$ eV

Fig. 4-49 Frontier orbitals for Diels-Alder reaction between a C-substituted diene and a C-substituted dienophile

Alston[36, 170] has pointed out that the regioselectivity in this and some other reactions may be better attributed, not to the primary interactions of the frontier orbitals that we have been using so far, but to the secondary interactions. 1-substituted dienes often show very small differences in the coefficients on C-1 and C-4 in the HOMO. (We have already seen, on p. 124, how some calculations make them come out the other way round from that shown on Fig. 4-38.) On the other hand, 1-substituted dienes regularly show a much larger difference between the coefficients on C-2 and C-3. The crude way we have handled the problem does not immediately demonstrate this: thus we can see in the example in Fig. 4-50 how the contribution of the triene-like character and the pentadienyl-cation-like

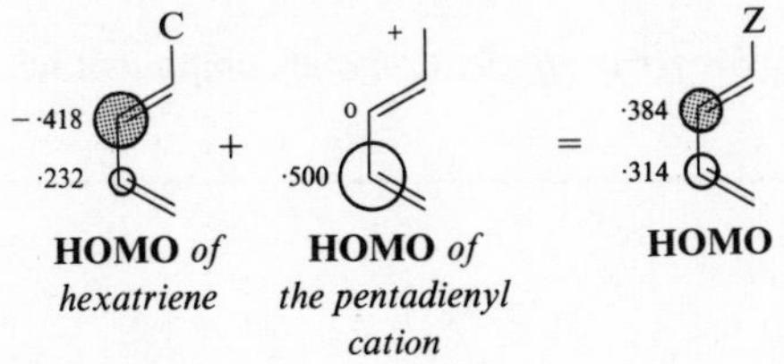

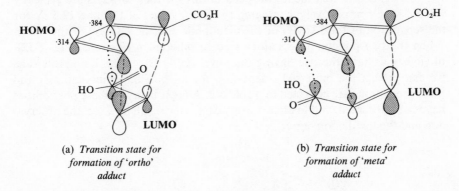

Fig. 4-50   Coefficients on C-2 and C-3 of a 1-Z-substituted diene

character have opposite effects on the coefficients on C-2 and C-3 of a Z-substituted diene. However, the differences in the coefficients on each component are noticeably large, and it is therefore easy to accept that C-2 and C-3 could have quite different coefficients if either contribution should dominate. Alston's calculation[170] gives the values shown; as usual, the triene-like character is evidently more important than the pentadienyl-cation-like character.

If we now look at the frontier orbital interactions for this case, we see that the arrangement giving the 'ortho' adduct (Fig. 4-51a) has a larger secondary interaction than that leading to the 'meta' adduct (Fig. 4-51b). It may be that the secondary interactions play

(a) *Transition state for formation of 'ortho' adduct*

(b) *Transition state for formation of 'meta' adduct*

Fig. 4-51   Secondary interactions as an influence on the regioselectivity of a Diels-Alder reaction

an important part in the regioselectivity shown by cycloaddition reactions. More work will have to be done before this can be demonstrated for certain, because most of the observations which have been made in this field are quite adequately explained by taking into account only the primary interactions of the frontier orbitals.

In summary, we predict the regioselectivity of a cycloaddition by the following sequence:

(*i*) *Estimate the energies of the HOMO and the LUMO of both components.*

(*ii*) *Identify which HOMO/LUMO pair is closer in energy.*

(*iii*) *Using this HOMO/LUMO pair, estimate the relative sizes of the coefficients of the atomic orbitals on the atoms at which bonding is to take place.*

(*iv*) *Match up the larger coefficient on one component with the larger on the other.*

4.4.1 Regioselectivity in Diels-Alder Reactions

Having established in the last section how easily frontier orbitals can be used to explain the regioselectivity of two rather special Diels-Alder reactions, we can now go on to see how well they explain the regioselectivity of many other Diels-Alder reactions.[171] Thus, if we use the rules summarized above together with the data in Fig. 4-46, we should predict that the Diels-Alder reaction of any C-, Z-, or $\ddot{X}$-substituted olefin with a 1-C- or 1-Z-substituted diene will give the 'ortho' adduct in greater amount. We must be careful, with electron rich dienophiles, to use the HOMO of the dienophile and the LUMO of the 1-C- or 1-Z-substituted diene: the separations for these orbitals are typically 9·5 and 8·5 eV respectively, compared to typical values of 11·2 and 12·5 eV for the separation of the LUMO of the dienophile and the HOMO of the diene. When the two possible interactions lead to different predictions (Fig. 4-52), obviously we take the one having the lower $(E_r - E_s)$ value. Using this rule, we can predict the preferred orientation for all possible combinations; and these predictions are collected in Table 4-2. A great many of these possibilities have been observed, and in almost every case, as we shall now see, the observation and the 'prediction' agree.

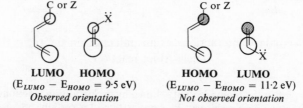

C or Z   C or Z

LUMO   HOMO   **HOMO   LUMO**
$(E_{LUMO} - E_{HOMO} = 9\cdot5 \text{ eV})$ $(E_{LUMO} - E_{HOMO} = 11\cdot2 \text{ eV})$
Observed orientation   *Not observed orientation*

Fig. 4-52   Regioselectivity for the Diels-Alder reaction of a C- or Z- substituted diene and an electron-rich dienophile

Table 4-2 Regioselectivity in Diels-Alder reactions predicted by considering only the frontier orbital contributions

Dienophile	Diene	Product expected	Case No.
 C (ethylene with C substituent)	 diene with C	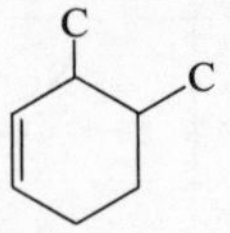 product with C, C	1
	 diene with Z	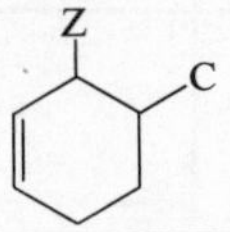 product with Z, C	2
	 diene with $\ddot{X}$	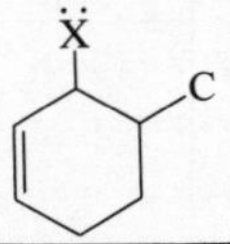 product with $\ddot{X}$, C	3^a
	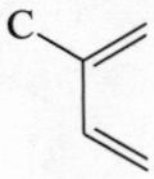 diene with C	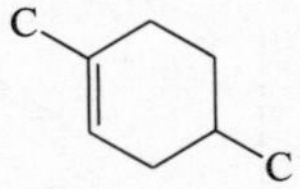 product with C, C	4
	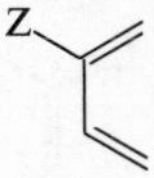 diene with Z	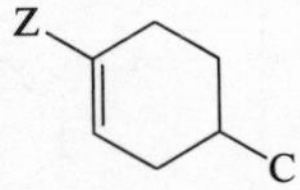 product with Z, C	5
	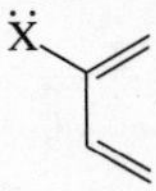 diene with $\ddot{X}$	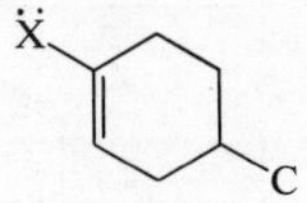 product with $\ddot{X}$, C	6^a
 Z (ethylene with Z substituent)	 diene with C	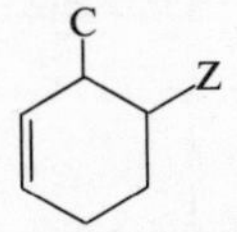 product with C, Z	7
	 diene with Z	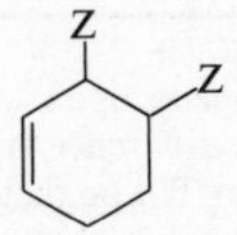 product with Z, Z	8
	 diene with $\ddot{X}$	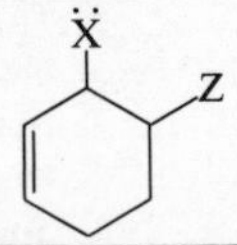 product with $\ddot{X}$, Z	9^a
	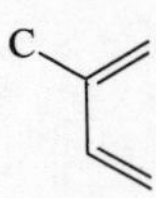 diene with C	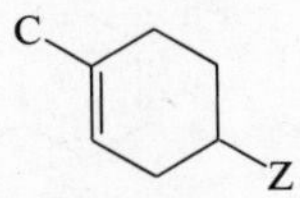 product with C, Z	10

Table 4-2 *continued*

	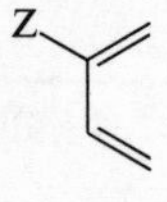 Z	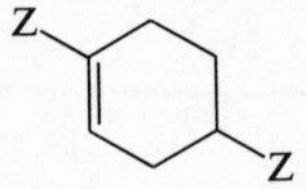 Z, Z	11
	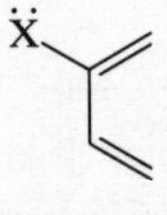 Ẍ	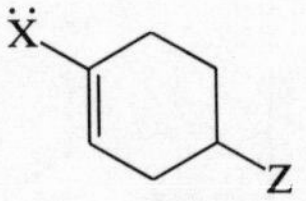 Ẍ, Z	12[a]
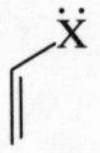 Ẍ	 C	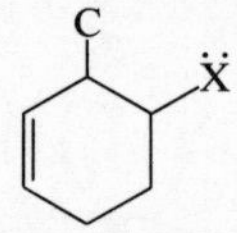 C, Ẍ	13[a]
	 Z	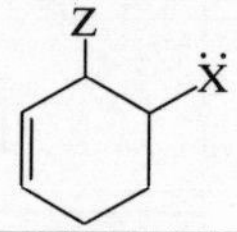 Z, Ẍ	14[a]
	 Ẍ	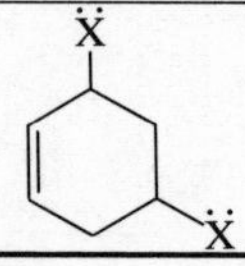 Ẍ, Ẍ	15
	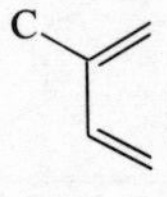 C	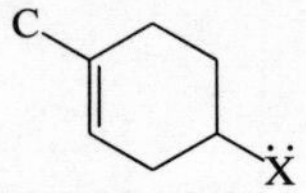 C, Ẍ	16[a]
	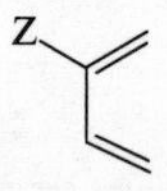 Z	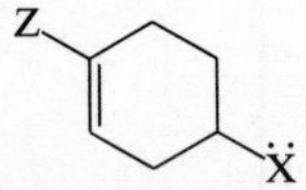 Z, Ẍ	17[a]
	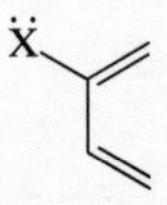 Ẍ	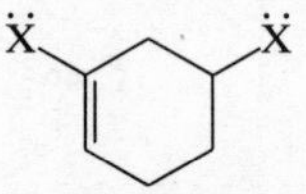 Ẍ, Ẍ	18

[a] These predictions depend upon which HOMO and which LUMO is taken. In each case the choice is an easy one, the difference in the two possible ($E_{HOMO} - E_{LUMO}$) values is always greater than $1·7$ eV (i.e. > 39 kcals). The prediction may, however, be upset by untypical dienes or dienophiles.

Case 1[172,173]

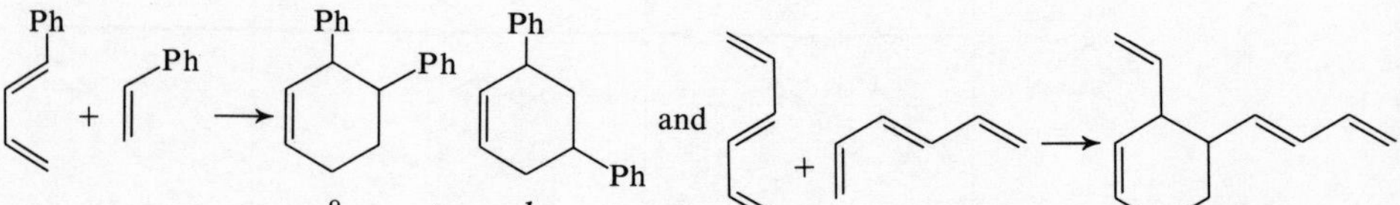

Ph + Ph → Ph, Ph   Ph, Ph   8 : 1   and   + →

Case 2[174,175]

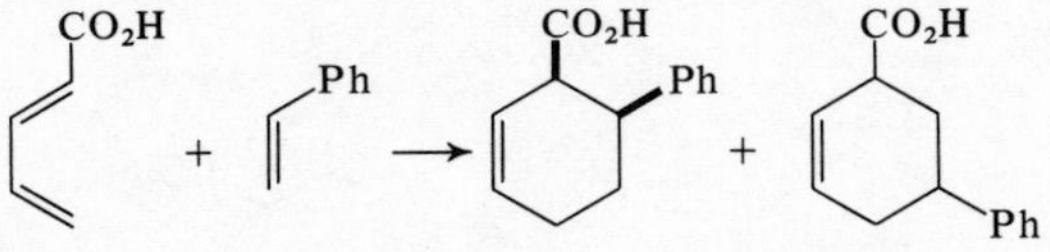

6 : 1

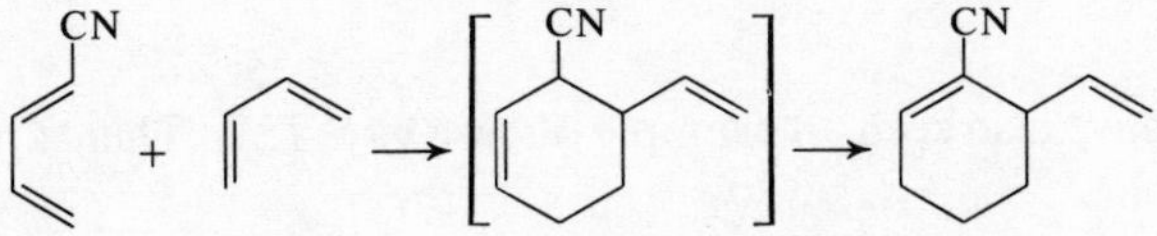

Case 3[176]

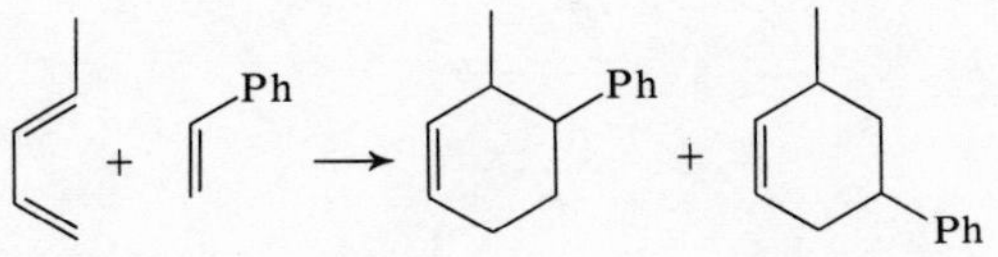

major *minor*

Case 4[177,178]

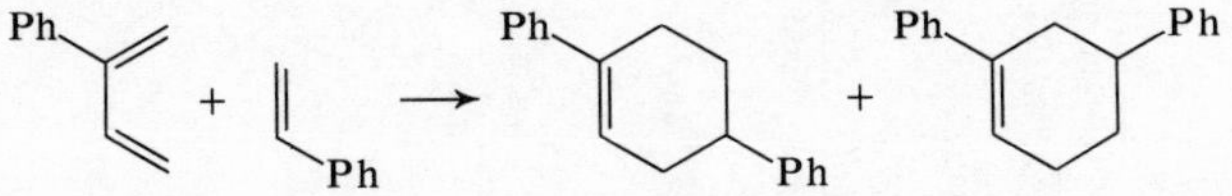

20 : 1

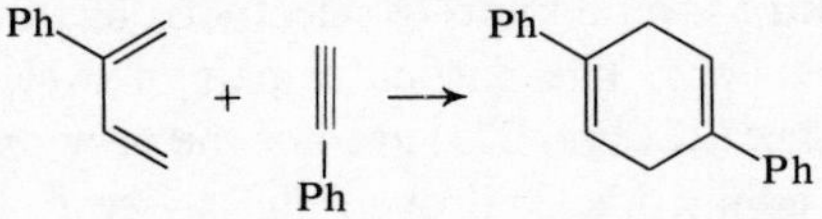

Case 5 and Case 11[179]

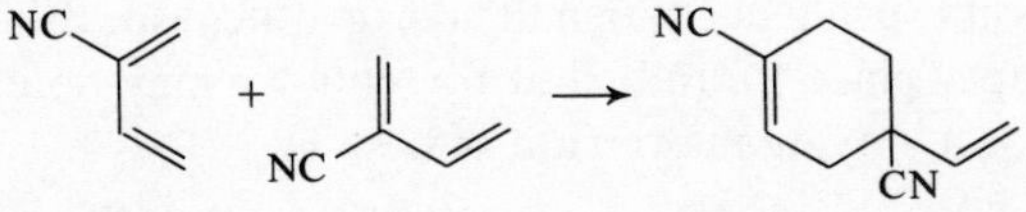

Case 6[180]

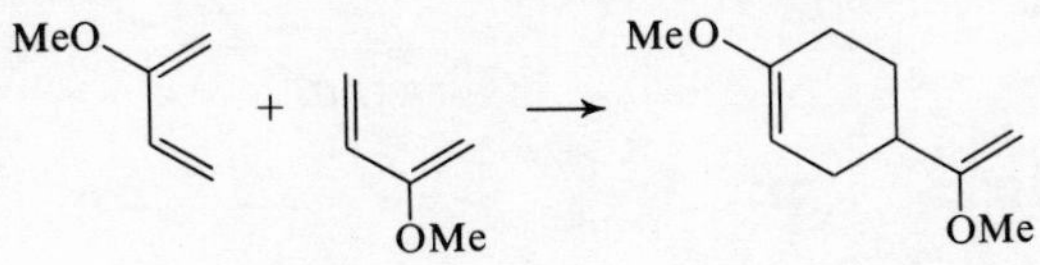

Case 7[181]

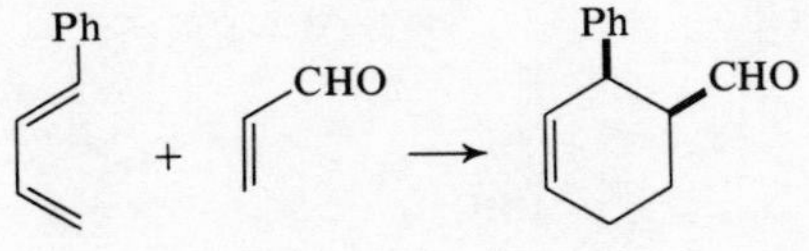

Case 8. We have already seen two examples of case 8 (p. 129). Here is another one.[182]

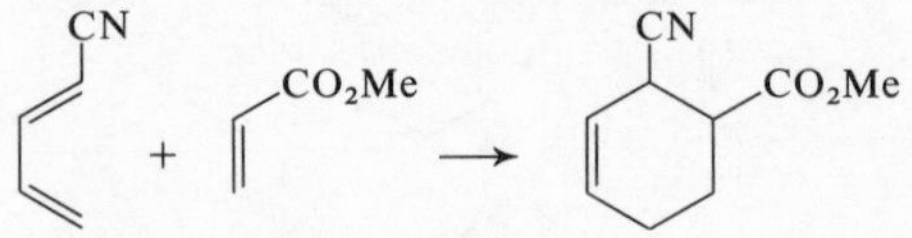

Case 9. We have also seen an example of case 9 (p. 121). There are many more; here is just one.[183]

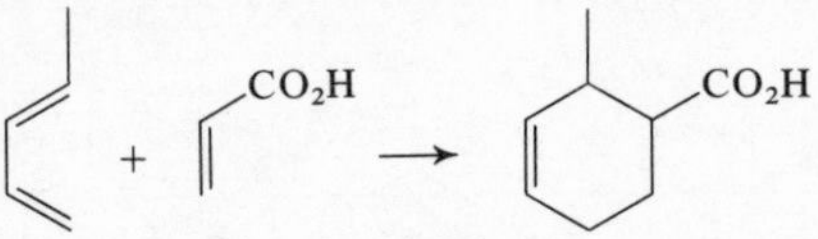

Case 10[178]

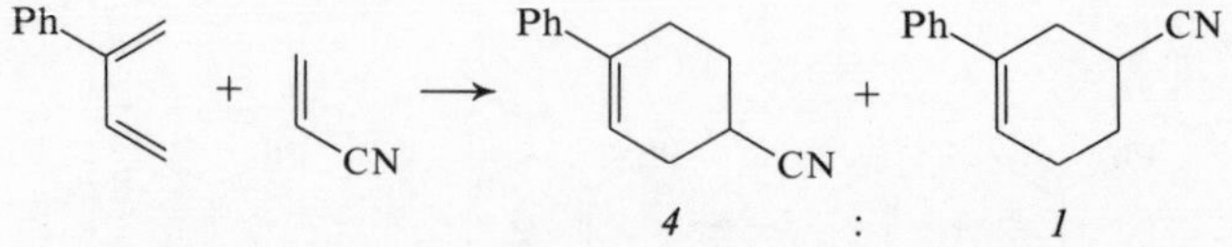

$$4 \quad : \quad 1$$

Case 11. (See also Case 5 above.) The formation of Thiele's ester (**223**)[184] is a remarkable example of several kinds of selectivity, all of which can be explained by frontier orbital theory. The particular pair of cyclopentadienes which do actually react together (**221** and **222**) are not the only ones present. As a result of very rapid 1,5-sigmatropic hydrogen shifts (see p. 99), all three isomeric cyclopentadiene carboxylic esters are present, and any combination of these is in principle possible. As each pair can combine in several different ways there are, in fact, 72 possible products. We shall take up other aspects of this selectivity later, in their proper place (p. 167), but for now we may note that the *regio-selectivity* shown is a vinylogous version of case 11.

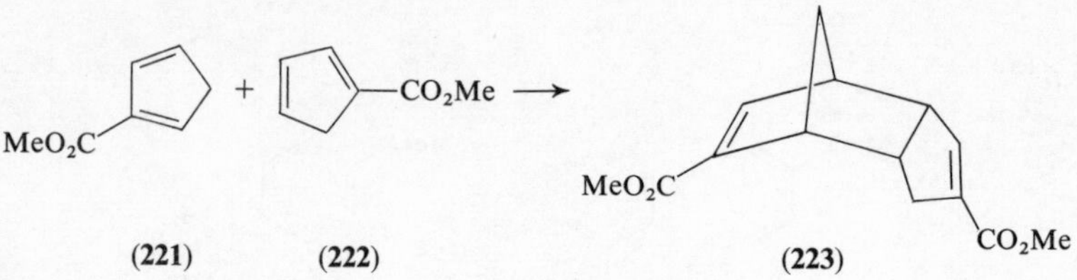

(**221**)   (**222**)   (**223**)

Case 12[185]

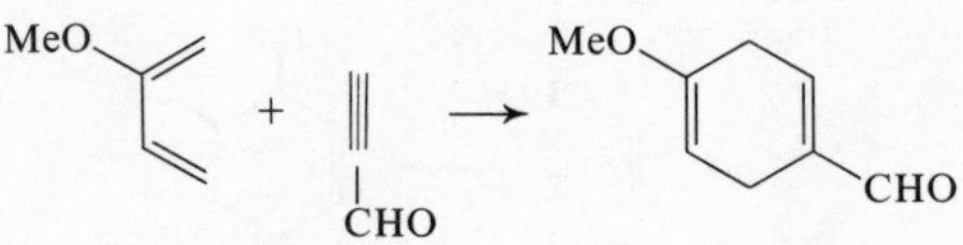

Case 13. The nearest situation to this reaction is the adduct of 1-phenylbutadiene (**224**) and citraconic anhydride (**225**).[186] However, appearances are misleading.

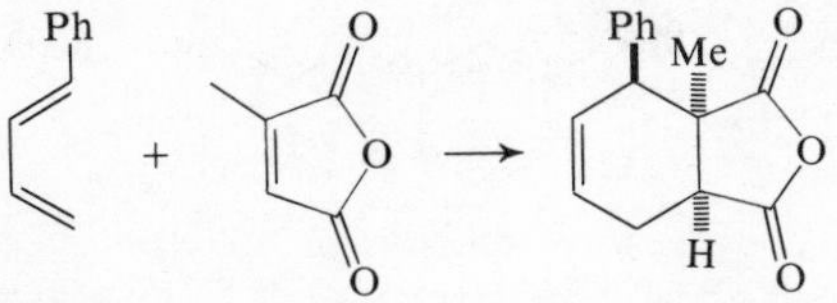

$$(224) \qquad (225)$$

With two carbonyl groups on the dienophile (**225**), the correct HOMO and LUMO to take in this case is HOMO(diene) and LUMO(dienophile). For this pair, if we ignore the carbonyl groups for the moment and treat the dienophile simply as an Ẍ-substituted olefin, we would predict the opposite of what is observed. But it is obviously unreasonable to ignore the carbonyl groups. One simple-minded way of looking at it is to say that the conjugation (**226**) of the methyl group, through the double bond, with the carbonyl group on C-3 will reduce the electron withdrawing effect of that carbonyl. The result is that the carbonyl group on C-2 is more important in guiding the reaction: the C=C orbital is more polarised by the C-2 carbonyl than by the C-3 carbonyl, and this reaction thus becomes an example of case 7.

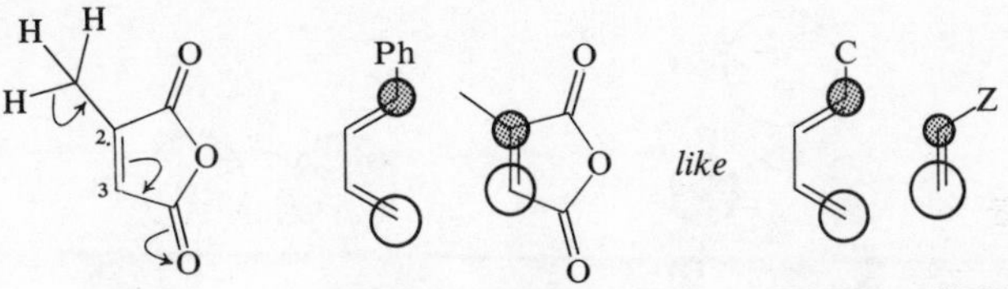

$$(226) \qquad \text{HOMO} \quad \text{LUMO} \qquad like \qquad \text{HOMO} \quad \text{LUMO}$$

Case 14. The examples of this type may only be formal ones, since most of the known reactions[187] could be stepwise ionic reactions.[188]

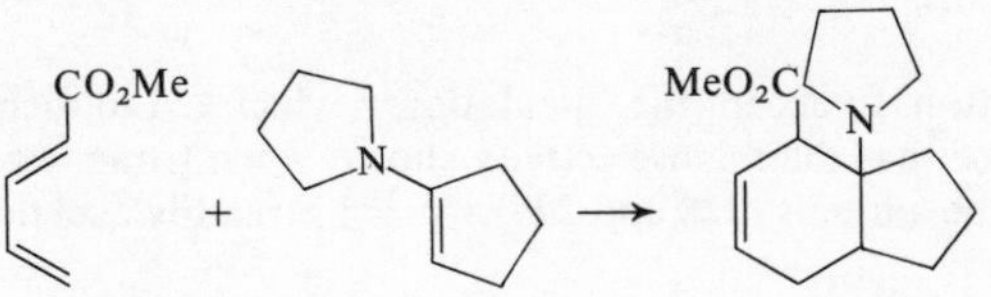

Case 15[189]

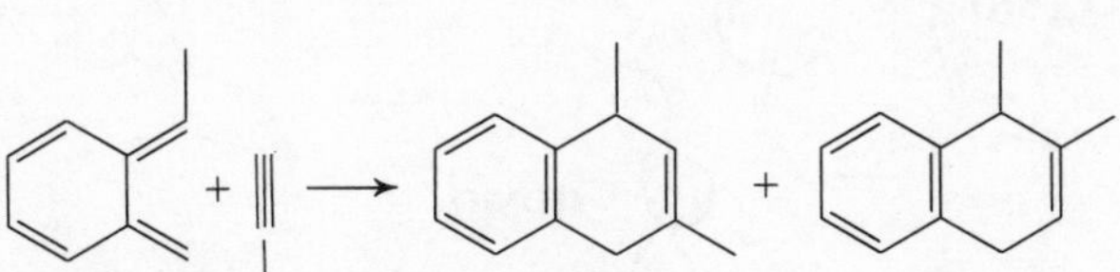

$$65\% \qquad 35\%$$

138

Cases 16 and 17 do not seem to be known.

Case 18[189]

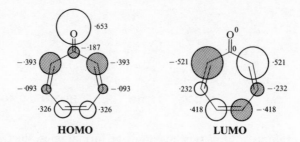

The examples of Cases 15 and 18 are particularly significant: they are the only cases where 'meta' adducts are predicted, and they are the only ones where the diradical theory[190] makes an opposite prediction to that of frontier orbital theory. Furthermore, there were virtually no examples of these two cases. The reactions shown above were actually done to test the predictive power of frontier orbital theory: it emerged triumphant.

Special Cases   There are a few special cases, in which the unsymmetrical diene is part of a conjugated system which cannot easily be placed in any of the categories, C-, Z- or Ẍ-substituted.

For example, tropone (**227**) occasionally reacts as a diene. Because of its symmetry, we have to work out the coefficients of the atomic orbitals by some other means than by the simple arguments used above. The coefficients of the HOMO and LUMO are shown in Fig. 4-53. These numbers are not easily guessed at, and we must be content, in this more

HOMO

·653
-·187
-·393   -·393
-·093   -·093
·326   ·326

LUMO

0
0
-·521   ·521
·232   -·232
·418   -·418

Fig. 4-53   The frontier orbital coefficients of tropone

complicated situation, to accept the calculation[15] which led to them. With this pattern in mind, we can see that the regioselectivity shown when tropone reacts as a diene, for example, to give the adducts (**228** and **229**),[191, 192] is readily explained.

LUMO

HOMO

(**227**)   Ph

(**228**)   Ph

(229)
major product

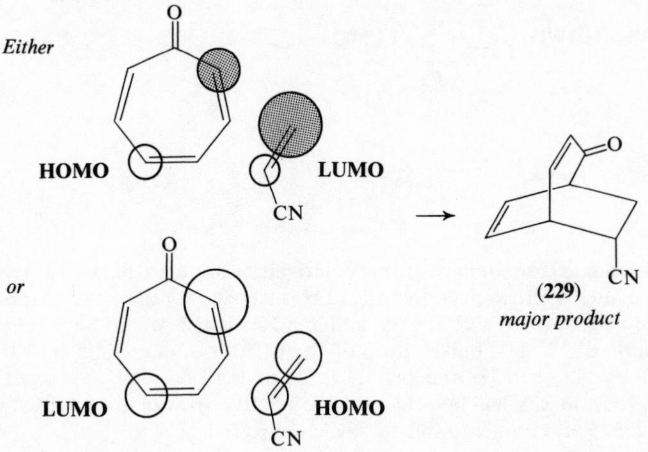

Another special case is that of styrene (**230**) when it reacts as a diene. The initial product (**231**) with maleic anhydride is not isolated, but instead reacts with more maleic anhydride to give **232** and **233**; it also rearranges to give the aromatic isomer (**234**).[193] Most of the

(230) **(231)**

(232) **(233)** **(234)**

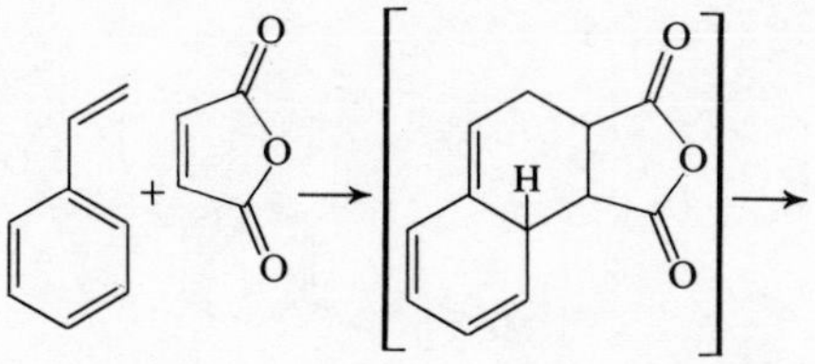

reactions of styrene as a diene are with maleic anhydride, where the presence of two carbonyl groups makes the dienophile reactive enough and also makes regioselectivity impossible. However, 1-vinylnaphthalene (**235**) is reactive enough to undergo addition to acrylic acid (**236**), and the product observed (**237**)[194] is exactly that which we should expect from the calculated[7] frontier orbitals.

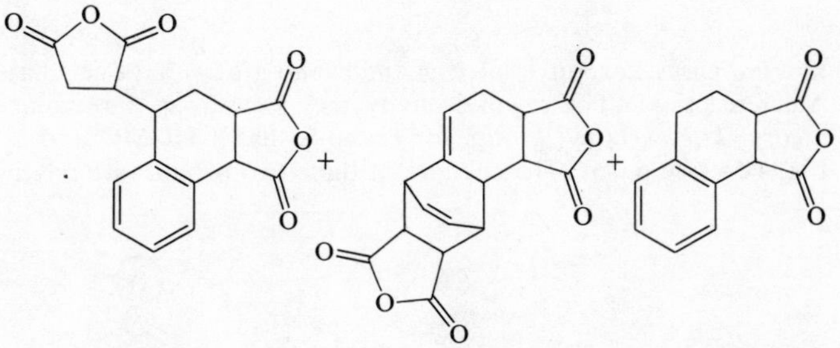

140

HOMO ⬭⬭ LUMO

CO₂H ⟶

CO₂H

(235)   (236)   (237)

An unfortunate consequence of this polarization was a set-back to a steroid synthesis along these lines. Citraconic anhydride (239) reacted with the 1-vinylnaphthalene (238) to give the adduct (240) and not the adduct (241) which would have been useful for a steroid synthesis.[195] The polarization of the LUMO of citraconic anhydride has been 'explained' earlier (p. 137) and plainly leads to the observed product. Evidently the methoxy group in 238 has not changed the relative sizes of the coefficients from that shown in 235. For a sequel to this set-back, see p. 164.

MeO

(238)   (239)

MeO   not   MeO

(240)   (241)

4.4.2 Regioselectivity with Heterodienophiles

In a few cases, carbonyl, nitrosyl and cyano groups have acted as dienophiles in Diels-Alder reactions. The regioselectivity they show is in agreement with frontier orbital theory. The carbonyl group, for example, has a HOMO and a LUMO as shown in Fig. 4-54 (see p. 16). The energies of these two orbitals are relatively low, and most of

LUMO   HOMO

Fig. 4-54 LUMO and HOMO of a carbonyl group

their Diels-Alder reactions will therefore be guided by the interaction between the HOMO of the diene and the LUMO of the dienophile. The following examples illustrate that this works.[196–198]

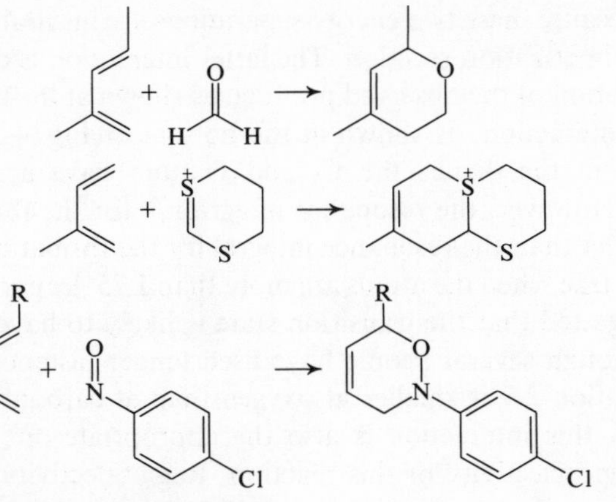

R = Ph, CO$_2$Me *or* OAc

4.4.3 Regioselectivity with Heterodienes

There are also Diels-Alder reactions in which the hetero-atom is part of the diene system.[199] The most notable of these is the dimerization of acrolein (**242**) giving the adduct (**243**).[200] This reaction has been a long-standing puzzle; as in

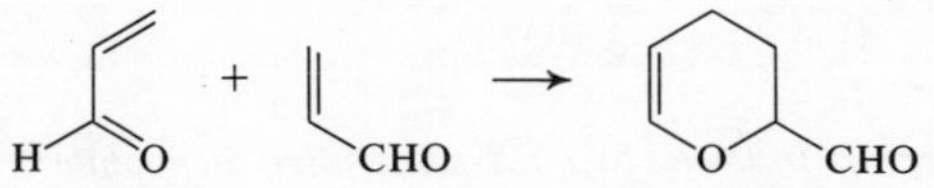

(**242**)   (**243**)

the reaction of butadiene carboxylic acid (**213**) with acrylic acid (**214**), the two 'electrophilic' carbon atoms are the ones which have become bonded. The first application of frontier orbital considerations also failed to solve the problem. In this work, Hückel theory was used to estimate the coefficients; and Hückel theory is notoriously weak in dealing with electron-distribution in heteroatom-containing system. More recent calculations[114,170,201,202] give a new set of coefficients, one of which[202] is shown in Fig. 4-55. With this refined set of values, we get the right answer: first of all, both HOMO(diene)/LUMO-(dienophile) and LUMO(diene)/HOMO(dienophile) interactions have to be

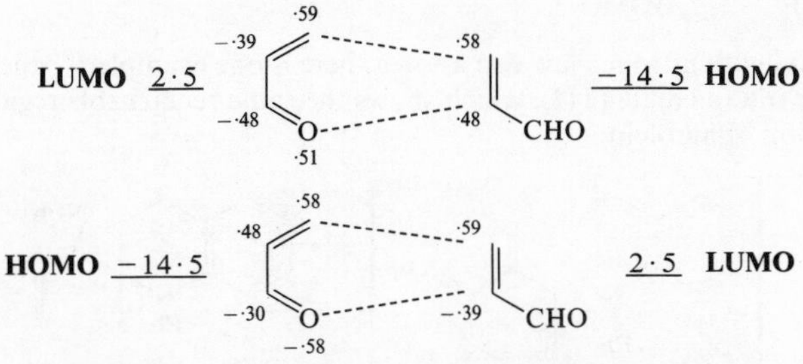

LUMO   2·5   −14·5 **HOMO**

HOMO −14·5   2·5 **LUMO**

Fig. 4-55   Energies and coefficients of acrolein

looked at, because these two energy-separations are inevitably the same as each other for a dimerization reaction. The latter interaction is directly appropriate for the formation of the observed product, as shown at the top of Fig. 4-55, but the former interaction, as shown at the bottom of Fig. 4-55, has no obvious polarization in the diene—the C- and O-atom have accidentally identical coefficients. However, the resonance integral, β, for the formation of a C—O bond is smaller than the resonance integral for the formation of a C—C bond. (This is only true when the atoms are more than 1·75 Å apart, see p. 150, but no one has suggested that the transition state is likely to have a shorter distance than this, though several people have used longer distances.) Thus the $(c\beta)^2$ term of equation 2-7 is smaller at oxygen than at carbon in this orbital, and consequently this interaction is also the appropriate one to explain the remarkable regioselectivity of this reaction. Regioselectivity fitting this picture is also shown in other cycloadditions of α,β-unsaturated aldehydes and ketones, as illustrated below, but the last of the examples, although naturally conforming to the pattern, may be a stepwise, ionic reaction.[199]

Another Z-substituted dienophile:[203]

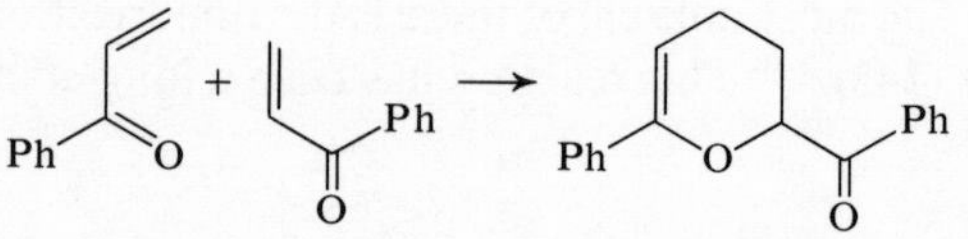

A C-substituted dienophile followed by an $\ddot{X}$-substituted dienophile:[204]

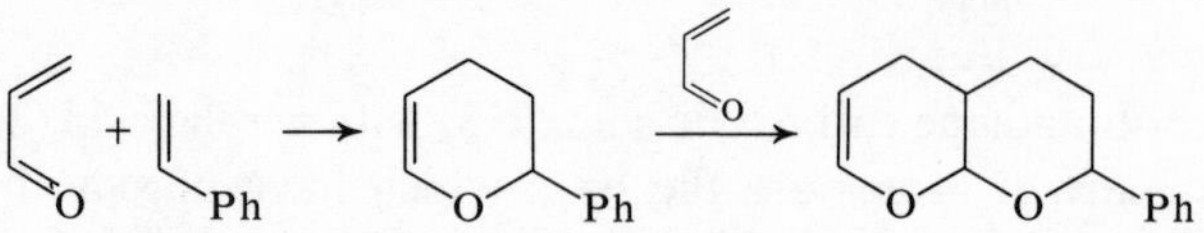

An $\ddot{X}$-substituted dienophile:[205]

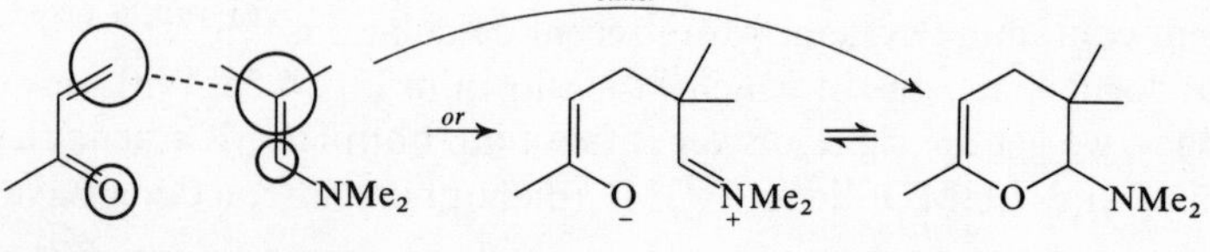

LUMO   **HOMO**

Heterodienes with nitrogen are also known; here is one example,[206] the dimerization of the reactive intermediate (244), which shows the same remarkable regioselectivity as the dimerization of acrolein.

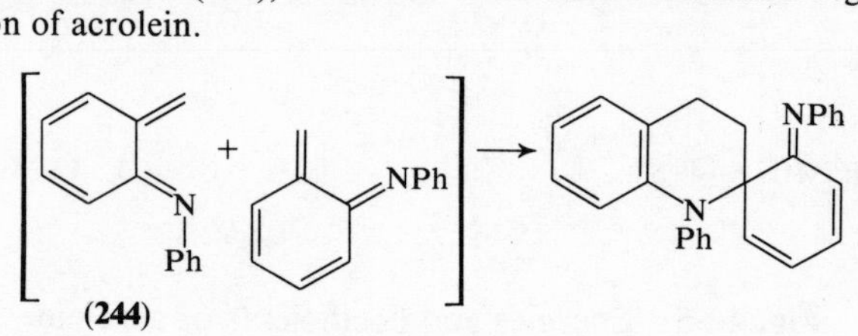

(244)

4.4.4 Regioselectivity in Ketene Cycloadditions and other 4-Membered Ring-Forming Reactions

Ketenes (e.g. **246**) undergo cycloadditions to double bonds (e.g. that of **245**) to give cyclobutanones (**247**).[207] The reaction is an easy one when the ketene has electron-withdrawing groups on it (like Cl) and when the olefin is relatively

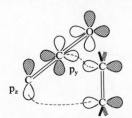

(245)   (246)   (247)

electron-rich. Already we can see that this follows the same pattern as the Diels-Alder reaction, and we can explain it in the same way. The electron-withdrawing groups lower the energy of the LUMO of the ketene, and the electron-rich olefins have a high-energy HOMO. The important interaction will therefore be HOMO(ketenophile)/LUMO(ketene). However, we must look more closely at this reaction, for at first sight it seems to be disobedient to the Woodward-Hoffmann rules. Nevertheless, there is a lot of evidence[208] that it is usually a concerted reaction: it is thought to find an allowed pathway in which a $[\pi 2s + \pi 2a]$ process occurs. One end of the π-bond of the olefin develops overlap to the p_z orbital at the terminal carbon atom of the ketene, while the other end of the π-bond of the olefin develops overlap to the p_y orbital of the central carbon atom of the ketene (Fig. 4-56). Regioselectivity is therefore determined

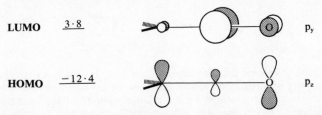

Fig. 4-56 Overlap in a cycloaddition of an olefin to a ketene

by the fact that the larger lobe of the HOMO of the olefin overlaps with the larger lobe of the LUMO of the ketene. The energies and coefficients of the frontier orbitals of ketenes[209] are shown in Fig. 4-57. These may be compared

LUMO   3·8 p_y

HOMO   −12·4 p_z

Fig. 4-57 Energies and coefficients of the frontier orbitals of ketene

with a simple carbonyl group (p. 16). Clearly the major interaction is that from the large LUMO coefficient on the central atom, and it is this atom, therefore, which will become bonded to the carbon atom having the larger coefficient in the HOMO of the olefin. This is exactly what has been observed with C-, Z- and Ẍ-substituted olefins, as the following examples show.

C-substituted ketenophiles:

$$(245) \quad + \quad (246) \rightarrow (247)$$

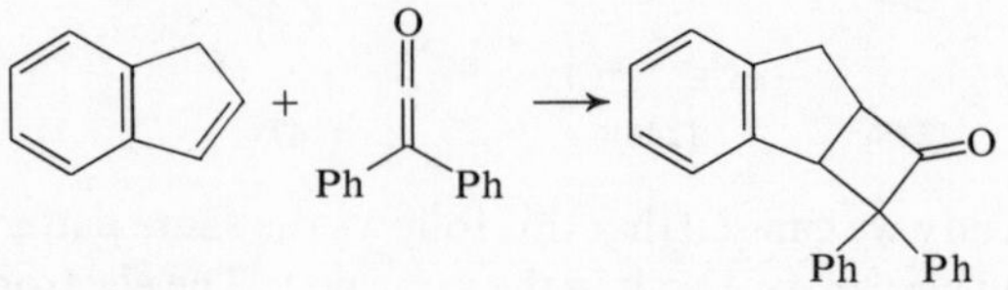

Z-substituted ketenophiles:[211,212]

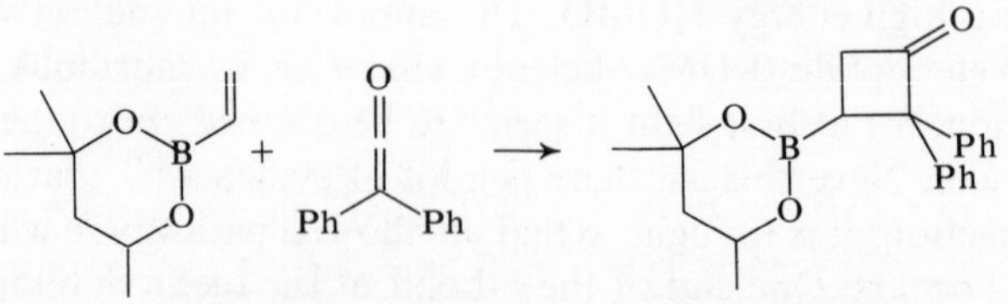

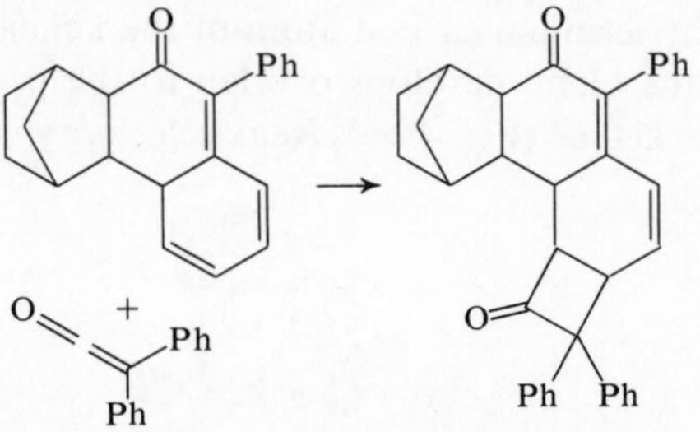

(This is a doubly vinylogous Z-substituted ketonophile.)

Ẍ-substituted ketenophiles:[213,214]

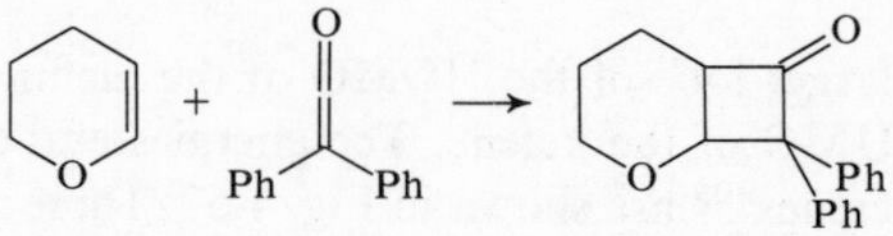

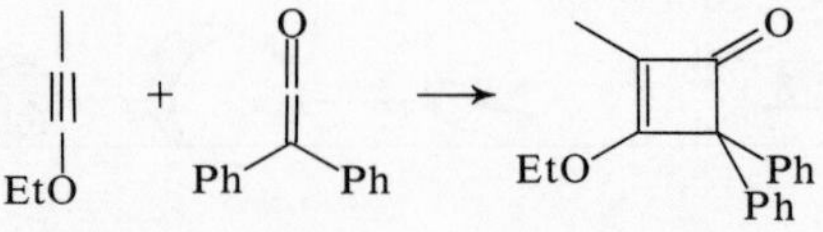

In the reaction of ketenes with enamines, the pathway is usually a stepwise one. Thus the regioselectivity of a reaction such as that between the enamine (**248**) and dimethyl ketene (**249**) is largely determined by the relative stability of the intermediate (**250**) and its regioisomer. The intermediate (**250**) is highly stabilized, and its regioisomer would not be. As in the case of the Diels-Alder reaction on p. 137, the formation of this intermediate is also consistent with frontier orbital control, in that the atoms with the large coefficients on the HOMO of the enamine and the LUMO of the ketene are the first ones to become bonded. However, some of this particular reaction[215] appears to be a concerted formation of the cyclobutanone (**251**), in which case this becomes another obedient case of the regioselectivity shown by an Ẍ-substituted olefin with a ketene, as illustrated above.

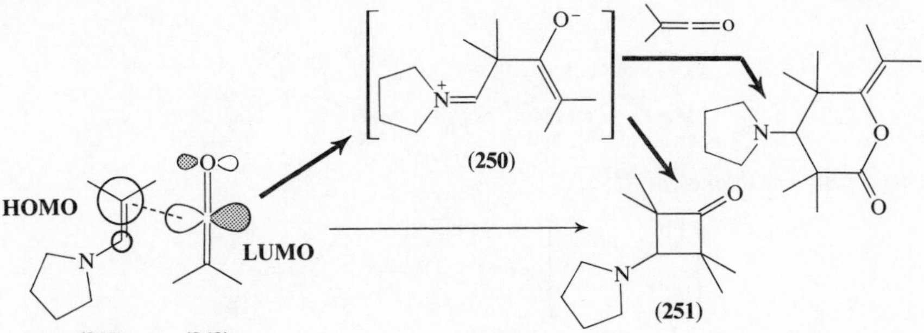

(**248**)　(**249**)　　(**250**)　　　　　　　(**251**)

The cycloaddition of ketenes to carbonyl compounds also shows the expected regioselectivity. In this case, both HOMO(ketone)/LUMO(ketene) and LUMO(ketone)/HOMO(ketene) interactions are likely to be important, but they lead to the same conclusions about regioselectivity. Lewis-acid catalysis is commonly employed in this case;

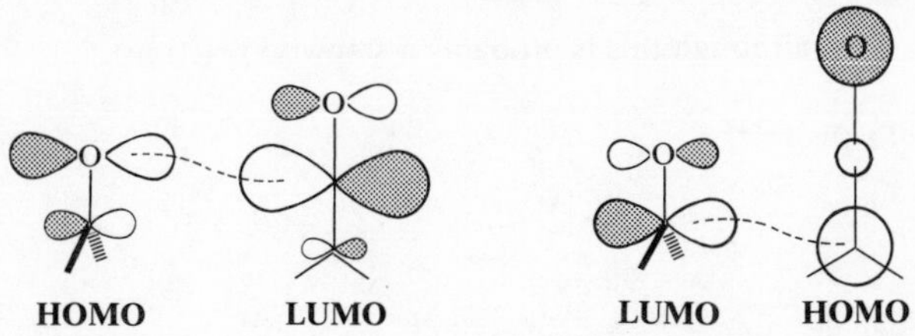

HOMO　　　LUMO　　　　　LUMO　　　HOMO

presumably it lowers the energy of the LUMO of the ketene (or that of the ketone) in the same way that it does with dienophiles (p. 161). Unsaturated ketones react with ketenes more easily than do saturated ketones, presumably because conjugation raises the HOMO of the ketones and lowers the energy of the LUMO of the ketone. Both of these factors will speed up the reaction. The following examples of heterocyclic 4-membered-ring-forming reactions, both of ketenes and of a number of related compounds, show in each case the expected regioselectivity.

Ketenes plus heteroketenophiles:[216-218]

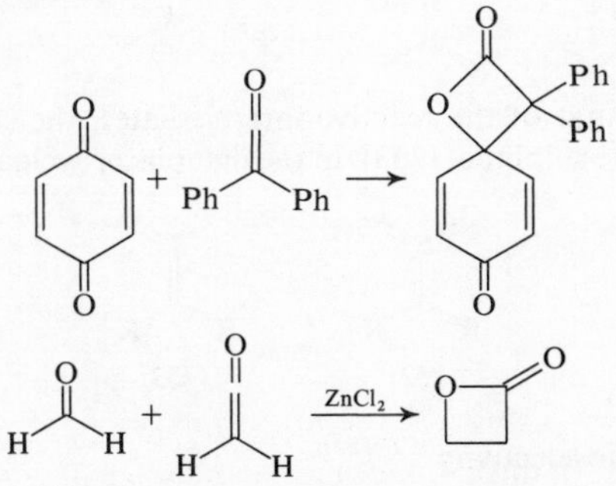

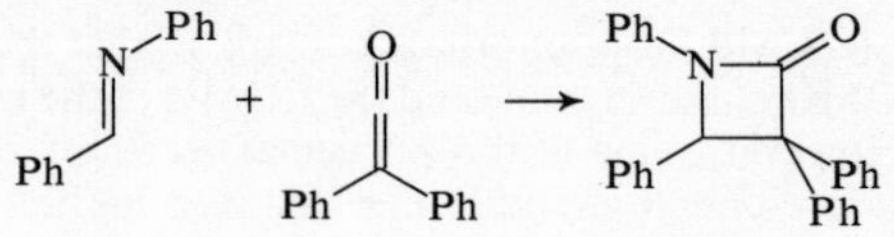

146

Isocyanates plus olefins:[219]

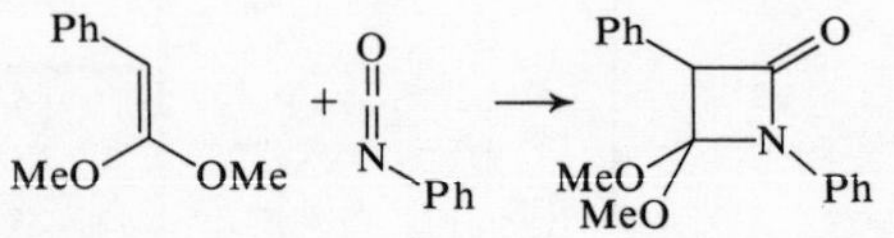

Dimerization of thioketones:[220]

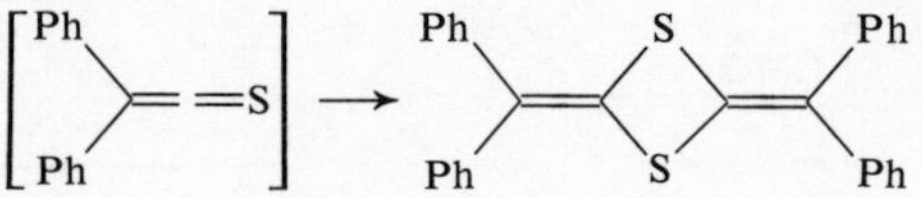

The Wittig reaction:[221]

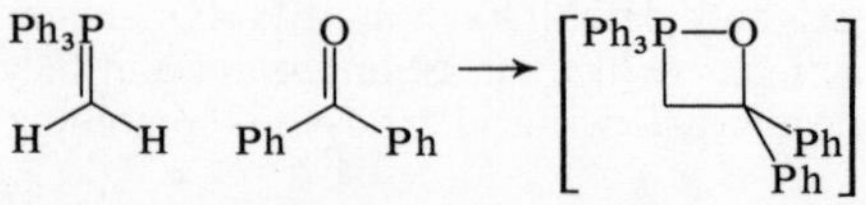

(although this is probably a stepwise reaction).

Dimerization of imines:[222]

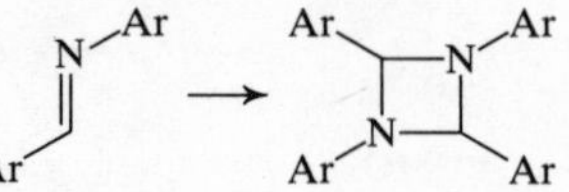

Dimerization of ketenes:[223, 224]

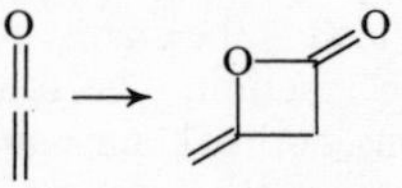

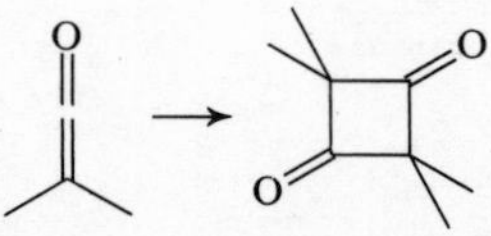

One special case is that of the reactive intermediates, the sulphenes (**252**), and the isolable compounds, the sulphines (**253**). In the first place, cycloadditions of these species

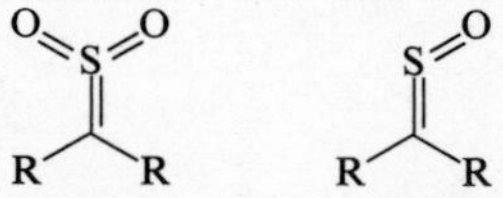

(**252**)   (**253**)

are normal in their regioselectivity:[225, 226]

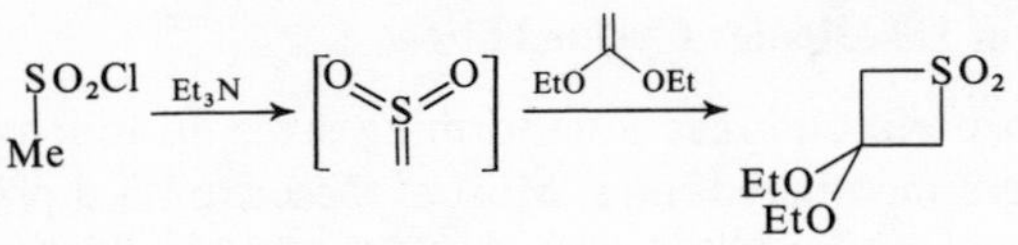

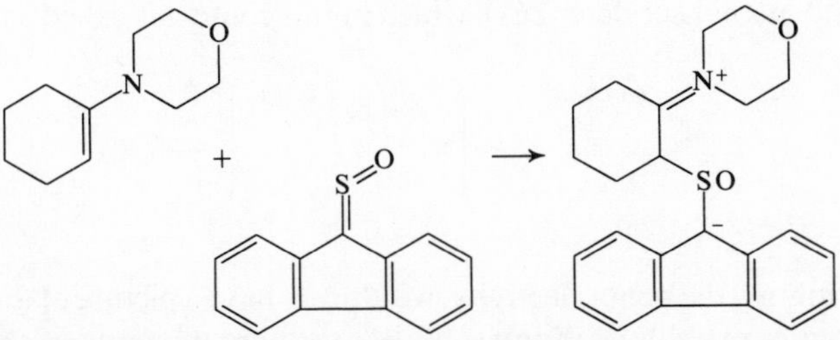

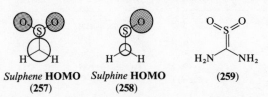

Calculations[227] indicate that these intermediates have very similar HOMO and LUMO energies and that their LUMO coefficients are very similar to each other; so we cannot look to these features to explain the striking difference in the reactivity of these two species. The former (252) cannot be isolated when sulphonyl chlorides are treated with base; instead the intermediate sulphene (e.g. 254) rapidly dimerizes.[228] By contrast, sulphines (e.g. 255 and 256) are an easily isolated class of compound.[226, 229] The same

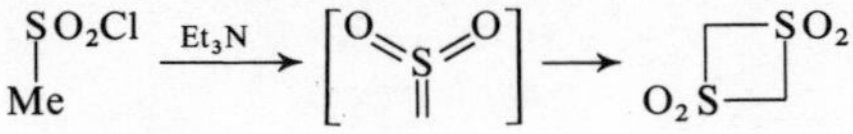

(254)

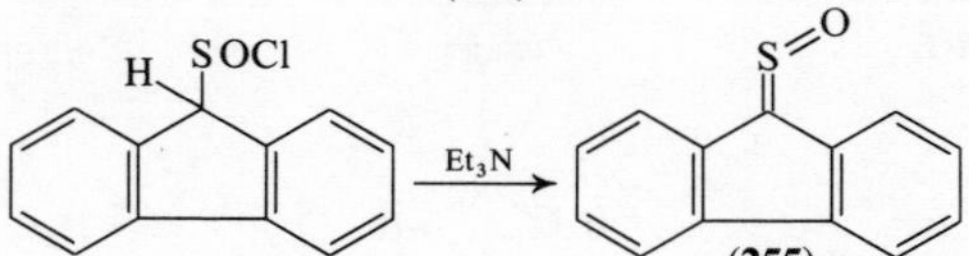

(255)

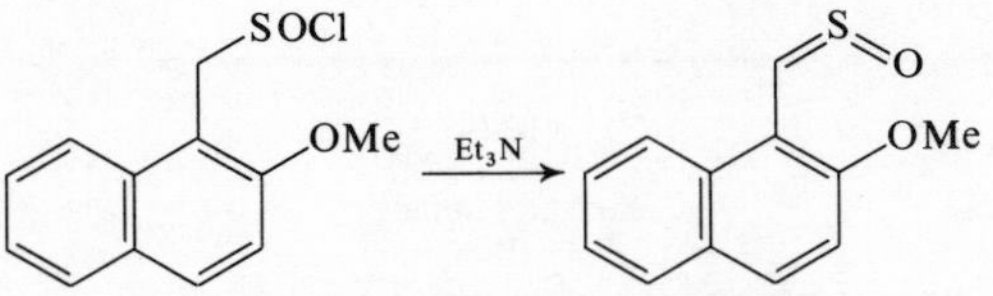

(256)

calculations, however, do reveal a difference in the coefficients of the HOMO: sulphenes (257) have a much larger coefficient on carbon and a much smaller coefficient on oxygen than do sulphines (258). The larger coefficient makes the $(c\beta)^2$ term of equation 2-7 larger and hence the dimerization reaction faster. The only sulphene which has been isolated is thiourea dioxide (259). The two $\ddot{\text{X}}$-substituents will polarize the HOMO of

Sulphene **HOMO**
(257)

Sulphine **HOMO**
(258)

(259)

the C=S in such a way as to reduce the coefficient on the carbon atom, and hence they decrease the rate of its dimerization.

4.4.5 Regioselectivity in 1,3-Dipolar Cycloadditions

We can now turn from pericyclic reactions forming six- and four-membered rings to those forming five-membered rings. Most of these are 1,3-dipolar cyclo-additions, typified by the reaction[230] of diazomethane (260), which is the 1,3-dipole, and methyl acrylate (261), which in this context is called a dipolarophile.

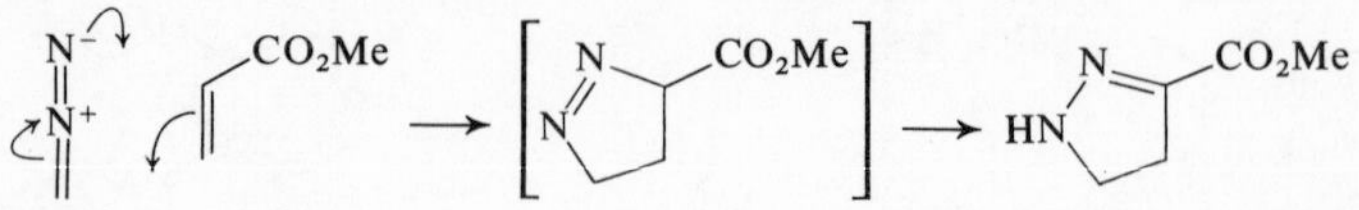

(260)   (261)

For determining the regioselectivity, we already have a picture of the polarization of the frontier orbitals of dipolarophiles: they are the same as the dienophiles with C-, Z-, and $\ddot{X}$-substituents, and for them we can continue to use Fig. 4-46. What we need now is a corresponding picture for dipoles, and this will not be nearly so easy to work out. To begin with, there are so many kinds of dipole. Even if we restrict ourselves to the elements carbon, nitrogen and oxygen, we still have many possible types of unsymmetrical dipoles (Fig. 4-58), and with

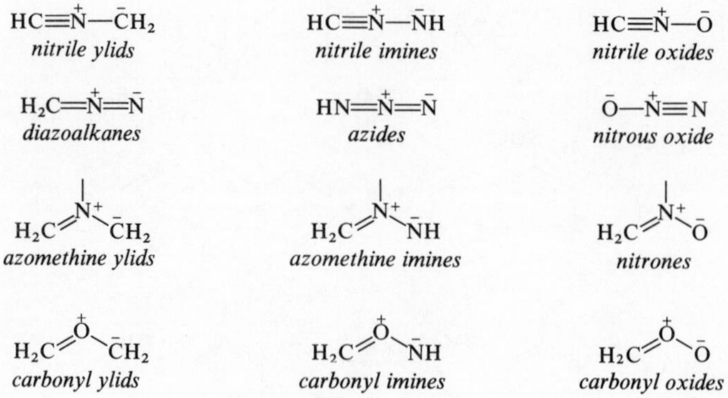

Fig. 4.58   The most important 1,3-dipoles

Table 4-3   Energies and 'coefficients' of 1,3-dipoles[231]

| Dipole | HOMO | | LUMO[a] | |
	Energy[b]	$(c\beta)^2/15^c$	Energy[b]	$(c\beta)^2/15^c$
Nitrile ylids	$-7\cdot7$	$HC\equiv\overset{+}{N}-\overset{-}{C}H_2$	$0\cdot9$	$HC\equiv\overset{+}{N}-\overset{-}{C}H_2$
$PhC\equiv\overset{+}{N}-CH_2^-$	$-6\cdot4$	$1\cdot07$ $1\cdot50$	$0\cdot6$	$0\cdot69$ $0\cdot64$
Nitrile imines	$-9\cdot2$	$HC\equiv\overset{+}{N}-\overset{-}{N}H$	$0\cdot1$	$HC\equiv\overset{+}{N}-\overset{-}{N}H$
$PhC\equiv\overset{+}{N}-\overset{-}{N}Ph$	$-7\cdot5$	$0\cdot90$ $1\cdot45$	$-0\cdot5$	$0\cdot92$ $0\cdot36$
Nitrile oxides	$-11\cdot0$	$HC\equiv\overset{+}{N}-\overset{-}{O}$	$-0\cdot5$	$HC\equiv\overset{+}{N}-\overset{-}{O}$
$PhC\equiv\overset{+}{N}-\overset{-}{O}$	$-10\cdot0$	$0\cdot81$ $1\cdot24$	$-1\cdot0$	$1\cdot18$ $0\cdot17$

Table 4-3 *continued*

Dipole	HOMO Energy[b]	HOMO $(c\beta)^2/15$[c]	LUMO[a] Energy[b]	LUMO[a] $(c\beta)^2/15$[c]
Diazoalkanes	$-9\cdot0$	$\overset{}{H_2C}=\overset{+}{N}=\overset{-}{N}$ 1·57 0·85	$1\cdot8$	$\overset{}{H_2C}=\overset{+}{N}=\overset{-}{N}$ 0·66 0·56
Azides	$-11\cdot5$	$HN=\overset{+}{N}=\overset{-}{N}$ 1·55 0·72	$0\cdot1$	$HN=\overset{+}{N}=\overset{-}{N}$ 0·37 0·76
$Ph\overset{+}{N}=N=\overset{-}{N}$	$-9\cdot5$		$-0\cdot2$	
Nitrous oxide	$-12\cdot9$	$\overset{-}{O}-\overset{+}{N}\equiv N$ 1·33 0·67	$-1\cdot1$	$\overset{-}{O}-N\equiv N$ 0·19 0·96
Azomethine ylids	$-6\cdot9$	$H_2C\overset{\overset{\overset{H}{+}}{N}}{\diagup\;\diagdown}CH_2$ 1·28 1·28	$1\cdot4$	$H_2C\overset{\overset{\overset{H}{+}}{N}}{\diagup\;\diagdown}CH_2$ 0·73 0·73
$ROOCCH\overset{\overset{\overset{Ar}{+}}{N}}{\diagup\;\diagdown}\overset{-}{C}HCOOR$	$-7\cdot7$		$-0\cdot6$	
Azomethine imines	$-8\cdot6$	$H_2C\overset{\overset{\overset{H}{+}}{N}}{\diagup\;\diagdown}\overset{-}{N}H$ 1·15 1·24	$-0\cdot3$	$H_2C\overset{\overset{\overset{H}{+}}{N}}{\diagup\;\diagdown}\overset{-}{N}H$ 0·87 0·49
$PhCH\overset{\overset{\overset{|}{+}}{N}}{\diagup\;\diagdown}\overset{-}{N}Ph$	$-5\cdot6$		$-1\cdot4$	
$H_2C\overset{\overset{\overset{|}{+}}{N}}{\diagup\;\diagdown}\overset{-}{N}COR$	$-9\cdot0$		$-0\cdot4$	
Nitrones	$-9\cdot7$	$H_2C\overset{\overset{\overset{H}{+}}{N}}{\diagup\;\diagdown}\overset{-}{O}$ 1·11 1·06	$-0\cdot5$	$H_2C\overset{\overset{\overset{H}{+}}{N}}{\diagup\;\diagdown}\overset{-}{O}$ 0·98 0·32
$H_2C\overset{\overset{\overset{R}{+}}{N}}{\diagup\;\diagdown}\overset{-}{O}$	$-8\cdot7$		$0\cdot3$	
$PhHC\overset{\overset{\overset{H}{+}}{N}}{\diagup\;\diagdown}\overset{-}{O}$	$-8\cdot0$		$-0\cdot4$	
Carbonyl ylids	$-7\cdot1$	$H_2C\overset{\overset{+}{O}}{\diagup\;\diagdown}CH_2$ 1·29 1·29	$0\cdot4$	$H_2C\overset{\overset{+}{O}}{\diagup\;\diagdown}CH_2$ 0·82 0·82
$Ar(CN)C\overset{\overset{+}{O}}{\diagup\;\diagdown}\overset{-}{C}(CN)Ar$	$-6\cdot5$		$-0\cdot6$	
$(NC)_2C\overset{\overset{+}{O}}{\diagup\;\diagdown}\overset{-}{C}(CN)_2$	$-9\cdot0$		$-1\cdot1$	
Carbonyl imines	$-8\cdot6$	$H_2C\overset{\overset{+}{O}}{\diagup\;\diagdown}\overset{-}{N}H$ 1·04 1·34	$-0\cdot2$	$H_2C\overset{\overset{+}{O}}{\diagup\;\diagdown}\overset{-}{N}H$ 1·06 0·49
Carbonyl oxides	$-10\cdot3$	$H_2C\overset{\overset{+}{O}}{\diagup\;\diagdown}\overset{-}{O}$ 0·82 1·25	$-0\cdot9$	$H_2C\overset{\overset{+}{O}}{\diagup\;\diagdown}\overset{-}{O}$ 1·30 0·24
Ozone	$-13\cdot5$		$-2\cdot2$	

[a] The important unoccupied orbital for cycloadditions is not, in a number of cases, strictly the lowest of the unoccupied orbitals; that is often an orbital at right angles to the reaction plane (see p. 93). The energies and coefficients in this column refer to the lowest of the unoccupied orbitals which is in the reaction plane.

[b] In eV; these values are estimates, wherever spectroscopic data ate available, and are otherwise calculated by CNDO/2.

[c] As explained in the text, both the coefficients and the β values are important in these cycloadditions. The $(c\beta)^2$ values have been divided by 15 to bring the numbers close to 1. The β values are calculated assuming that the new bonds are being made from *carbon* atoms in the dipolarophile to the carbon, nitrogen or oxygen atoms of the dipole.

very few of them can we find simple arguments from which to deduce the relative sizes of the coefficients at the ends of the conjugated systems or the energies of the frontier orbitals. Furthermore, in all the unsymmetrical cases, the bonds being made are no longer always C—C bonds, as they are in the common Diels-Alder reactions. Just as with heterodienes and heterodienophiles, we must include an estimate of the appropriate resonance integral, β, as well as the co-efficients of the atomic orbitals, c. Fortunately, all this work has been done for us by Houk,[231] and we shall take his figures on trust. They are summarized in Table 4-3, which gives the frontier orbital energies for the unsubstituted cases and also gives the relative values of $(c\beta)^2$ at each end of the dipole for both frontier orbitals. Because β contains S, the overlap integral, it is a distance-dependent function, so that the values chosen by Houk involved a guess about the distance apart of the atoms in the transition state. In Table 4-4 you can see that the choice of distance is critical: it was made at 1·75 Å on the basis that cycloaddition reactions show large negative activation volumes and sizeable steric effects.

Table 4-4 β-values in eV calculated by CNDO/2 for σ overlap between 2p orbitals of C, N, and O

R, Å	β_{CC}	β_{CN}	β_{CO}	β_{NN}	β_{NO}	β_{OO}
1·50	6·97	7·20	7·05	7·18	6·92	6·63
1·75	6·22	5·83	5·38	5·35	4·81	4·19
2·00	5·00	4·35	3·77	3·65	3·02	2·45
2·50	2·63	2·14	1·53	1·40	1·04	0·68
3·00	1·20	0·78	0·55	0·45	0·28	0·16

The first thing we must do to account for the regioselectivity of 1,3-dipolar cycloadditions is to assess whether a particular reaction that we are looking at has a smaller separation between HOMO(dipole) and LUMO(dipolarophile)—as will be common for electron-deficient dipolarophiles (see Fig. 4-26)—or between the LUMO(dipole) and the HOMO(dipolarophile)—as will be common for electron-rich dipolarophiles (see Fig. 4-26). The former we shall call *dipole-HO controlled* and the latter *dipole-LU controlled* reactions. We can do this simply by looking at the energies of the dipoles in Table 4-3, and at the energies of the representative dipolarophiles in Fig. 4-46. Often our reactions are not done with the simple unsubstituted dipoles, and we therefore have to assess the effect on the energies of any substituents which may be present. To do this we can take advantage of our qualitative understanding of the effects of C-, Z-, and Ẍ-substituents on olefins. Table 4-3 shows the effect of some of the commonly found substituents on the energy of the frontier orbitals; you will see there that phenyl groups do indeed raise the HOMO and lower the LUMO energy, and that ester and cyano groups likewise lower both the HOMO and the LUMO energy.

Having discovered which interaction is of primary importance, we next look at the 'coefficients' of the relevant orbitals. Let us say that the reaction is a dipole-HO-controlled one: in this case they will be the coefficients of the HOMO of the dipole (Table 4-3) and of the LUMO of the dipolarophile (Fig. 4-46). Regioselectivity should follow in the usual way from the large-large/small-small interaction. Again, substituents present on the dipole will modify the coefficients shown for the unsubstituted cases in Table 4-3, but it should be an easy matter, at least qualitatively, to predict how these substituents will affect the coefficients.

The following discussion is limited to a few representative cases which illustrate the way all this works; there is more discussion of these and all the other cases in Houk's papers.[231] The cases discussed below, and in Houk's papers, need a much more elaborate discussion than was necessary for the Diels-Alder reactions earlier in this chapter. The balance of factors leading to a particular regioselectivity is often a very close one: the choice of which pair of frontier orbitals to take is sometimes difficult, and the fact that some frontier orbitals are not strongly polarized forces us to judge each case carefully on its merits. Generalizations are not as simple and predictions not as firm as they were for Diels-Alder reactions. Nevertheless, it should be remembered that before frontier orbital theory was applied to this problem there appeared to be no easily comprehended pattern at all. The best effort at explaining it was Huisgen's theory[125] of the maximum gain in σ-bonding energy, where the effect of the nature of the σ-bonds being made is in fact related to the β-term in the frontier orbital treatment. The suggestion[232,190] that diradical intermediates were involved has been severely criticized in a paper[233] which, coming before frontier orbital theory had been developed, described the orientation of 1,3-dipolar cycloadditions as the biggest unsolved problem in the field.

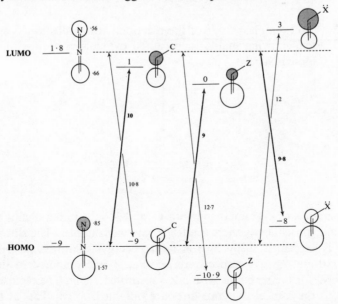

Fig. 4-59 Frontier orbitals for diazomethane and dipolarophiles

4.4.5.1   An Unsubstituted Dipole: Diazomethane. Diazomethane has frontier orbital energies and $(c\beta)^2$ values as shown in Fig. 4-59, which also includes the energies and coefficients of the common types of dipolarophile. We can see immediately that the smallest separation in energy of all of the possible frontier orbital interactions (the double-headed arrows) is for the reaction between diazomethane and a Z-substituted olefin: $E_{LUMO(dipolarophile)}-E_{HOMO(dipole)}$ in this case is 9 eV. Reactions of diazomethane with electron-deficient olefins are by far the fastest and most often encountered of the cycloaddition reactions of diazoalkanes, and we can now see that the reason for this is the strong frontier orbital term (in equation 2-7) for this particular combination. This reaction is therefore dipole-HO-controlled, so we can now look at the coefficients, where we see that the orientation will be that (Fig. 4-60) in which the carbon end of the dipole becomes bonded to the β-carbon of the Z-substituted olefin. This is exactly the orientation which has so often been observed, and which we used as an example at the beginning of this book in posing the problem of how we should explain such selectivity.

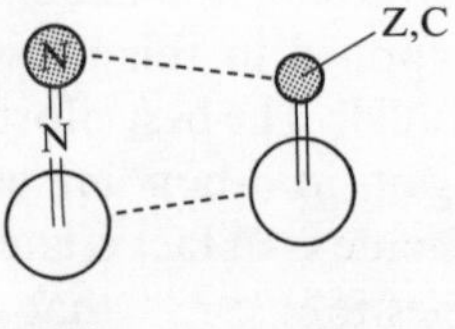

HOMO   LUMO

Fig. 4-60   Regioselectivity for diazomethane reacting with electron-deficient and conjugated olefins

Here are two examples[234, 235] of the reaction with Z-substituted olefins, the second of them showing that Δ^1-substituted pyrazolines are the first-formed products in these reactions.

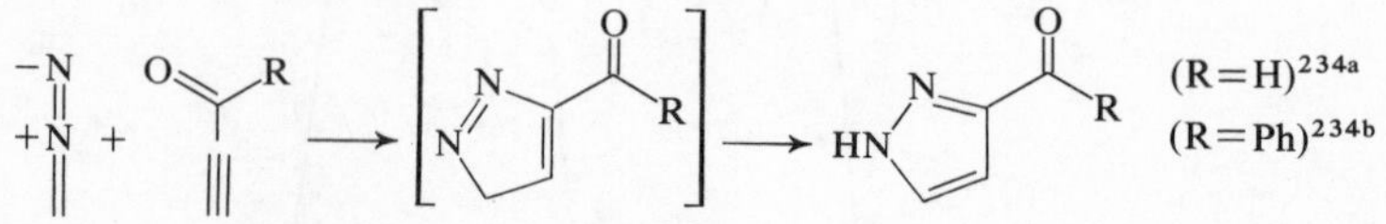

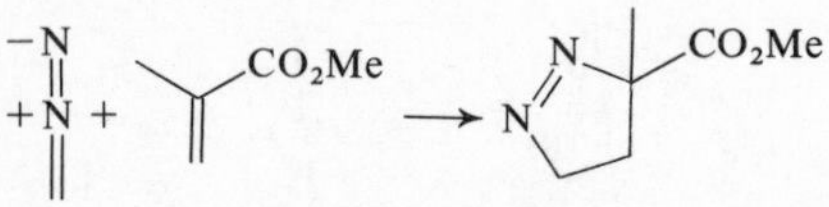

The reactions of diazomethane with C- and $\ddot{X}$-substituted olefins are much slower, and consequently there are fewer known examples. The slower rate of reaction is explained easily by the larger energy-separation in the frontier orbitals (10 and 9·8 eV, respectively, see Fig. 4-59) compared to the typical value (9 eV) for the reactions with a Z-substituted olefin. The regioselectivity, however, is the same: Δ^1-pyrazolines are obtained.[236, 237] This at first sight surprising observation, is clearly explained by the change from dipole HO control in the cases of the C- and Z-substituted olefins (supported, incidentally, by the positive ρ-value for the former[236]) to dipole LUcontrol in the case of $\ddot{\text{X}}$-substituted olefins.

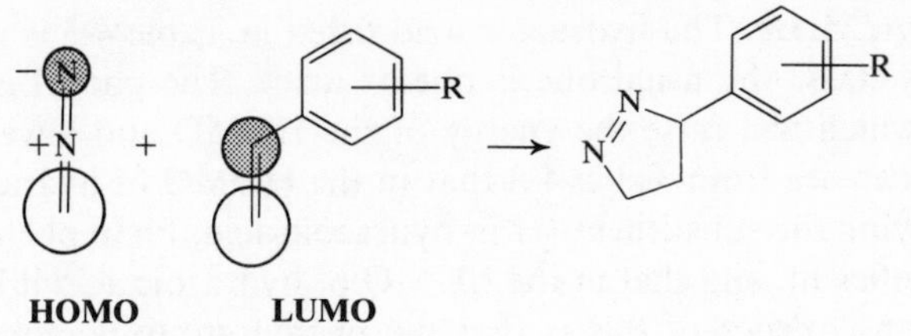

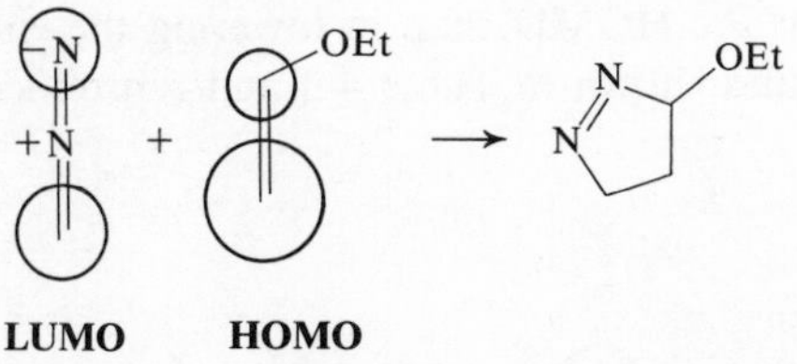

4.4.5.2 Substituted Diazomethanes. The effect of substituents on the diazoalkane can readily be predicted. Electron-donating substituents will raise both the HOMO and the LUMO and will speed up reactions with electron-deficient olefins. This is confirmed by the greater reactivity of alkyl diazomethanes in cycloadditions.[238] Electron-withdrawing substituents, such as the keto group in a diazoketone, will lower both the HOMO and the LUMO energies of the dipole and will speed up reactions only with electron-rich dipolarophiles. Diazoketones do react easily with enamines. Furthermore, with a Z-substituent on the carbon atom of the diazomethane, the coefficient on the carbon atom will be reduced in the LUMO, just as it is in the LUMO of an olefin with a Z-substituent. Since the $(c\beta)^2$ terms for the LUMO of diazomethane are so similar, the Z-substituent should polarize them decisively. This reaction will now be decidedly a dipole-LU-controlled one, and the regioselectivity will be that shown in Fig. 4-61. This regioselectivity has been observed[239] with an enamine (**263**) and diazoacetic ester (**262**).

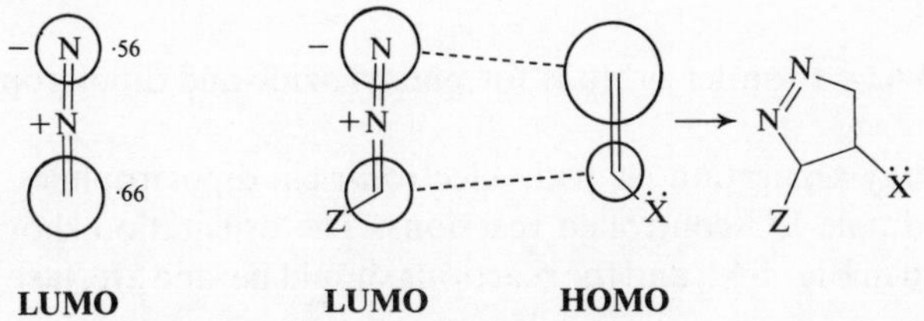

Fig. 4-61   Regioselectivity for a diazoketone reacting with an electron rich olefin

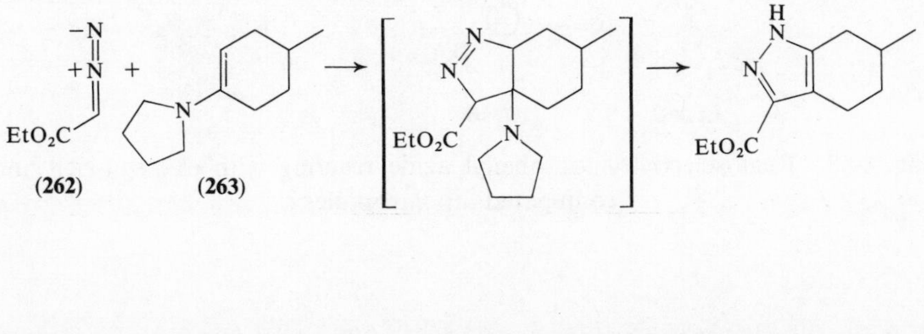

154

4.4.5.3 *Phenyl Azide.* The hydrazoic acid listed in Table 4-3 is not the usual dipole of this class; the usual one is phenyl azide. The phenyl group is a C-substituent, which will raise the energy of the HOMO and lower that of the LUMO. We can see from Table 4-3 that in the HOMO of hydrazoic acid the nitrogen carrying the substituent (H in hydrazoic acid, Ph in phenyl azide) has the larger coefficient, and that in the LUMO of hydrazoic acid it has a smaller value. The consequence of this is that the phenyl group is more effective in raising the energy of the HOMO than in lowering the energy of the LUMO. The result is the values shown in Table 4-3 and reproduced in Fig. 4-62. The

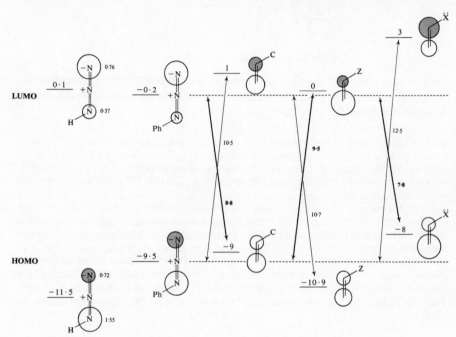

Fig. 4-62 Frontier orbitals for phenyl azide and dipolarophiles

smallest energy-separation is with electron-rich dipolarophiles, which will clearly give dipole-LU-controlled reactions. The orientation should therefore be that shown in Fig. 4-63, and the reactions should be, and are, fast. Orientation

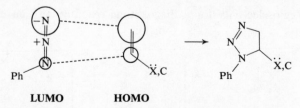

Fig. 4-63 Regioselectivity for phenyl azide reacting with electron-rich and conjugated dipolarophiles

like this has often been observed. Here are just two examples:[240, 241]

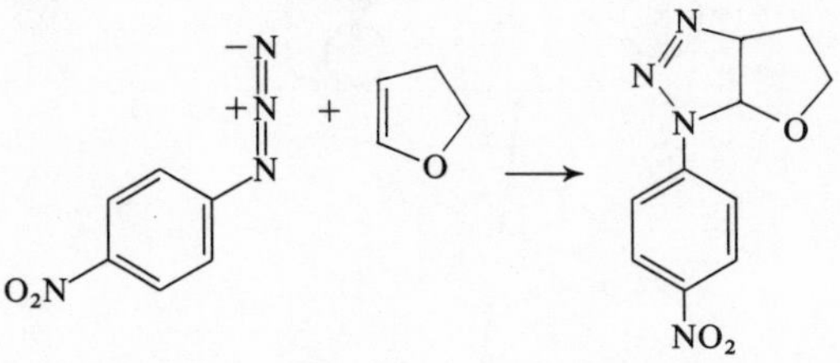

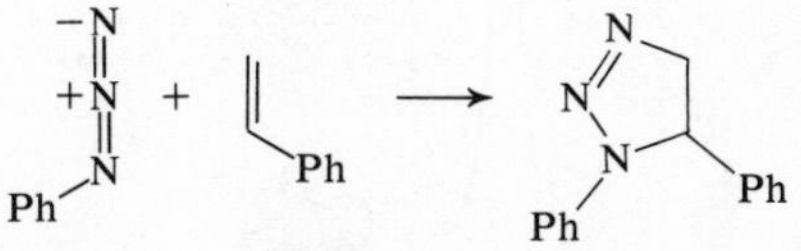

(264)

When phenylacetylene (265) replaced styrene (264) in the last reaction, the regio-selectivity was sharply reduced, and nearly equal amounts of the 1,5-diphenyltriazole (266) and the 1,4-diphenyltriazole (267) were obtained.[242]

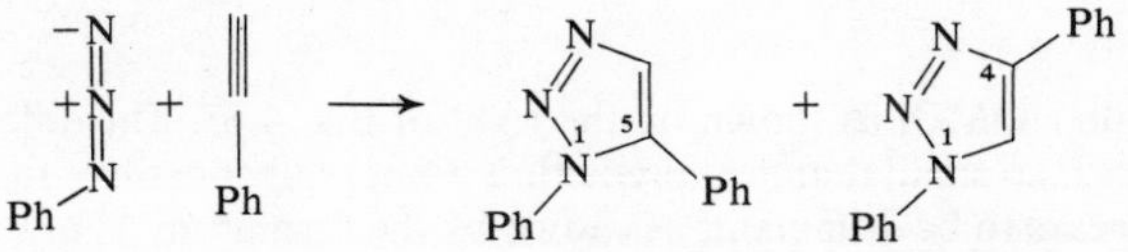

(265)　　　　(266) 52%　　　　(267) 43%

This puzzling observation can readily be explained, using frontier orbital theory. The second π-bond of an acetylene is stronger than the first, because it is made between two atoms held close together by the first π-bond. The overlap of the p orbitals on carbon is therefore stronger, and an acetylene has a lower-energy HOMO than ethylene. This is shown to be true by photo-electron spectroscopy, where the HO level is generally found to be 0·4 to 0·9 eV lower than that of the corresponding alkene. We can also relate this observation to the familiar notion that alkynes are *less* reactive towards electrophiles like bromine than are the corresponding alkenes. Curiously, the LUMO is not raised for alkynes relative to alkenes. This is shown by UV spectroscopy, where phenylacetylene (λ_{max} 245 nm) and styrene (λ_{max} 248 nm) can be seen to have rather similar *separations* of their HOMOs and LUMOs. Thus, with a LUMO not much, if at all, raised, it is not unexpected to find that acetylenes with electron-withdrawing substituents are more reactive towards nucleophiles than are the corresponding alkenes.

The effect of going from styrene to phenylacetylene is therefore to lower the HOMO and the LUMO by about, say, 0·5 eV. This makes what was clearly a dipole-LU-con-trolled reaction into one which is affected by both interactions (Fig. 4-64). Since dipole-HO control leads to the opposite regioselectivity, it is not so surprising that both orientations are now observed.

With electron-deficient dipolarophiles and phenyl azide, the situation is again deli-cately balanced. The reaction is only just a dipole-HO-controlled one (9·5 eV against 10·7 eV). For the dipole-HO-controlled reaction, we should expect to get adducts orientated as shown on the left of Fig. 4-65. However, a phenyl group reduces the co-efficient at the neighbouring atom both for the HOMO and for the LUMO, and this will reduce the polarization of the HOMO. Conversely, it will increase the polarization for the LUMO and hence increase the effectiveness of the dipole LUMO's interaction with

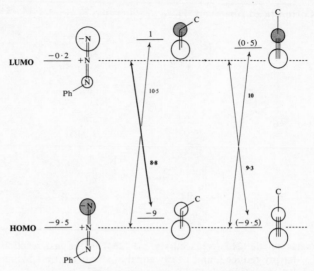

Fig. 4-64 Frontier orbitals for phenyl azide, styrene and phenylacetylene

the dipolarophile HOMO, as shown on the right of Fig. 4-65. The difference in energy for the two cases is so small that firm prediction is not really possible. In practice, dipole HO control appears to be dominant, as shown by the formation[243] of the adduct (**268**),

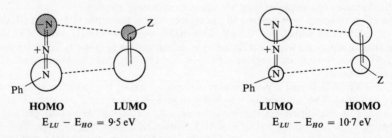

but it only needs a small change in the structure of the dipolarophile, such as the addition of an α-methyl group (**269**), for some dipole-LU control to become evident. The methyl group in **269** will, of course, raise both the LUMO and the HOMO of the dipolarophile, making the HO/LU separations still more nearly equal.

Fig. 4-65 Regioselectivity for phenyl azide reacting with electron-deficient olefins

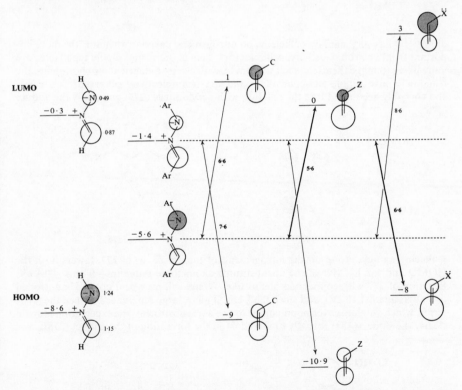

4.4.5.4 *Azomethine Imines*. The commonly used azomethine imine (**270**) has phenyl groups at both ends and hence has a raised HOMO relative to the unsubstituted system. Because the coefficients at the terminal atoms of the dipole are smaller in the LUMO than they are in the HOMO, the phenyl groups do not lower the energy of the LUMO as much as they raise the energy of the HOMO. These effects on the energy are recorded in Table 4-3 and reproduced in Fig. 4-66.

Fig. 4-66 Frontier orbitals for **270** and dipolarophiles

With conjugated dipolarophiles like styrene, the reaction is only just a dipole-HO-controlled one, and mixtures can be expected. Styrene does, in fact, give both regio-isomers (**271** and **272**) in nearly equal amounts.[125] With an acetylenic dipolarophile,

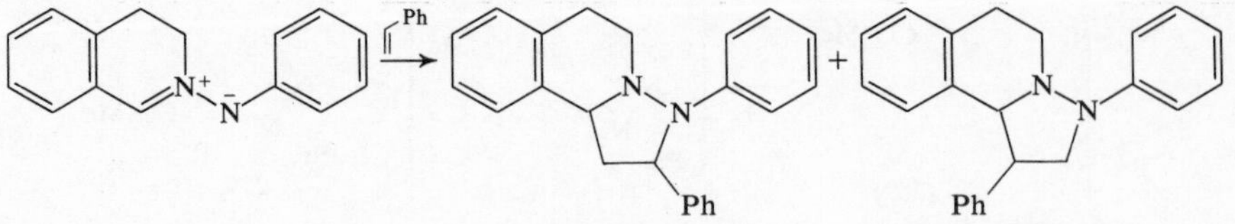

(270)   (271) *31%*   (272) *55%*

dipole-LU control   *dipole-HO control*

phenylacetylene, the lowering of the HOMO of an acetylene relative to that of the corresponding olefin should make the reaction more predominantly dipole-HO-controlled. The experimental observation[125] is, in fact, the opposite of what we would expect: phenylacetylene (273) with the azomethine imine (270) gives only the adduct (274). The

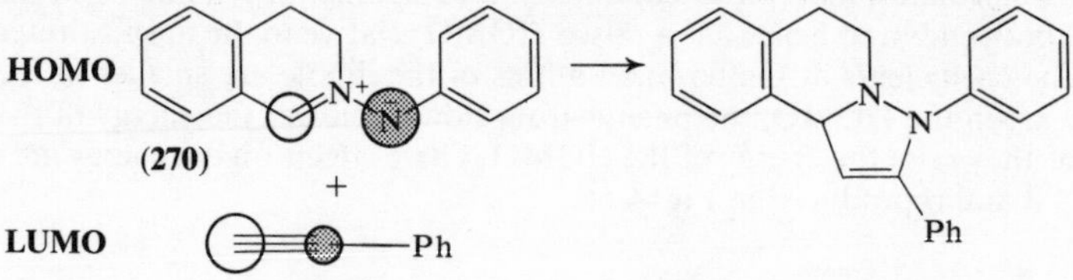

dipole-HO has very similar coefficients on nitrogen and carbon; when it is the more important frontier orbital, as here, other factors, such as steric and dipole repulsions, are more likely to make themselves felt. However, with electron-deficient dipolarophiles, the reaction is also dipole-HO-controlled, and the regioselectivity observed[125] is easily and correctly accounted for: the reaction with acrylonitrile (275) gives only the adduct (276).

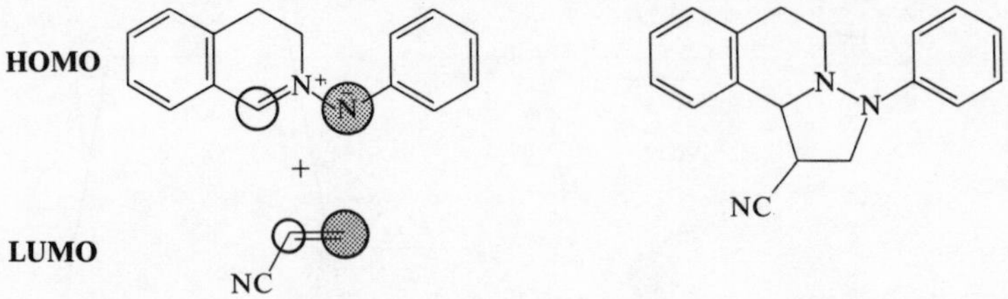

Placing an acyl group on the nitrogen end of the dipole, as in 277, lowers both the HOMO and the LUMO of the unsubstituted azomethine imine to -9 and -0.4 eV. Reaction of 277 with conjugated olefins like styrene will now be dipole-LU-controlled (8.6 eV as against 10 eV), and since the LUMO has a large difference between the $(c\beta)^2$ terms, which will moreover be enhanced by the acyl substituent, the expectation is again clearly the same as that actually observed[244] in the formation of the adduct (278).

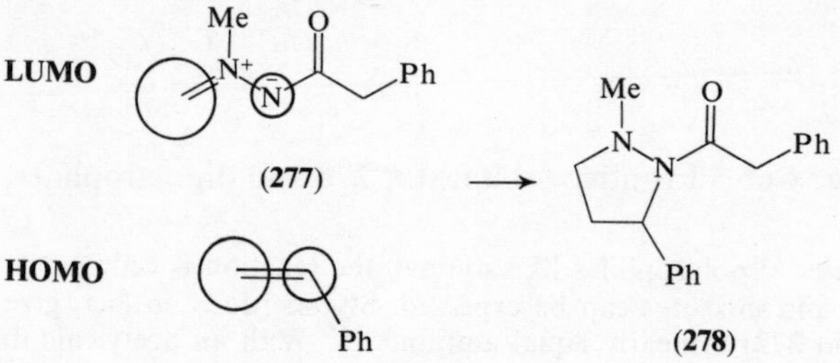

Placing an acyl group on the carbon atom end of the dipole, as in **279**, again leads to dipole-LU control, but this time the acyl group will reduce the difference between the coefficients. Overwhelming preference for one orientation is not to be expected, but all three kinds of dipolarophile should give adducts of the type (**280**). This is exactly what

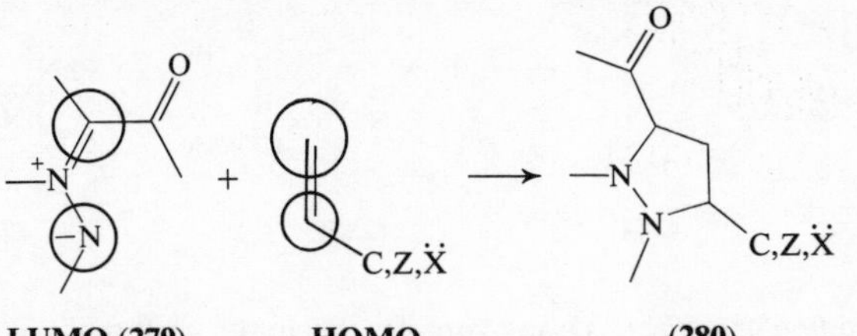

LUMO (279)   **HOMO**   (**280**)

has been observed for the cycloaddition reactions of sydnones (**281**). An intermediate of the general formula (**282**) is produced in the first instance, and this loses carbon dioxide in a retro 1,3-dipolar cycloaddition. With olefinic dipolarophiles, a further tautomerism

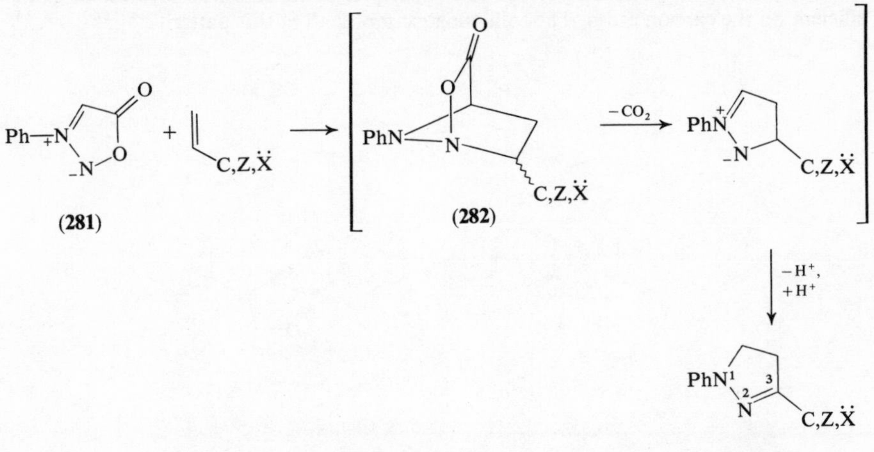

(**281**)   (**282**)

$-CO_2$

$-H^+,$
$+H^+$

(**283**)

takes place, and the major product is always the 3-substituted pyrazoline (**283**). Here is an example of each kind of dipolarophile showing this regioselectivity:[245]

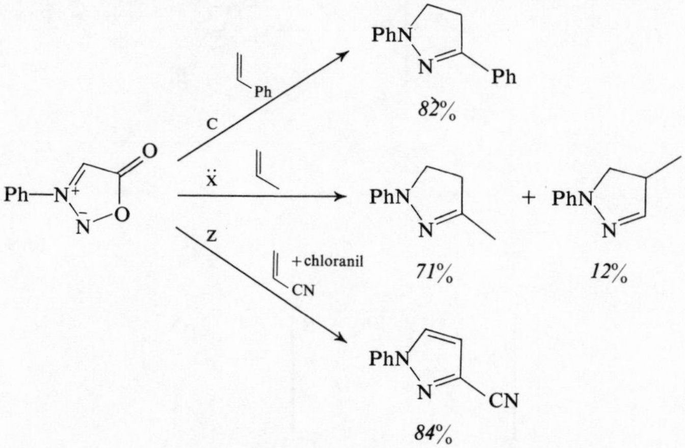

82%

71%   12%

+ chloranil

84%

With an acetylenic dipolarophile, slightly more dipole-HO control can be expected, and this has been observed. Propiolic ester (284) gives[246] a substantial, but still minor, amount of the 4-substituted pyrazoline (286).

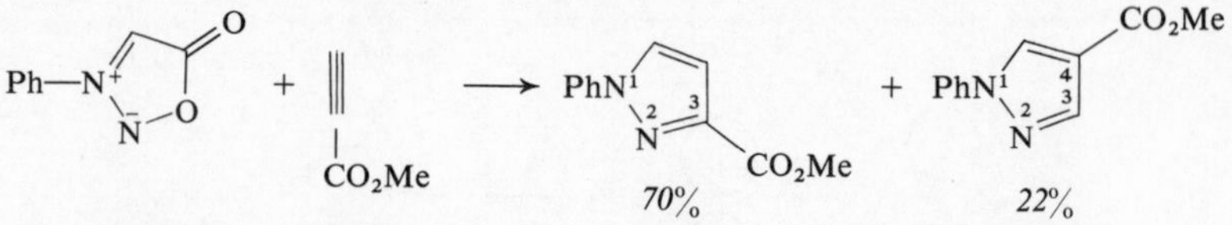

Ph—N$^+$ ... CO_2Me (281) + CO_2Me (284) ⟶ PhN ... CO_2Me 70% (285) + PhN ... CO_2Me 22% (286)

4.4.5.5 Heterodipolarophiles. Dipolarophiles such as carbonyl groups and cyano groups (for the frontier orbitals see p. 16) also show an orientation in agreement with frontier orbital theory. These dipolarophiles will usually be involved in dipole-HO-controlled reactions, and the orientation observed will therefore depend upon the LUMO of the dipolarophile. For a C=X double (or triple) bond, this will have the large co-efficient on the carbon atom. The following examples all fit this pattern:[241,247,125,248]

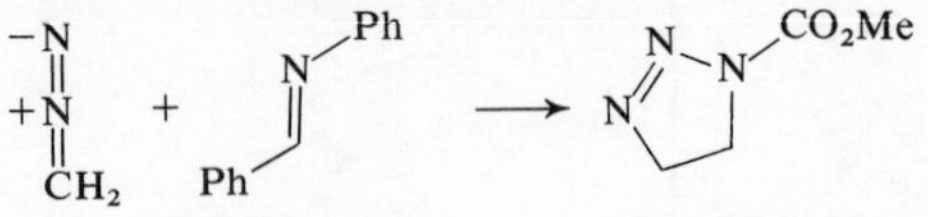

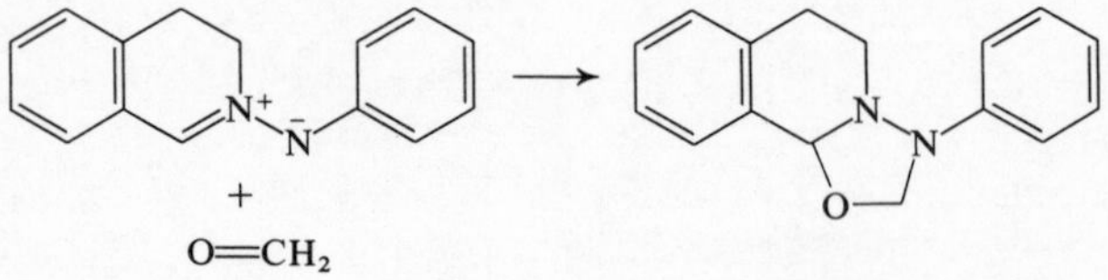

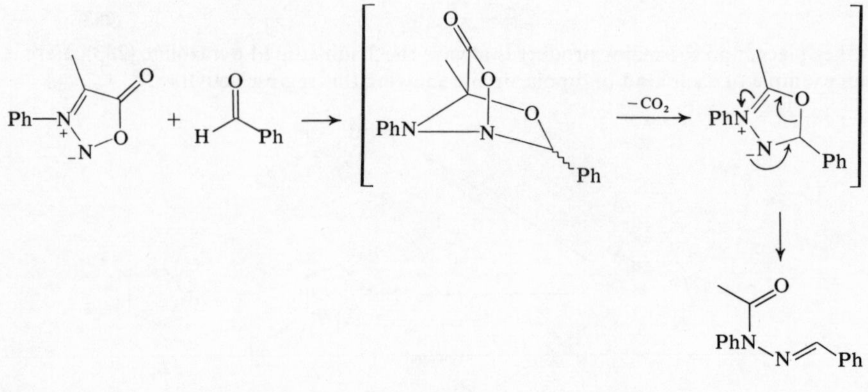

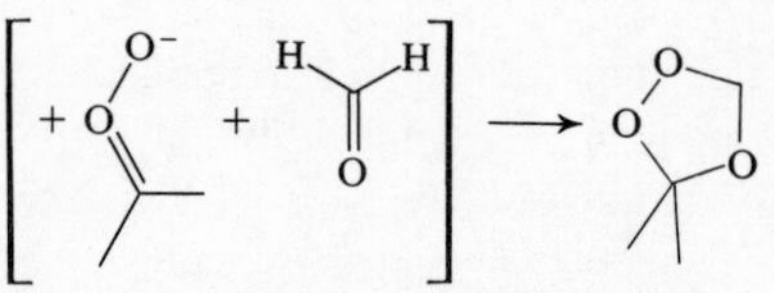

We are far from exhausting the subject of regioselectivity in dipolar cyclo-additions with these few examples. Frontier orbital theory has been successful in accounting for most of the otherwise bewildering trends in regioselectivity. No other theory, whether based on polar or steric factors, or on the possibility of diradical intermediates, has had anything like such success. It is plain that, as so often happens in science, a very large body of data has at last been reduced to an amenable set of principles.

4.5 Lewis Acid Catalysis of Diels-Alder Reactions

Diels-Alder reactions are, as we know, little influenced by polar factors, such as changing the solvent from a non-polar to a polar one. Yet Lewis acids exert a strong catalysing effect. Furthermore, Lewis-acid-catalysed Diels-Alder reactions are not only faster but also more stereoselective and more regio-selective than the uncatalysed reactions. For this reason, the catalysed reactions are of great synthetic importance. Thus, cupric ion has been used[249] to catalyse the Diels-Alder reaction of the 5-substituted cyclopentadiene (287), in order to

(287)

k_1 is unaffected by catalysis, k_2 is greatly increased in the presence of Cu^{2+}

make the Diels-Alder reaction compete more favourably with the 1,5-hydrogen transfer reaction, which isomerizes the cyclopentadienes. This is an example in which advantage is taken of the increase in rate of the catalysed reaction. Piperylene (288) and methyl acrylate (289) give mainly the 'ortho' product (290),

(288)	(289)		(290)		
		without AlCl₃	*90%*		*10%*
		with AlCl₃	*98%*		*2%*

as we already know, but this preference is increased with Lewis acid catalysis.[250] Similarly, in a synthesis of a natural product, isoprene (291) and 3-methylbut-3-ene-2-one (292) gave a mixture of ketones in which the "para" isomer (293) was the major one, but it could only with great difficulty be separated from the unwanted "meta" isomer. The synthesis of the "para" isomer alone was achieved by adding stannic chloride to the reaction mixture.[251] Finally, let us add an example of enhanced stereoselectivity to these examples of enhanced regioselectivity: the reaction of cyclopentadiene (294) with methyl acrylate (295). The *endo* adduct (296) is the major one in the uncatalysed reaction, but

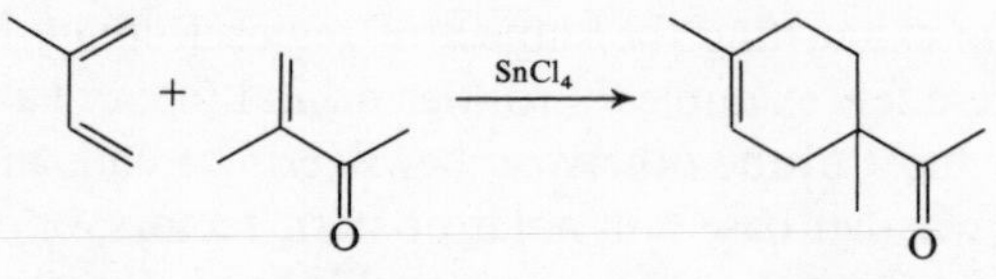

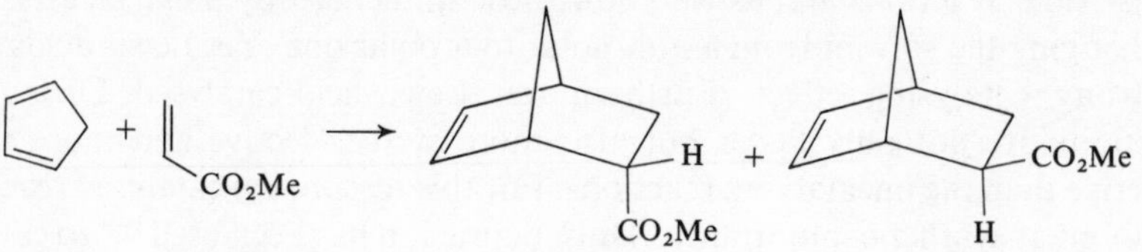

(294)	(295)	(296)	(297)
	without AlCl₃ at 0°	*88%*	*12%*
	with AlCl₃ at 0°	*96%*	*4%*
	with AlCl₃ at −80°	*99%*	*1%*

the proportion of this isomer is much higher when aluminium chloride is present.[252]

All these features of Lewis acid catalysis can be explained by the effect the Lewis acid has on the LUMO of the dienophile.[36, 202, 253] We shall take acrolein (298) as the simplest dienophile. The Lewis acid forms a salt (299) with the

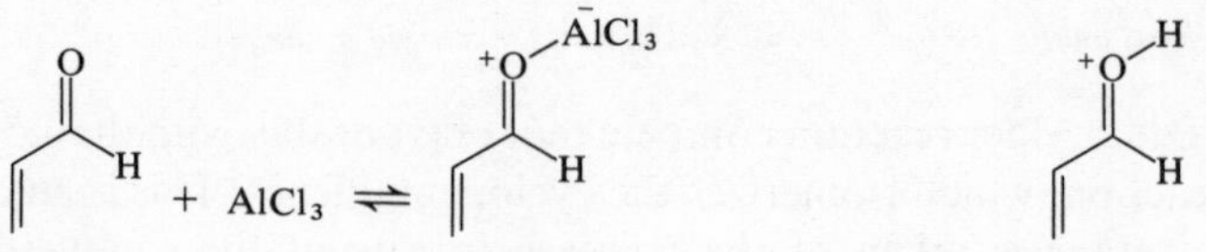

dienophile, and it is this salt which is the more active and selective dienophile. For simplicity in the calculation and the discussion, protonated acrolein (300) is used instead of the Lewis salt (299).

When we were trying to estimate the energies and polarities of the frontier orbitals of acrolein itself (pp. 115 and 124), we added to the orbitals of a simple diene a contribution from the allyl-cation-like nature of acrolein. The effect of adding a proton to acrolein is to enhance its allyl-cation-like nature. For the frontier orbitals of acrolein, therefore, we must add a larger contribution from the allyl cation. The results are: (i) both HOMO and LUMO are even lower in energy, (ii) the HOMO will have the opposite polarity at the C=C double bond, the contribution from the allyl cation now outweighing the contribution from butadiene, and (iii) the LUMO will have even greater polarization, the β-carbon carrying an orbital with an even larger coefficient, and the α-carbon carrying an

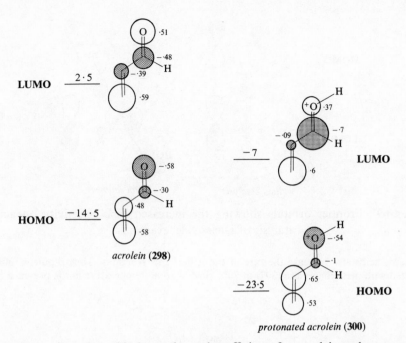

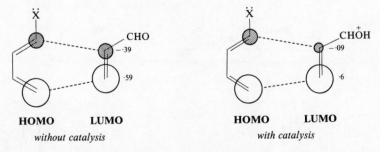

Fig. 4-67 Frontier orbital energies and coefficients for acrolein and protonated acrolein

orbital with an even smaller coefficient. A calculation[202] sets values on these energies and coefficients, as shown in Fig. 4-67.

The lowering in energy of the LUMO makes the $E_{LUMO(dienophile)} - E_{HOMO(diene)}$ a smaller number and therefore increases the rate. The increased polarisation of the LUMO of the C=C double bond increases regioselectivity (Fig. 4-68). Finally the increased LUMO coefficient on the carbonyl carbon makes the secondary interaction (Fig. 4-69) greater than in the uncatalysed case, and accounts for the greater *endo* selectivity.

Fig. 4-68 Frontier orbitals showing increased regioselectivity for acid-catalysed Diels-Alder reactions

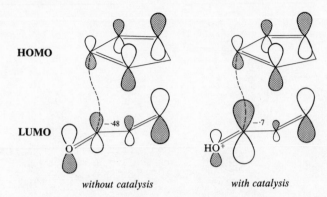

HOMO

LUMO

without catalysis　　　　*with catalysis*

Fig. 4-69　Frontier orbitals showing the increased *endo* selectivity on acid catalysis of Diels-Alder reactions

Lewis acids also increase the rate of the Alder 'ene' reactions. Thus β-pinene (**301**) reacted with methyl acrylate (**302**) in 70% yield at room temperature in the presence of

aluminium chloride,[254] a reaction which would probably not have been possible without the Lewis acid. The lowering of the LUMO energy of the methyl acrylate again accounts for the increase in the rate of reaction. Nothing is known of the effect of Lewis acids on the regioselectivity or stereoselectivity of 'ene' reactions.

We can conclude this section on Lewis acid catalysis with a striking example of its use in solving a long-standing problem in steroid synthesis. We saw, on p. 140, how citraconic anhydride added to 1-vinylnaphthalene with inappropriate regioselectivity for steroid synthesis. The same is true of the uncatalysed reaction of 2,6-xyloquinone (**304**) with the diene (**305**). But, when boron trifluoride was added to the reaction mixture, it formed a salt (**306**) at the more basic (and, as it happens, less hindered) carbonyl group. The result was that the polarization of the LUMO of the C=C double bond was reversed, the major adduct was the appropriate one (**307**), and a steroid synthesis could be completed.[255]

4.6 Site Selectivity in Cycloadditions

Site selectivity is that selectivity shown by a reagent towards one site (or more) of a polyfunctional molecule, when several sites are, in principle, available. The preference for electrophilic attack at the *ortho* and *para* positions of $\ddot{X}$-substituted benzenes is just one of many examples discussed in Chapter 3. In cycloadditions, site selectivity always involves a pair of sites; thus, butadiene reacts faster with the quinone (**308**) at C-2 and C-3 than at C-5 and C-6,[256] and Diels-Alder reactions of anthracene (**309**) generally take place[257] across the

9,10-positions rather than across the 1,4 or 3,9a. The familiar explanation for this example of site selectivity is that reaction at the 9,10-position creates two isolated benzene rings, whereas reaction at the 1,4-position would create a naphthalene nucleus, which is a less stable arrangement of two benzene rings. This explanation relies on the influence of product-like character in the transition state, but we may also note that the same product is accounted for by looking at the frontier orbital coefficients of the starting materials: the largest coefficients in the HOMO of **309** are at the 9,10-positions (p. 58). Most examples of site selectivity are similar, in that the more stable product is obtained and the influence of the frontier orbitals is also such as to favour the formation of these products. In the following discussion, therefore, we shall concentrate on those cases where the two approaches make different predictions.

In cycloadditions, the frontier orbital interactions are almost always between orbitals well separated in energy, and consequently they are second order and follow the form of the third term of equation 2-7. As long as only the frontier orbitals are being considered, we can ignore the $E_r - E_s$ term, because, for any particular pair of reagents, it is always the same, whichever way they combine. Furthermore, as long as the atoms at the reaction-sites in each component are the same (for instance, if they are both carbon atoms in one reagent, and both carbon atoms or both oxygen atoms in the other) then we can, as a first approximation, also ignore the β^2 term. Thus all we have left is the Σc^2 term. The dimerization of hexatriene (310) will serve as an example. The dimer formed[258] is 311 and not 312. Clearly the former, which retains a conjugated diene system,

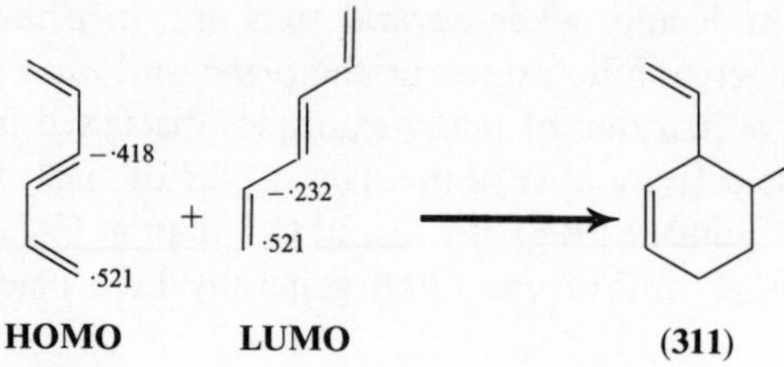

HOMO   LUMO

(310)

−·418   ·521   −·232   ·521

(311)

HOMO   LUMO

−·418   ·521   −·418   ·418

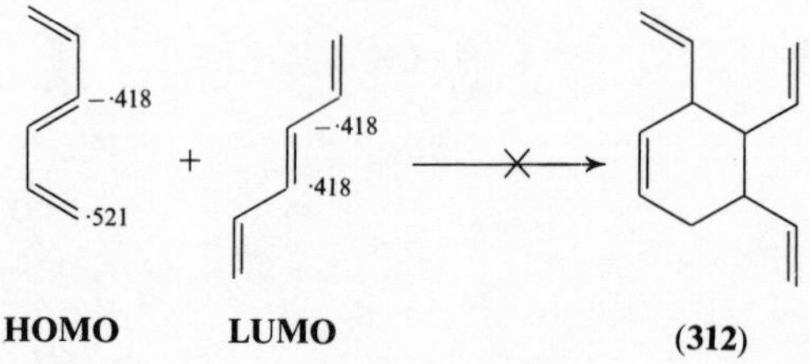

(312)

will be lower in energy than the latter, which has lost all conjugation; 311 is therefore the thermodynamically favoured product. In this example, the frontier orbitals, unusually, make the opposite prediction, by a small margin. Thus we have the coefficients for the HOMO and LUMO from Table 4-1. The Σc^2 term for the observed reaction is given by:

$$\Sigma c^2 = [(0\cdot418 \times 0\cdot232) + (0\cdot521 \times 0\cdot521)]^2 = 0\cdot136$$

and for the reaction which is not observed by:

$$\Sigma c^2 = [(0\cdot418 \times 0\cdot418) + (0\cdot521 \times 0\cdot418)]^2 = 0\cdot154$$

The former reaction is unsymmetrical, and it is likely that the transition state is also unsymmetrical. This will make the β^2 term, which is distance-dependent, different for each of the C—C bonds being formed, and will therefore modify the already delicate balance in the values of Σc^2 calculated above. The inclusion of other than frontier orbital terms will also adjust this balance. Clearly, however, a simple FO treatment is inadequate in this instance.

With substituted conjugated systems, such as heptatriene (**314**) and butadiene nitrile (**319**), a similar kind of selectivity is shown: with maleic anhydride (**315**)[259] and isoprene (**318**)[260] respectively, they give the lower energy adducts (**316** and **320**), rather than the alternatives (**313** and **317**). Once again a frontier orbital treatment, in its simplest form, has little to offer. We need the coefficients in the two compounds (**314** and **319**), but in neither case do we have them to hand. To avoid doing a calculation for the triene (**314**), we could add a little of the coefficients of the HOMO of the heptatrienyl anion (p. 123) to those of the HOMO of hexatriene. Similarly, for butadiene nitrile (**319**), we could

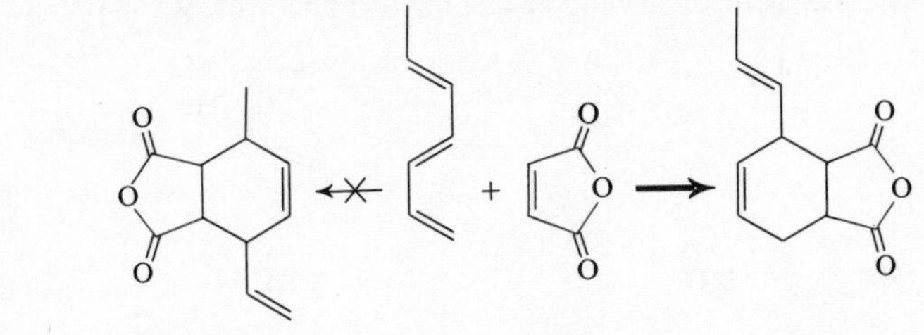

(313)	(314)	(315)	(316)
23%			68%

| (317) | (318) | (319) | (320) |

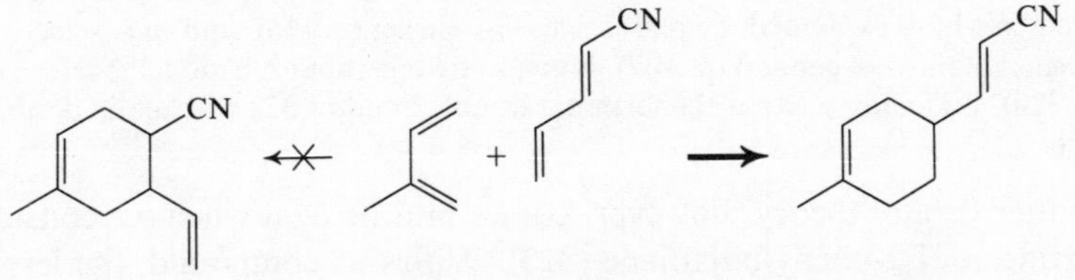

add a little of the coefficients of the LUMO of the pentadienyl cation to those of the LUMO of hexatriene. (In both cases, the HOMO (diene)/LUMO (dienophile) inter-action is the important one.) If you try this for yourself you will find that it makes no difference: the delicate balance in the values of Σc^2 remains slightly in favour of attack at the central double bond of **319**, and the two diene systems of **314** are expected to be equally reactive. Thus all these reactions are probably best thought of as being governed largely by the nature of the products.

Similarly when we come to complete the account of why Thiele's ester[184] should have the structure (**324** = **223**) of the 72 possible isomers (see p. 136).

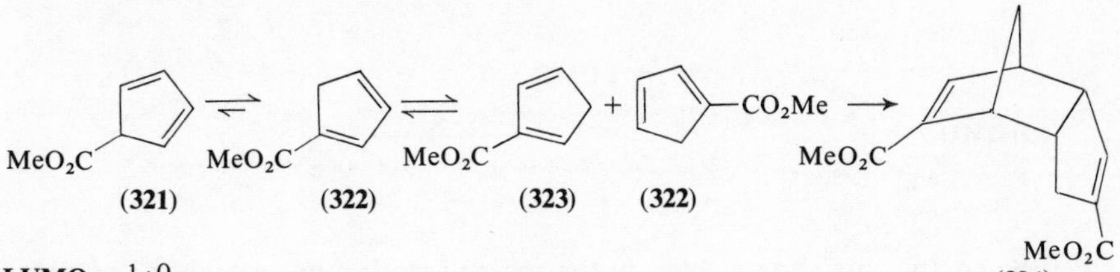

(321)	(322)	(323)	(322)	(324)

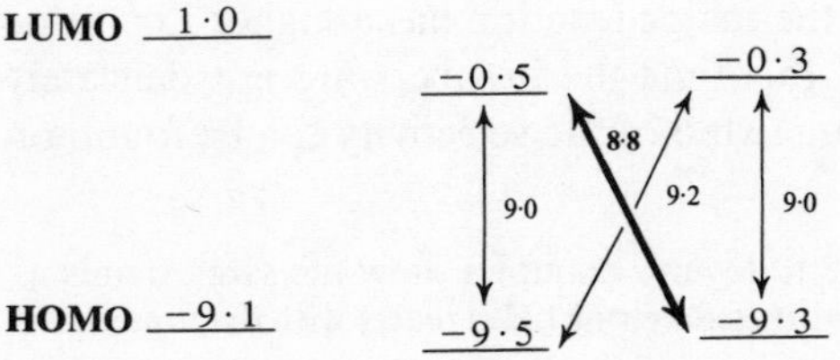

LUMO __1·0__

−0·5   −0·3

8·8

9·0   9·2   9·0

HOMO __−9·1__   −9·5   −9·3

168

The three isomers of cyclopentadiene carboxylic ester (**321**, **322** and **323**) are all present and rapidly equilibrating. The HOMO and LUMO energies of these isomers will be close to the 'typical' values used in Fig. 4-46, and these are repeated under the structures above. The isomer (**321**) ought to be the most reactive diene, because it has the highest-energy HOMO, but it is known to be present only to a very small extent—evidently too low a concentration to be a noticeable source of products. Leaving this isomer (**321**) out of account, the smallest energy-separation is between the HOMO of **323** and the LUMO of **322**. These isomers, therefore, will be the ones to combine, and they can react in either of two main ways: **323** as the diene and **322** as the dienophile or **323** as the dienophile

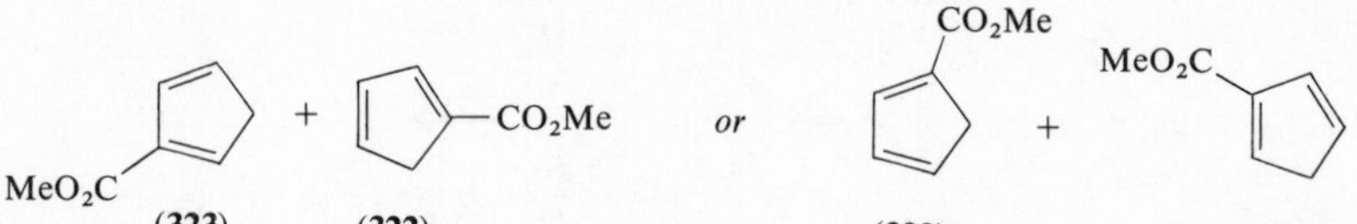

MeO$_2$C (**323**)   +   (**322**) —CO$_2$Me   *or*   CO$_2$Me (**322**)   +   MeO$_2$C (**323**)

and **322** as the diene. The second of these suffers severe steric hindrance, because two fully bonded carbon atoms would have to be joined, always a difficult feat. This leaves the combination of **323** as the diene and **322** as the dienophile, the reaction actually observed. We have explained the regioselectivity earlier (p. 136), and *endo* selectivity has also been explained in general (p. 107). Now, knowing about the model reaction (**319** + **318** → **320**), we finally see that the terminal double bond of **322** will be the double bond to react.

Frontier orbital theory, however, comes into its own when we consider the dimerization of 2-phenylbutadiene (**325**). With this compound, the less stable product (**326**) is the major one.[177] Furthermore, its formation involves attack at the more crowded double bond, so that neither a product-stability nor a steric argument explains why **326** is the major product. The frontier orbitals

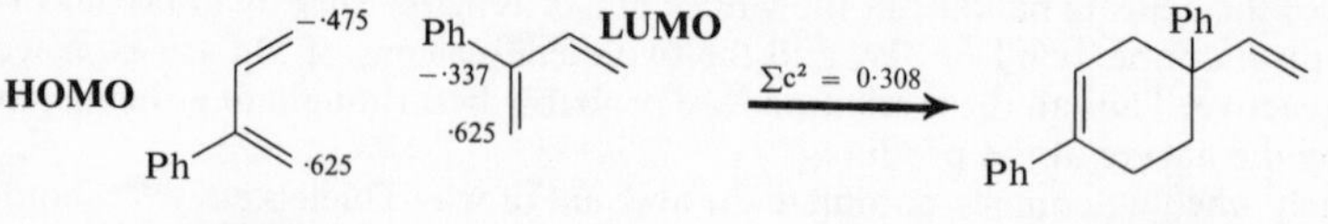

HOMO −·475 ·625 Ph (**325**)   Ph −·337 ·625 **LUMO** $\Sigma c^2 = 0.308$ → Ph Ph (**326**)

HOMO −·475 ·625 Ph   Ph −·256 ·475 **LUMO** $\Sigma c^2 = 0.175$ → Ph Ph

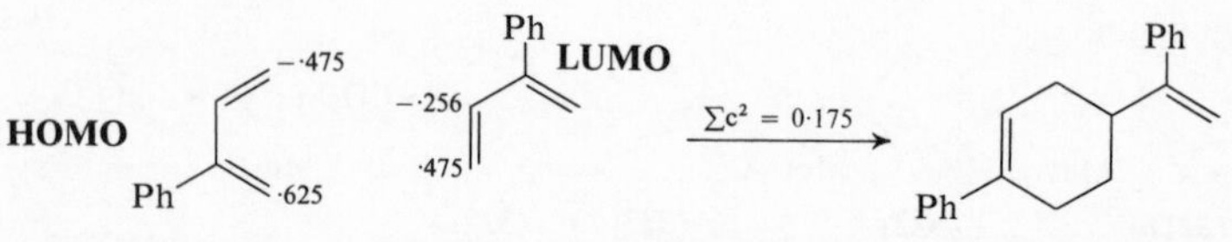

have the coefficients shown: clearly the major reaction has a higher Σc^2-value (0·308) than the minor (0·175), and this time the numbers are not delicately balanced. Another example of the same kind of site selectivity can be found on p. 135.

Among 1,3-dipolar cycloadditions, the following examples show the same trends as the Diels-Alder reactions above. The unsaturated ketone (**327**) reacts with diazomethane at the $\gamma\delta$-double bond, to give the adduct (**328**),[261] and diazoacetic ester (**329**) adds to the terminal double bond of 1-phenylbutadiene (**330**) (the product actually isolated[262] was the cyclopropane, **331**).

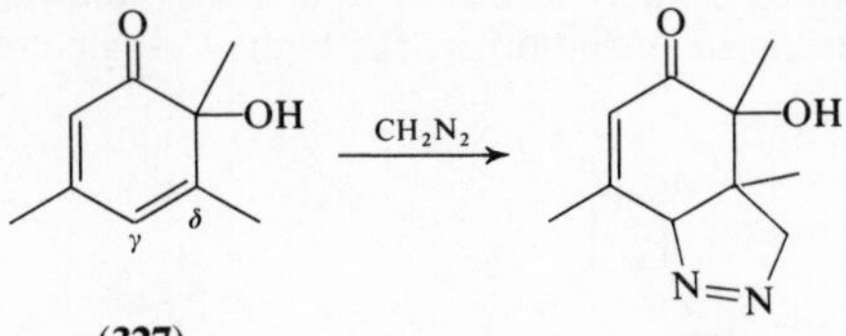

(**327**) (**328**)

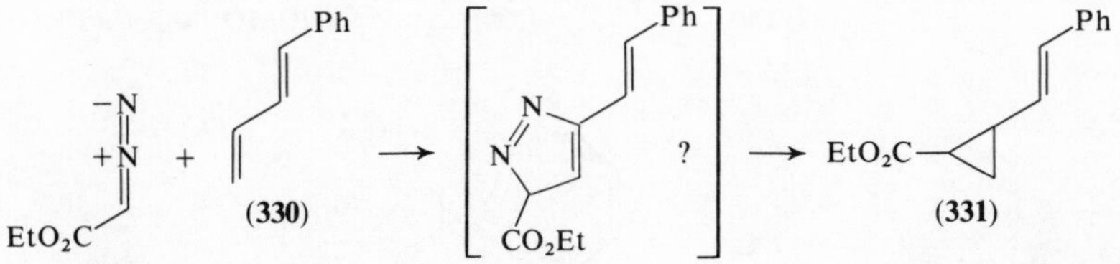

(**329**)

Site selectivity in a number of other concerted cycloadditions which are not [4 + 2] cycloadditions is also explained by frontier orbital control. Thus diphenylketene (**332**) reacts with isoprene (**333**) mostly at the more substituted double bond, and with *cis*-butadiene-1-nitrile (**334**) at the terminal double bond.[263] Dichlorocarbene reacts at the terminal double bond of cycloheptatriene (**335**),[264] and the Simmons-Smith reaction (**336** + **337**)[265] also takes place at the site with the higher coefficients in the HOMO.

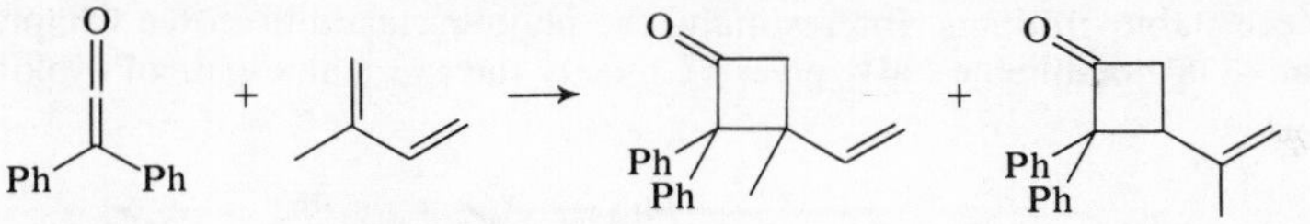

(**332**) (**333**) *70%* *30%*

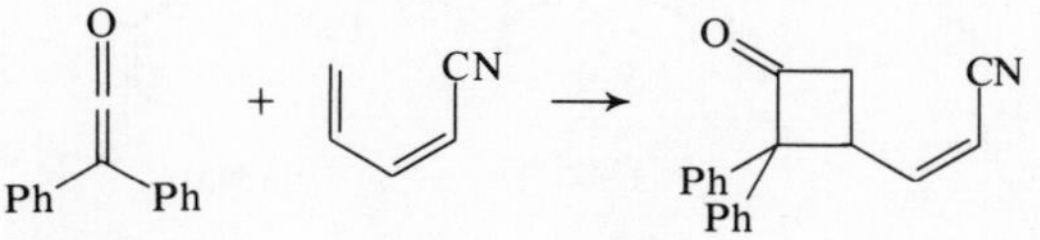

(**332**) (**334**)

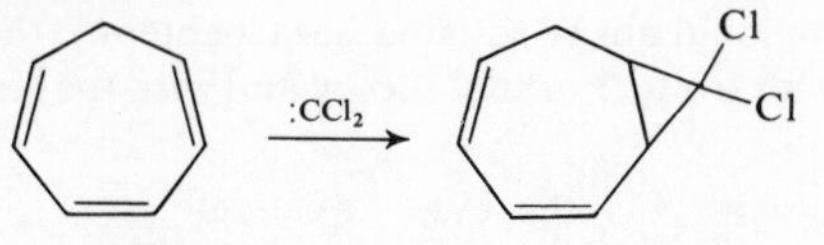

(**335**)

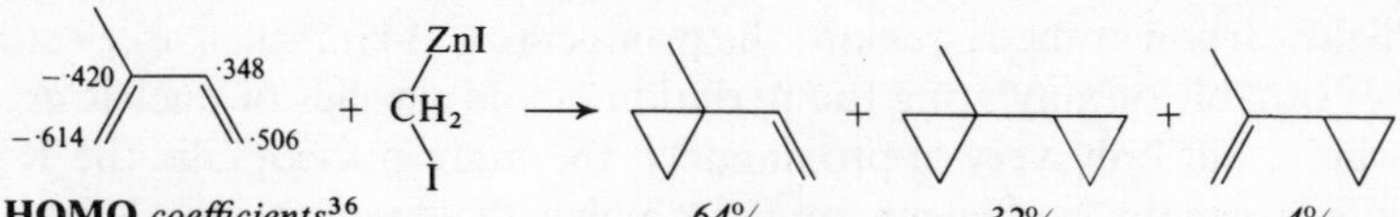

HOMO *coefficients*[36] *64%* *32%* *4%*

(**336**) (**337**)

Cycloadditions of ketenes and carbenes to piperylene (338) generally take place at the less substituted double bond.[266,267] The 'highest adjacent pair' of coefficients (p. 171) is at C-3 and C-4 in the HOMO of the diene, but the highest single coefficient is at C-4. Since both of these reagents are likely to attack in a highly unsymmetrical fashion, with one bond forming well ahead of the other, the highest single coefficient

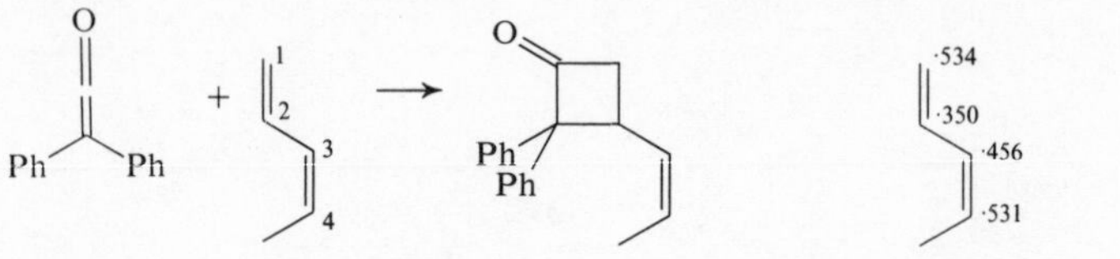

(338)   (339)   HOMO *coefficients*[36]

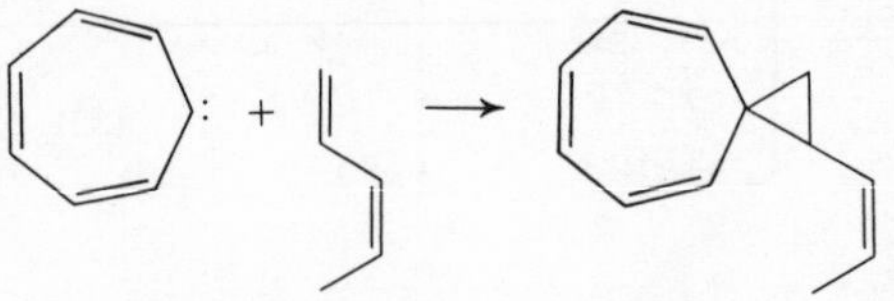

(338)   (340)

(at C-1) effectively becomes more important than the 'highest adjacent pair'. In piperylene (338), the coefficient on C-1 is only very slightly larger than the coefficient on C-4, so it may well be that a major factor in these reactions is that the products observed (339 and 340) have retained the more substituted double bond; they are, therefore, likely to be the more stable products. Interestingly, the photocycloaddition (see Chapter 6) of piperylene (338) to stilbene (341) gives[268] mostly the *less* stable pair of cyclobutanes

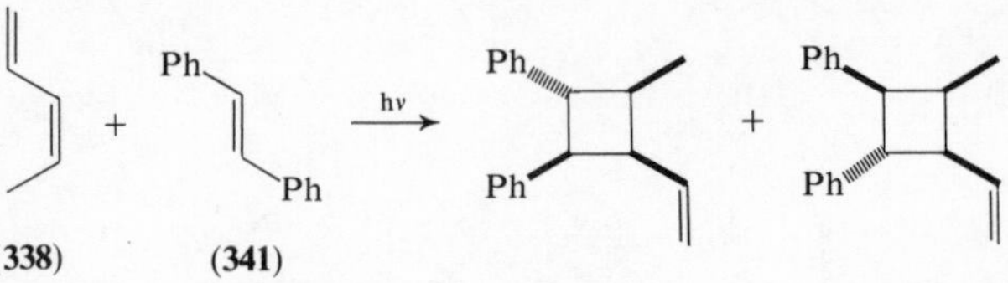

(338)   (341)   (342)   (343)

(342 and 343). Both the HOMO and the LUMO of piperylene will have the 'highest adjacent pair' at C-3 and C-4 (both the HOMO and the LUMO are important in photochemical reactions, and the coefficients should not be squared, see Chapter 6); therefore, this observation is in agreement both with frontier orbital theory and with the reaction's being a concerted one.

The so-called K-region of polycyclic hydrocarbons is implicated in the carcinogenicity of these compounds. It is believed that the body epoxidizes the hydrocarbon at the K-region; the product (e.g. 345) is then an electrophilic species capable of alkylating the pyrimidines and purines of nucleic acids. On the whole, but only very approximately, the more nucleophilic the K-region (i.e. the larger the coefficients and the higher the energy of the HOMO), the more carcinogenic the hydrocarbon proves to be, presumably because it is

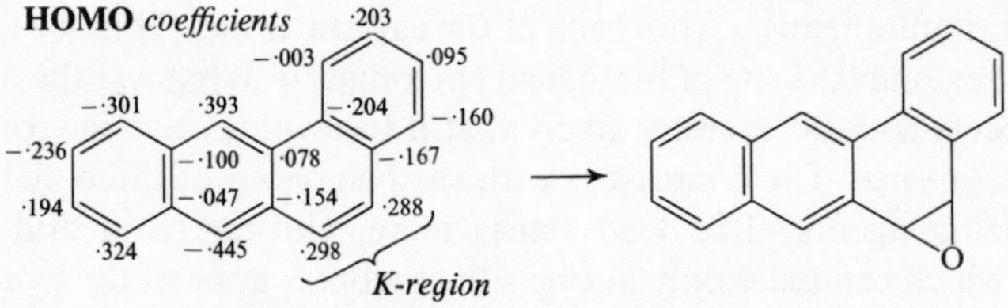

HOMO *coefficients*

(344)

(345)

epoxidized more readily. We shall now look at some cycloaddition reactions of polycyclic hydrocarbons, where we shall find that, although the K-region may not have the highest coefficients in the molecule, it often has the *highest adjacent pair*. Thus if benzanthracene (344) is to take part in a cycloaddition reaction in which it provides the 2-electron component, a high value of Σc^2 is got from the K-region. (0·298 + 0·288 is larger than 0·194 + 0·324 or any other sum of adjacent coefficients not involving the angular carbons; reaction at the angular carbon atoms presumably involves the loss of too much conjugation,

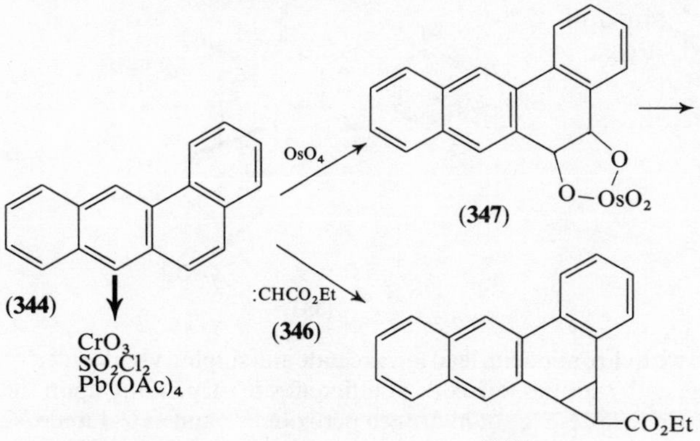

(344)

(347)

(348)

(346)

(349)

see however, p. 172.) This is the site of attack by osmium tetroxide,[269] which probably reacts in a cyclic process to give the ester (347) as an intermediate, and it is also the site of attack by the carbene (346) derived from diazoacetic ester.[270] The same idea accounts for the site-selectivity of the hydrocarbons (350 and 351) both of which react with osmium tetroxide in the K-region. Epoxidation is like a cycloaddition, in that the new bonds to the oxygen atom

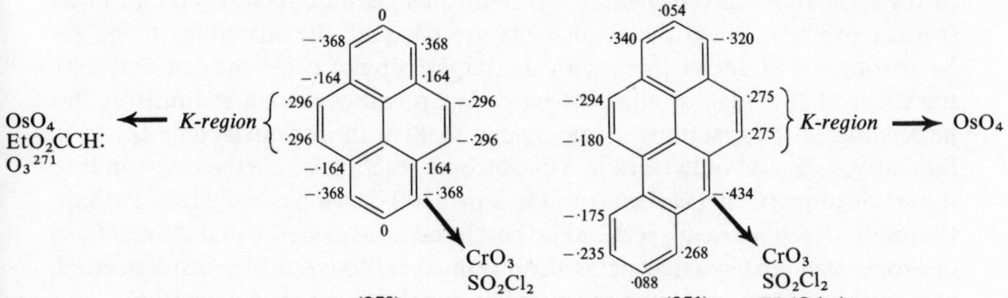

(350)

(351)

are formed simultaneously from each of the carbon atoms. Thus we can explain why the K-region is the site of biological epoxidation. Whatever the mechanism of that process may be, it seems likely that it too will involve the formation of both bonds at once. The contrast is with the behaviour of these hydrocarbons with oxidizing agents—like lead tetraacetate, chromic acid and sulphuryl chloride—which can react only at one site at a time: none of the hydrocarbons (**344**, **350** or **351**) reacts in the K-region with these reagents.[272] Instead, reaction takes place at the site with the highest single coefficient in the HOMO, just as we would expect for an electrophilic substitution.

Anthracene (**352**) and chrysene (**353**) show less of a dichotomy of behaviour, because the highest single coefficient is part of the site with the highest adjacent pair. Thus anthracene reacts with dichlorocarbene and peracid at C-9 (and possibly C-9a), and it

also reacts at C-9 (and C-10) with chromic acid, lead tetraacetate and sulphuryl chloride.[272] However, anthracene reacts with osmium tetroxide at different sites depending upon the conditions: at C-9 and C-10 in the presence of hydrogen peroxide,[273] and at C-1 and C-2 on its own;[274] and anthracene reacts with the carbene (**346**) at C-1 and C-2.[270] Plainly, other factors are at work: attack at C-9 and C-9a by the carbene (**346**) may be reversible, for example, or a 'product-stability' argument may apply, attack at C-9 and C-9a being inhibited by the high energy of the product of such attack.

Note: The use in this section of the 'highest adjacent pair' taken from only one component of a cycloaddition was permissible because the other component of the cycloaddition (OsO_4 and O_3) had *equal* coefficients at each end in its frontier orbitals. When the coefficients are unequal, the advantage to be got by pairing the larger of them with the largest single coefficient can outweigh the effect of the 'highest adjacent pair'. We possibly saw an example of this happening in the reactions of piperylene (**338**) with diphenylketene (p. 170). Similarly, it should only work in a cheletropic reaction where the two bonds to the single atom (C in a carbene or O in a peracid) develop equally fast. Perhaps the results described in this section can best be taken as evidence that the reactions of ozone, osmium tetroxide, peracid, and ethoxycarbonyl carbene are concerted, with both new bonds formed more or less equally in the transition state.

4.7 Periselectivity in Pericyclic Reactions

There is a special kind of site-selectivity which has been called periselectivity. When a conjugated system enters into a reaction, a cycloaddition for example, the whole of the conjugated array of electrons may be mobilized, or a large part of them, or only a small part of them. The Woodward-Hoffmann rules limit the total number of electrons (to 6, 10, 14 etc. in all-suprafacial reactions, for example), but they do not tell us which of 6 or 10 electrons would be preferred if both were feasible. Thus in the reaction of cyclopentadiene (**355**) and tropone (**356**), mentioned at the beginning of this book, there is a possibility of a Diels-Alder reaction, leading to **354**, but, in fact, an equally allowed, ten-electron reaction is actually observed,[121] namely the one leading to the adduct (**357**). The product is probably not thermodynamically much preferred to the

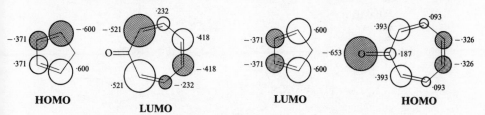

alternative, if at all, so that will not be a very compelling argument to account for this example of periselectivity. The frontier orbitals, however, are clearly set up to make the longer conjugated system of the tropone more reactive than the shorter. The coefficients of the frontier orbitals of tropone were given on p. 138, and are reproduced in Fig. 4-70. The largest coefficients of the LUMO of

Fig. 4-70 Frontier orbitals of cyclopentadiene and tropone

tropone are at C-2 and C-7, with the result that bonding to these sites is easier than to C-2 and C-3, which would have led to the other adduct (**354**).

In general, the ends of conjugated systems carry the largest coefficients in the frontier orbitals, and we should therefore expect pericyclic reactions to use the longest part of a conjugated system compatible with the Woodward-Hoffmann rules. This proves to be true up to a point, with the provision that the reactions have also to be geometrically reasonable. The following cases are all ones where the largest possible number of electrons have been mobilized, when smaller, but equally allowed numbers might have been instead. Frontier orbital considerations, including orbital-symmetry allowedness, account for all of them.

Electrocyclic Reactions:[275,276]

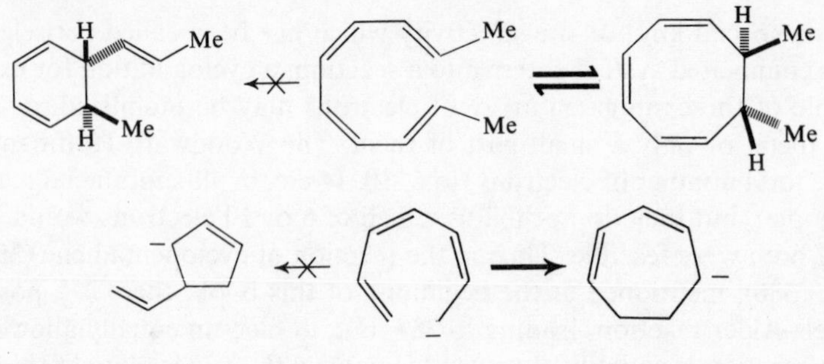

Cycloadditions:[277-280]

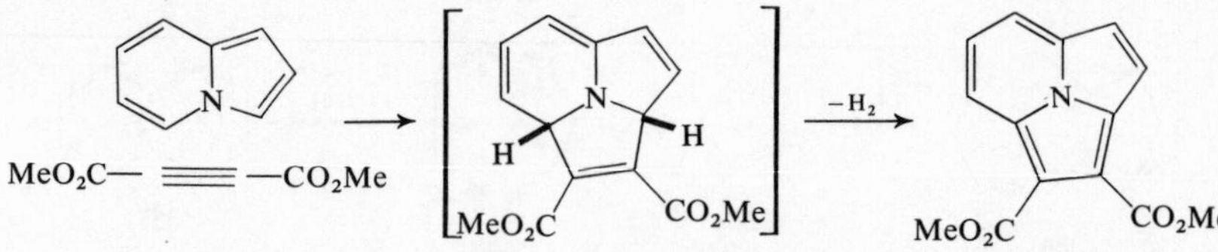

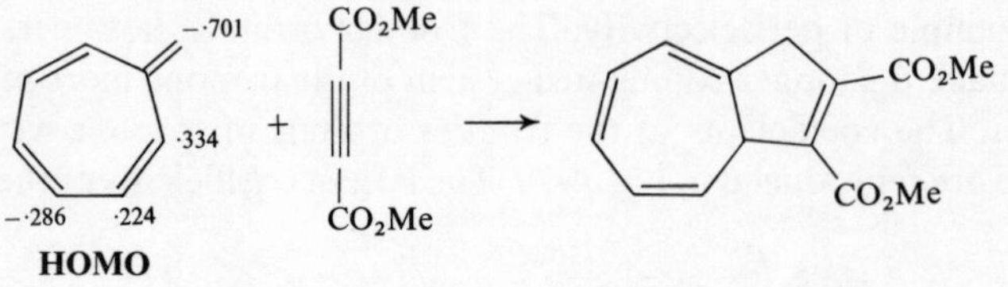

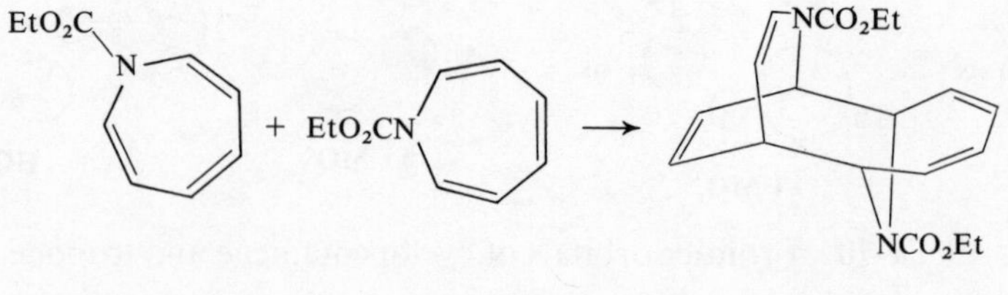

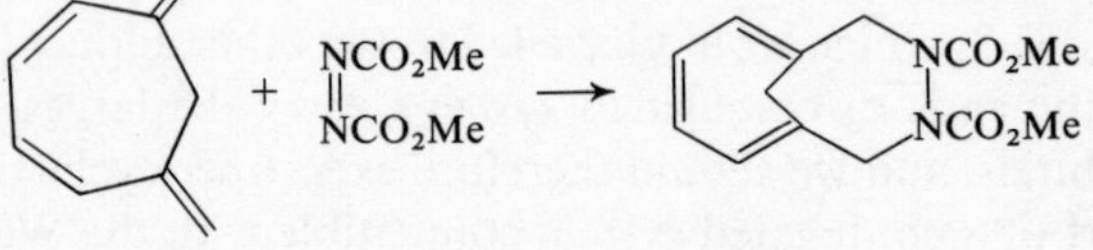

$$MeO_2C— \equiv —CO_2Me$$

HOMO

A Cheletropic Reaction:[281]

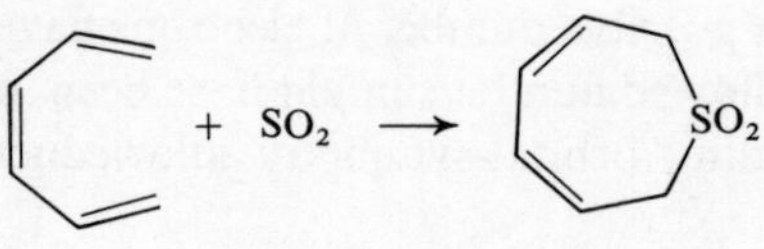

Sigmatropic Rearrangements:[282-287]

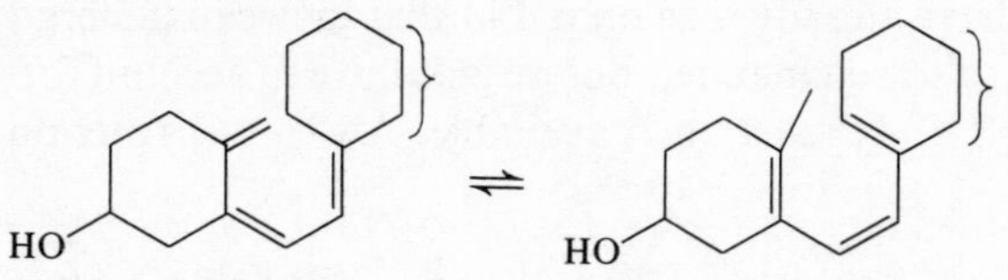

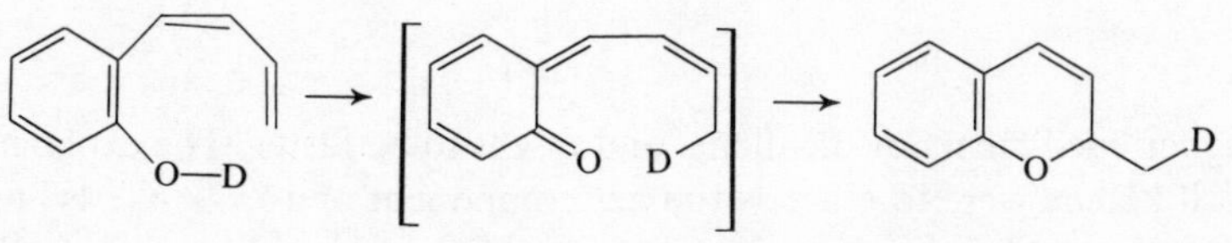

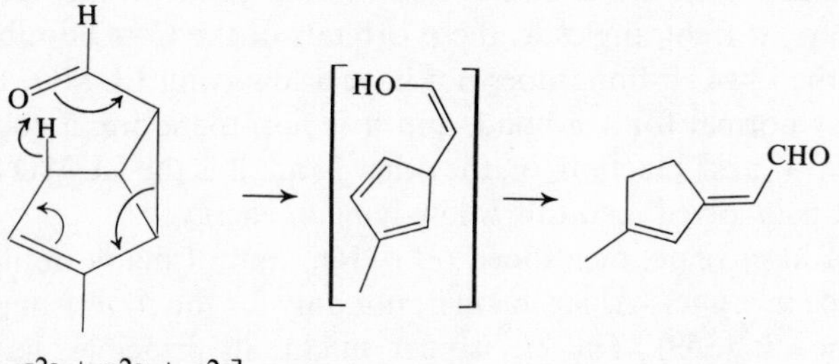

$[\pi 2s + \sigma 2s + \pi 2a + \sigma 2s]$

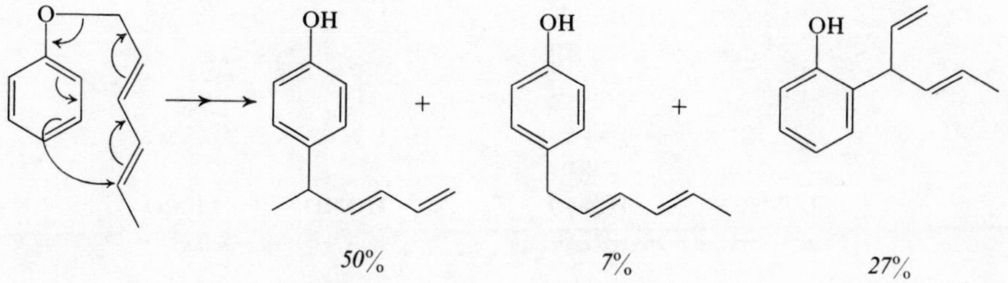

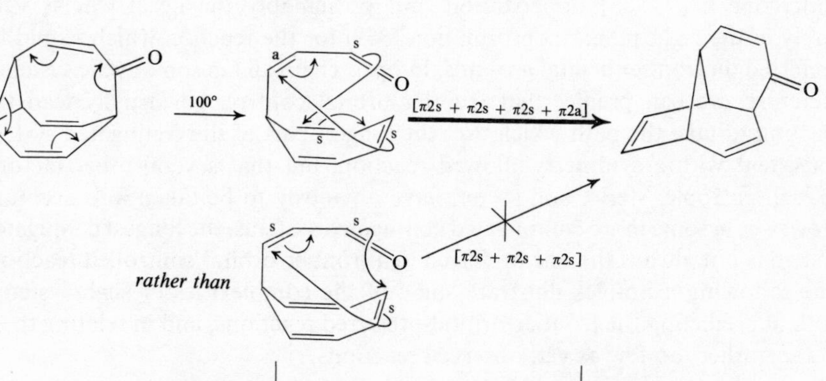

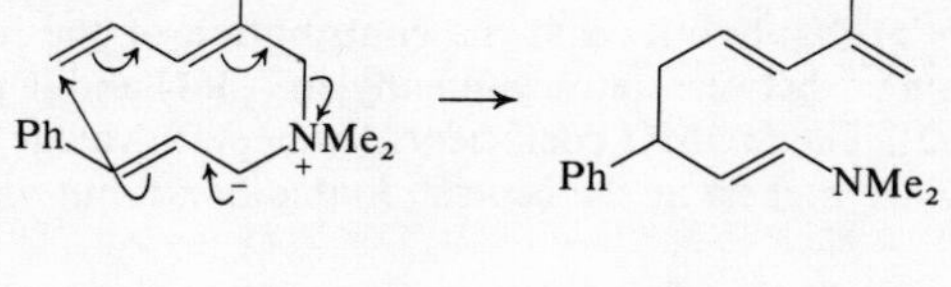

50%   7%   27%

$[\pi 2s + \pi 2s + \pi 2s + \pi 2a]$

$[\pi 2s + \pi 2s + \pi 2s]$

rather than

However, at first sight, ketenes (**358**) seem to be avoiding the higher co-efficients. We have already seen on p. 143 that they can undergo [2 + 2] cyclo-additions in an allowed manner, but we also have to account for why they do so, even when a [2 + 4] reaction is available. A [2 + 4] reaction would involve

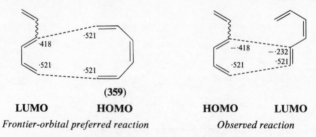

(**358**)

the higher coefficients of the diene, and ought to be faster. The explanation[288] is that if ketene were to react as the $\pi 2s$ component of a [$\pi 2s + \pi 4s$] reaction, the orbital most localized on the C=C double bond would be involved. The p orbitals of this bond are at right angles to the p orbitals of the C=O double bond. Consequently, the C=C π-bond does not have a low-lying LUMO. Its LUMO is more or less normal for a π-bond, and it is not, therefore, a good dienophile. In the [$\pi 2a + \pi 2s$] reaction, on the other hand, it is the LUMO of the C=O π-bond that is involved, and this *is* low-lying in energy.

The dimerization of hexatriene, mentioned on p. 166, and of many similar compounds, takes place in a Diels-Alder manner, not only for the *trans* isomer (**310**) but also for the *cis* (**359**). The *cis* isomer might, in principle, have

(**359**)

LUMO   **HOMO**   **HOMO**   **LUMO**

Frontier-orbital preferred reaction   *Observed reaction*

undergone a [4 + 6] dimerization, but presumably the hexatriene is very rarely in the right planar conformation (**359**) for the reaction which would be preferred on frontier orbital grounds. In open-chain and in some cyclic systems, therefore, we can predict that frontier orbital control will usually lead the reaction to take the path which uses the longest part of the conjugated system consistent with a symmetry-allowed reaction, but that several other factors, spatial, entropic, steric, and so on, have obviously to be taken into account. However, in some more complicated conjugated systems, the longest conjugated system is not always the one preferred in a frontier-orbital controlled reaction. The following examples illustrate some of the complexities of such systems, both in predicting the frontier-orbital-preferred reactions, and in relating them to the, rather too few as yet, observed reactions.

Geometrical factors probably outweigh the contribution of the frontier orbitals in the remarkable reaction[128] between tetracyanoethylene (**361**) and heptafulvalene (**360**) to give the adduct (**362**). The HOMO coefficients for heptafulvalene are shown on the diagram. Clearly they are highest at the central double bond, but any reaction at this site would have to be, by the Woodward-Hoffmann rules, antarafacial on one of the components, and this is geometrically unreasonable. (In a carbene cycloaddition, in which an antarafacial element can be taken up by the carbene, p. 95, it *is* the central double bond which is attacked.[289]) The best possibility, from the frontier orbital point

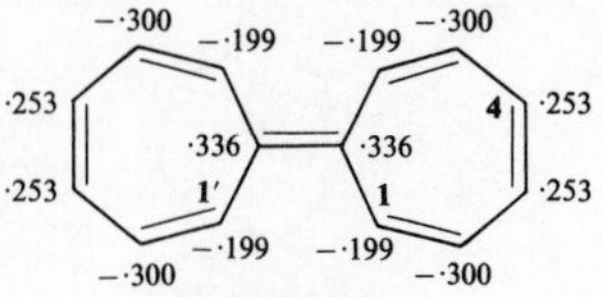

(360)

HOMO *coefficients*[7]

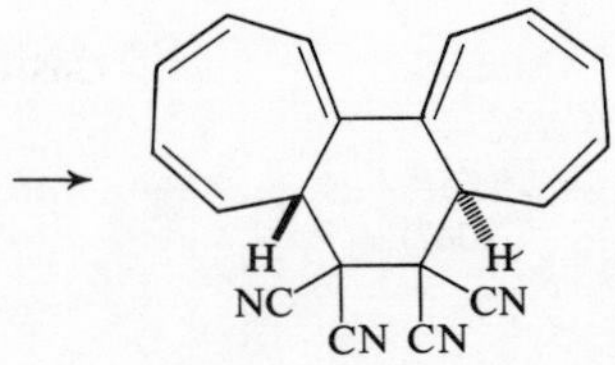

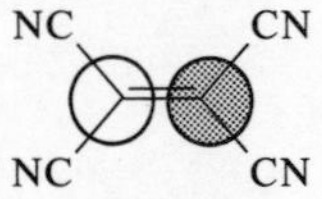

LUMO (361)

(362)

of view, would be a Diels-Alder reaction across the 1,4-positions, but this evidently does not occur, probably because the carbon atoms are held too far apart. This is well-known to influence the rates of Diels-Alder reactions—cyclopentadiene reacts much faster than cyclohexadiene, which reacts much faster than cycloheptatriene, see p. 113. The only remaining reaction is at the site of lowest frontier-orbital electron population, the antarafacial reaction across the 1,1'-positions.

Sesquifulvalene (**363**) presents another case where frontier orbital control is not the only factor that governs periselectivity. The sequifulvalene (**364**) does give the adduct

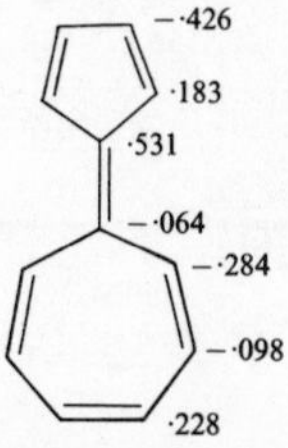

HOMO *coefficients*[7]

(363)

(**366**)[290] with tetracyanoethylene (**365**), as expected from the coefficients of the HOMO of the unsubstituted system. However, the sesquifulvalene (**367**) gives a [4 + 2] adduct (**368**)[290] instead, and the sesquifulvalene (**369**) gives a [12 + 2] adduct (**370**).[291]

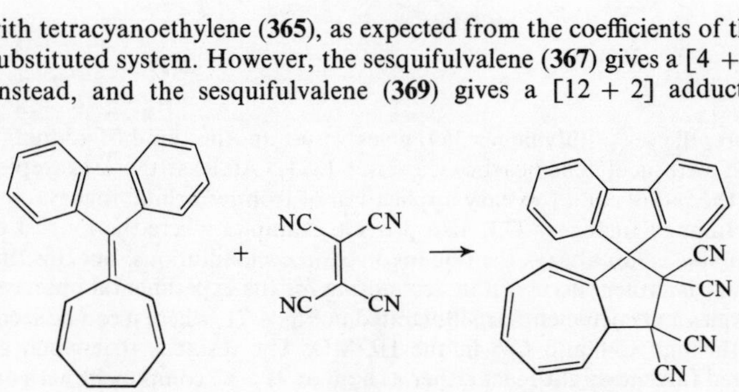

(364)   **(365)**   **(366)**

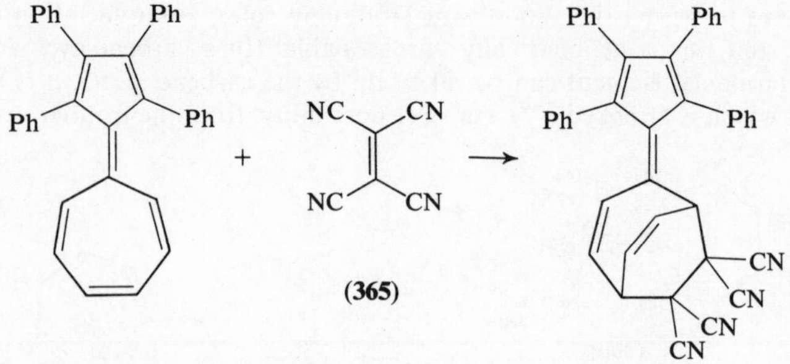

(367)   (365)   (368)

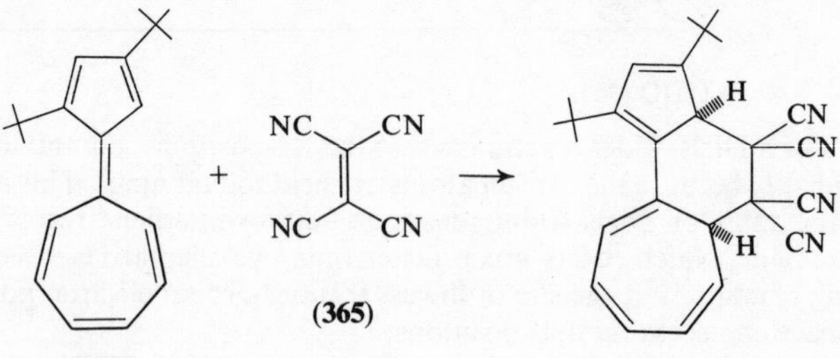

(369)   (365)   (370)

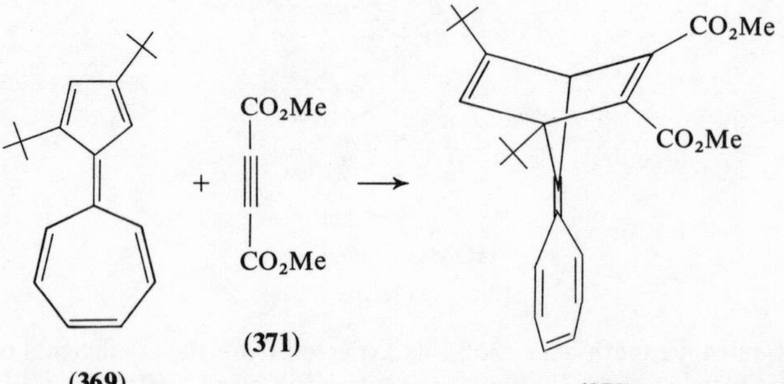

(369)   (371)   (372)

Furthermore, the sesquifulvalene (369) gives[292] yet another kind of adduct (372) when it is treated with acetylenedicarboxylic ester (371). At least these examples serve to emphasize the pitfalls of a too easy application of frontier orbital theory.

The reactions of fulvenes (373) also provide examples where the longest conjugated system available is not always the one involved in cycloadditions, but this time frontier orbital theory is rather successful in accounting for the experimental observations. The orbital energies and coefficients are illustrated in Fig. 4-71, where it can be seen that there is a node through C-1 and C-6 in the HOMO. The result is that when a relatively unsubstituted fulvene might react either as a $\pi6$ or as a $\pi2$ component with an electron-deficient (low-energy LUMO) diene or dipole, it should react as a $\pi2$ component because

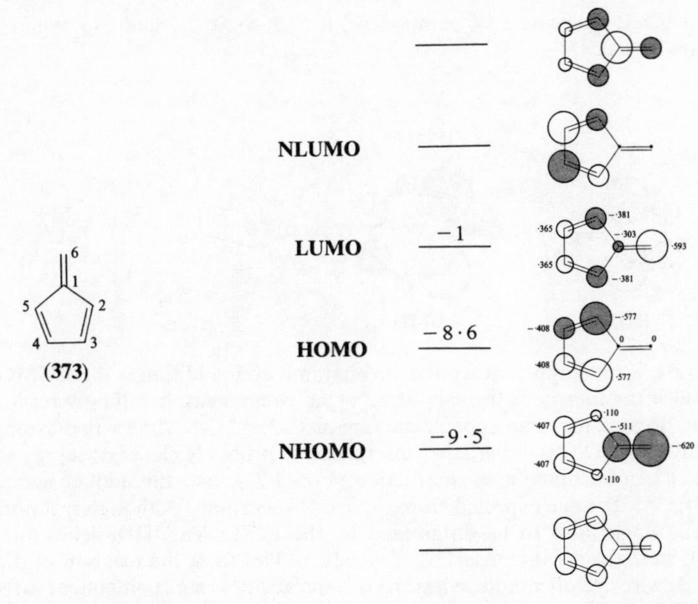

Fig. 4-71 Energies and coefficients of the orbitals of fulvene (373)[293]

of the zero coefficient on C-6 in the HOMO of the fulvene. This is the usual reaction observed:[294-296]

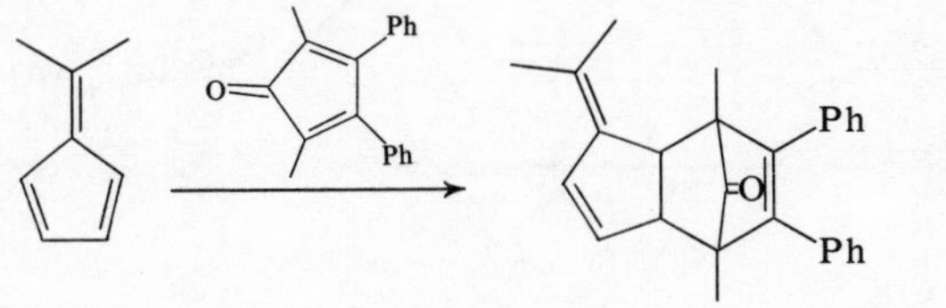

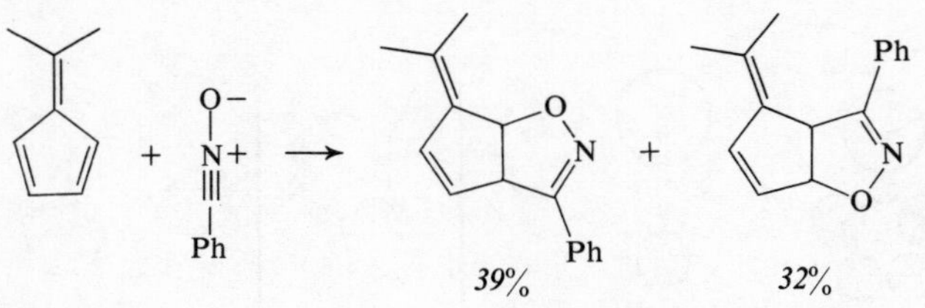

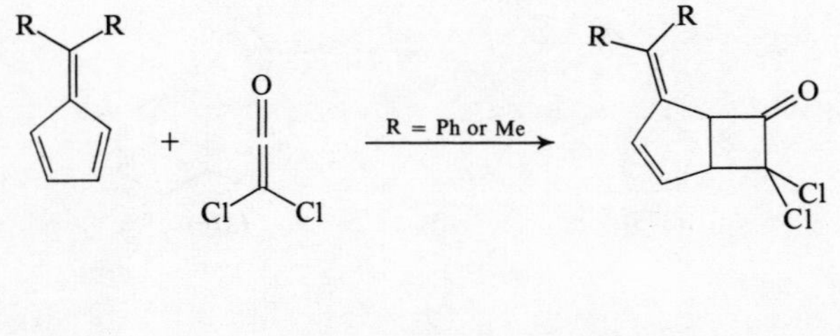

Similarly, when it reacts as a π4 component, it does so at C-2 and C-5, where the coefficients are largest:[297]

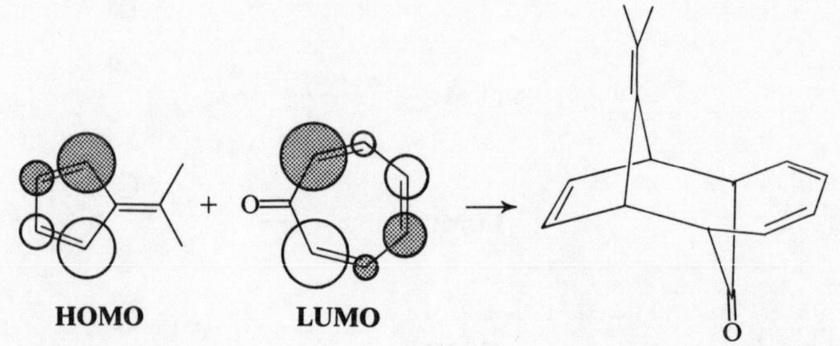

HOMO   LUMO

By contrast, if the important frontier orbital in a cycloaddition is the LUMO of the fulvene, and if the fulvene is to react as π2 or π6 component, it will now react as a π6 component, because the largest coefficients are on C-2 and C-6. Thus with diazomethane, the LU (fulvene)/HO (diazomethane) interaction is probably closer in energy than the HO(fulvene)/LU(diazomethane) interaction (Fig. 4-72), and the adduct actually obtained (374)[298] is the one expected from these considerations. Although reaction with a simple diene ought also to be dominated by the LU(fulvene)/HO(diene) interaction (Fig. 4-72), nevertheless the observed[294] product (376) from the reaction of dimethylfulvene (375) with cyclopentadiene has the fulvene acting as a π2 component rather than

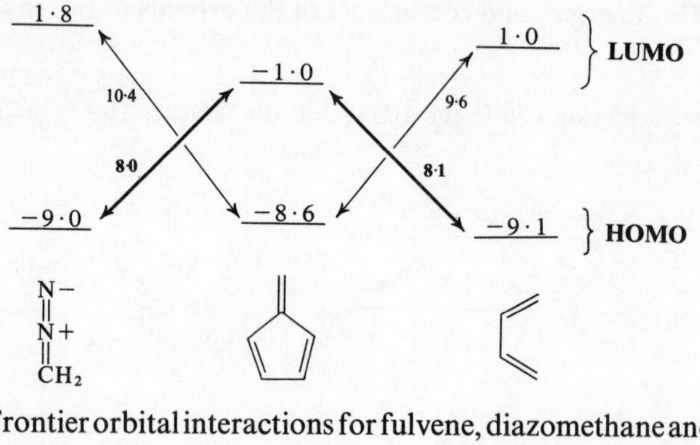

Fig. 4-72   Frontier orbital interactions for fulvene, diazomethane and butadiene

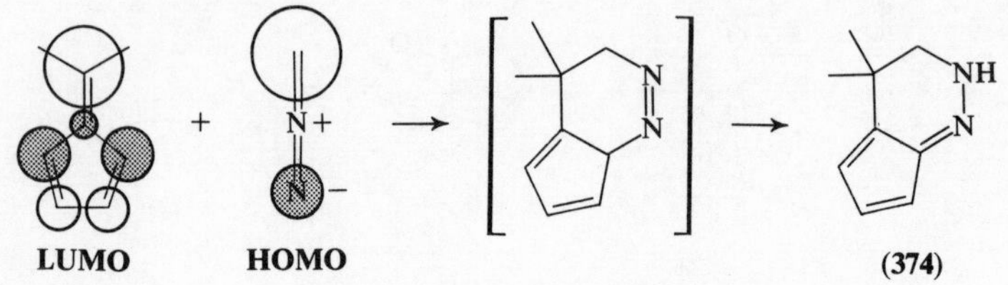

LUMO   HOMO   (374)

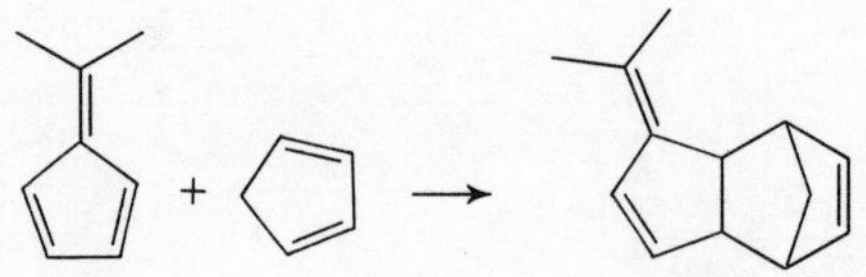

(375)   (376)

as a π6 component. This anomaly has been explained[299] by invoking the next-lowest unoccupied orbital (NLUMO) of fulvene. This orbital has zero coefficients on C-1 and C-6, and hence relatively large coefficients on C-2 and C-3. The interaction of this orbital with the HOMO of cyclopentadiene is apparently large enough to tip the balance in favour of the [2 + 4] adduct (376). This example comes as a useful reminder that the *frontier* orbitals are not the only ones to be interacting as the reaction proceeds. They are usually the most important orbitals, but, in a delicately balanced situation like this, they may not be decisive. The delicacy of the balance is illustrated by the reaction of the same fulvene (375) with a different diene (377), where the expected [6 + 4] adduct (378) is in

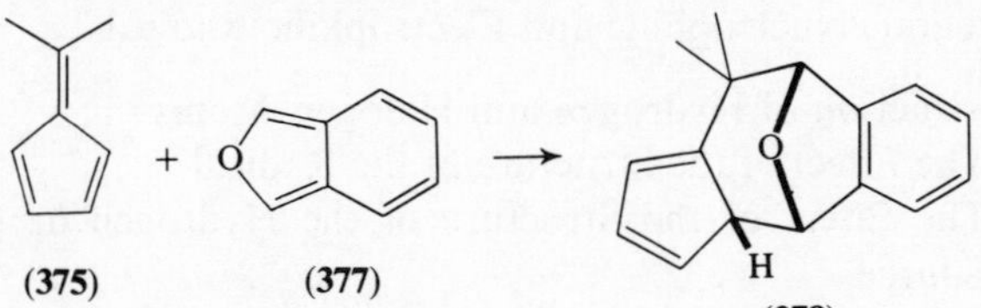

(375)   (377)   (378)

fact obtained.[300] The HOMO of this diene will be higher in energy than that of cyclopentadiene. The result is that the interaction of this orbital with the LUMO of the fulvene is proportionately greater than that with the NLUMO.

This section on periselectivity has been unusually long. It is one of those subjects which frontier orbital theory has rationalized particularly well. It is a fitting close to this chapter to reflect upon the bewildering variety of cyclo-additions shown by fulvenes, and to consider how difficult it would have been to explain before frontier orbital theory came along.

CHAPTER 5

Radical Reactions

5.1 Introduction: Nucleophilic and Electrophilic Radicals

Radicals should be very soft entities—most of them do not have a charge, and in most chemical reactions they react with uncharged molecules. Thus the Coulombic forces are usually small while the frontier orbital interactions remain large. In a sense this is borne out by such well-known reactions as the attack of radicals at the conjugate position of α,β-unsaturated carbonyl compounds,

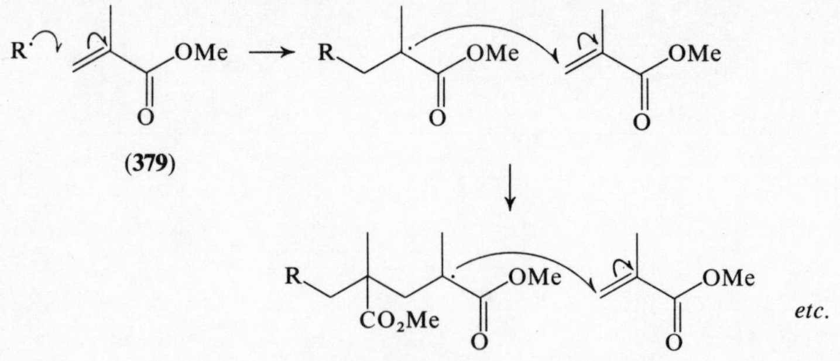

(379)

etc.

rather than at the carbonyl group, and the attack by α-carbonylmethyl radicals from the carbon atom, not from the oxygen atom. The clean polymerization of methyl methacrylate (**379**) demonstrates both these typically soft patterns of behaviour. But the story is more complicated with radicals than it was with ions. Plainly, the frontier orbital of the radical is the singly occupied one (SOMO). This orbital will interact with both the HOMO and the LUMO of the molecule it is reacting with, as shown in Fig. 5-1.[301] Plainly the interaction with the

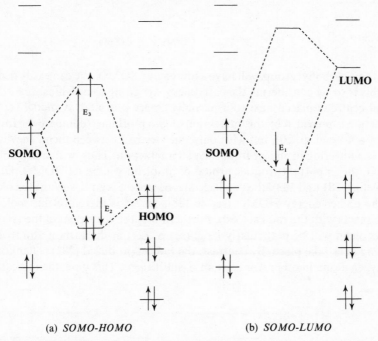

      (a) *SOMO-HOMO*   (b) *SOMO-LUMO*

Fig. 5-1 The interaction of the singly occupied molecular orbital (SOMO) of a radical with (a) the HOMO and (b) the LUMO of a molecule

LUMO will lead to a drop in energy (E_1 on Fig. 5-1b); but so does the interaction with the HOMO. Because there are two electrons in the lower orbital and only one in the upper, there will be a drop in energy ($2E_2 - E_3$) from this interaction. Radicals with a high-energy SOMO (Fig. 5-2a) will react fast with molecules having a low-energy LUMO, and radicals with a low-energy SOMO (Fig. 5-2b) will react fast with molecules having a high-energy HOMO.

This conclusion is strikingly confirmed by the observation of alternating co-polymerization.[302] Thus, the radical-initiated polymerization of a mixture of vinyl acetate (**381**) and dimethyl fumarate (**383**) takes place largely[303] to give a polymer in which the fragments derived from the two monomers alternate along the chain. In this case it is evident that a growing radical such as **380** attacks vinyl acetate rather than fumarate; but the new radical (**382**), so produced, attacks fumarate rather than vinyl acetate. The radical (**380**), because it

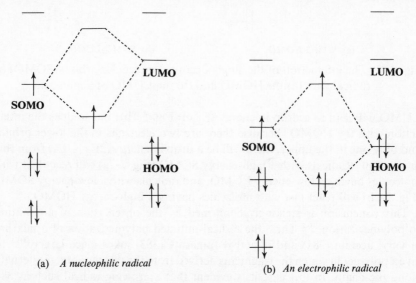

is next to a carbonyl group, will have a low energy SOMO. We can easily deduce that this is so by considering the carbomethoxyl group as a Z-substituent. The radical centre is placed next to a partially empty p orbital. A model for this interaction is provided by the interaction of two p-orbitals (compare the formation of a π-bond, p. 20), but with only one electron between them. Inevitably there is a lowering in energy from such an interaction. Thus, with a low-energy SOMO, the important frontier orbital of an olefin will be its HOMO. Of the two olefins (**381** and **383**), the vinyl acetate, because it is an $\ddot{\text{X}}$-substituted olefin, has the higher energy HOMO (see p. 128), and it is therefore this molecule which reacts with the radical (**380**). Furthermore, the coefficient in the HOMO of this olefin will be particularly large (see p. 124) at the carbon atom where bonding does take place. By contrast, the radical produced (**382**) is flanked by an oxygen atom, in other words by an $\ddot{\text{X}}$-substituent. This time, the radical will

(a) *A nucleophilic radical*

(b) *An electrophilic radical*

Fig. 5-2 Important frontier orbital interactions for radicals (*a*) with high-energy SOMOs, and (*b*) with low-energy SOMOs

have a high-energy SOMO (Fig. 5-3), because the interaction with the lone pair on the oxygen atom is like that which we saw in connection with the α-effect (p. 77), except that there is only one electron in the upper orbital. (We may note, in passing, that because two electrons go into the lower and only one into the upper orbital, there is an overall drop in energy from the interaction of a radical centre with a lone pair (Fig. 5-3). This accounts for the stabilization

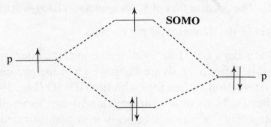

Fig. 5-3 The interaction of a lone pair with a radical centre

of such radicals as **384** and **385**, and for the existence of such isolable radicals as **386** and **387**.) The upper orbital is the SOMO of the radical (**382**). Because of its high energy, it reacts faster with the molecule having the lower energy

(384)	**(385)**	**(386)**	**(387)**

LUMO, namely the Z-substituted olefin (**383**)—and so on, as the polymerization proceeds. This explanation for alternating polymerization avoids the vague terms, such as "polar factors", which have often been used in the past.

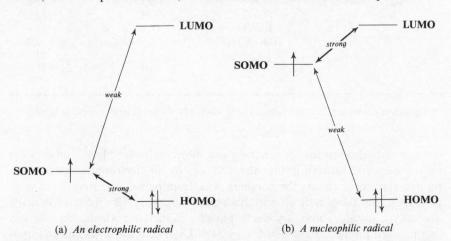

(a) *An electrophilic radical*   (b) *A nucleophilic radical*

Fig. 5-4 Frontier orbital interactions for an electrophilic and a nucleophilic radical

> *In general: radicals which have a low-energy SOMO show electrophilic properties, and radicals which have a high-energy SOMO show nucleophilic properties.*

5.2   The Abstraction of Hydrogen and Halogen Atoms

5.2.1   The Effect of the Structure of the Radical

When a radical attacks a C—H or C-halogen bond, the interactions are with σ and σ^* orbitals. The latter orbitals are usually high in energy, and we can expect that the major interaction is therefore with the HOMO (Fig. 5-4a), namely the σ orbital. Radicals abstracting hydrogen atoms are generally regarded as electrophilic. Reactions of various radicals with p-substituted toluenes have been studied and Hammett plots made (Table 5-1). The ρ-values are small,

Table 5-1 ρ-Values for hydrogen-abstraction from p-substituted toluenes (**388**)[304]

R·	ρ	SOMO-energy[a]
Bu^t·	$+1\cdot0$	$-6\cdot9$
C_5H_{11} $\diagdown$ CH· $\diagup$ Me	$0\cdot7$	$-7\cdot8$
$C_{10}H_{23}CH_2$·	$0\cdot5$	$-8\cdot7$
Et_3Si·	$+0\cdot3$	~ -7
Ph·	$-0\cdot1$	$-9\cdot2$
Me·	$-0\cdot1$	$-9\cdot8$
H·	$-0\cdot1$	$-13\cdot6$
Bu^tO·	$-0\cdot4$	$? -12$
Bu^tOO·	$-0\cdot6$	$\sim -11\cdot5$
HO_2CCH_2·	$-0\cdot7$	$\sim -10\cdot9$
Cl·	$-0\cdot7$	-13
Br·	$-1\cdot4$	$-11\cdot8$
Cl_3C·	$-1\cdot5$	$-8\cdot8$

(structure **388**: R·↷H, a CH with H H on a ring bearing Y)

[a] Ionisation potential in electron volts. (The higher the IP is, the lower in energy the SOMO.)

because radical reactions are generally not subject to polar effects, but they are mostly negative, indicating that the attack is by an electrophilic species. Although agreement among the numbers is far from perfect, the trend does seem to suggest that those radicals with high-energy SOMOs, like the trimethylsilyl and alkyl radicals, show less electrophilicity than those which, like the oxy and halogen radicals, have low-energy SOMOs. The alkyl series shows a particularly good correlation between SOMO-energy and the ρ-value.

When the SOMO/HOMO interaction is the more important one, and assuming, as is usually true for hydrogen-abstraction reactions, that the SOMO-energy lies *between* that of the HOMO and the LUMO, the radical with the higher-energy SOMO will be less reactive than the one with the lower-energy SOMO (because $2E_2 - E_3$ in Fig. 5-1 will be smaller). This explains why the Bu^{t}OO· radical is 10,000 times *less* reactive in hydrogen abstraction than the Bu^{t}O· radical.[305] Here we see the α-effect making an electrophilic radical less reactive, whereas it made a nucleophile more reactive (see p. 77); the cause is the same, namely the raising of the energy of the HOMO. It may be that the lower reactivity of the Bu^{t}OO· radical makes it more selective than the Bu^{t}O· radical (see Table 5-1), and similar factors may explain the other anomalous entries in Table 5-1.

5.2.2 The Effect of the Structure of the Hydrogen or Halogen Source

Radicals are well known to attack hydrogen atoms in the order: allylic faster than tertiary, faster than secondary, faster than primary. The more neighbouring groups a C—H bond has, the more overlap (hyperconjugation) can take place. Since such overlap is between filled orbitals and filled orbitals, the effect is to raise the energy of the HOMO. This effect therefore puts the energy of the HOMOs of the C—H bonds in the same order as their ease of abstraction.

The ease of the abstraction reactions:

$$R \overset{\frown}{} halogen \overset{\frown}{} R'$$

is in the order I > Br > Cl. Again, the energies of the HOMO of the σ-bonds fall in this order. Clearly both these series are compatible with the major involvement of the SOMO of the radical with the HOMO of the other molecule.[306]

Methyl radicals preferentially attack the hydrogen atoms on C-2 of propionic acid (**389**). On the other hand, chlorine atoms preferentially attack the hydrogen

atoms on C-3.[307] Frontier orbital theory explains this extraordinary pattern of reactivity.[308] The C—Hs on C-2 are conjugated to a Z-substituent, and will therefore have a relatively low-energy HOMO and LUMO. The C—Hs on C-3 are conjugated to a CH_2CO_2H group (a very mild Ẍ-substituent), and will have a higher-energy HOMO and LUMO. The methyl radical has a much higher energy SOMO than the chlorine atom (Table 5-1). The orbital interactions are therefore those shown in Fig. 5-5, where we can see how the interaction *A* can be more effective than *B* for a chlorine atom, and *C* more effective

188

than *D* for a methyl radical. If *A* and *C* are the dominant interactions, then the observed pattern of reactivity is explained.

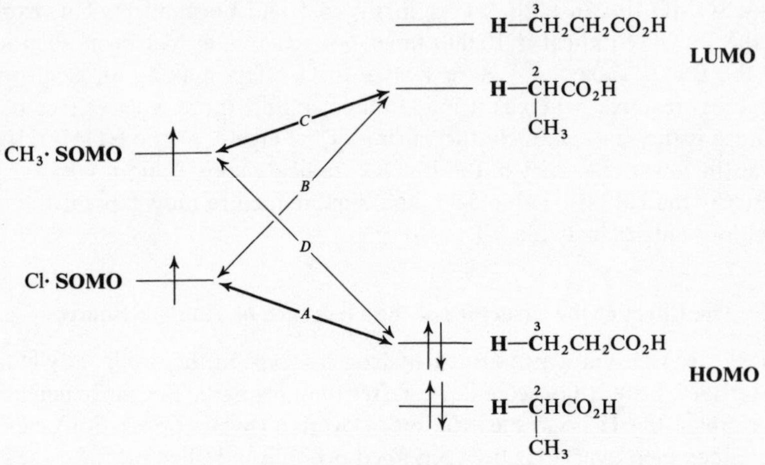

$$H-\overset{3}{C}H_2CH_2CO_2H \quad \textbf{LUMO}$$

$$H-\overset{2}{C}HCO_2H$$
$$|$$
$$CH_3$$

CH₃· SOMO

Cl· SOMO

$$H-\overset{3}{C}H_2CH_2CO_2H \quad \textbf{HOMO}$$

$$H-\overset{2}{C}HCO_2H$$
$$|$$
$$CH_3$$

Fig. 5-5 Interactions for the attack of methyl and chlorine radicals on propionic acid

5.3   The Addition of Radicals to π-Bonds

5.3.1   Attack on Substituted Alkenes

There is a great deal of information available about the addition of radicals to π-bonds, since it is such an important subject in the study of polymerization. A lot of this information is straightforward—the more stable 'product' (e.g. **390**,[309] **391**[310] or **392**[311]) is almost always obtained, and the site of attack usually has the highest coefficient in the appropriate frontier orbital. With C- and Z-substituted olefins (Fig. 5-6), the site of attack will be the same

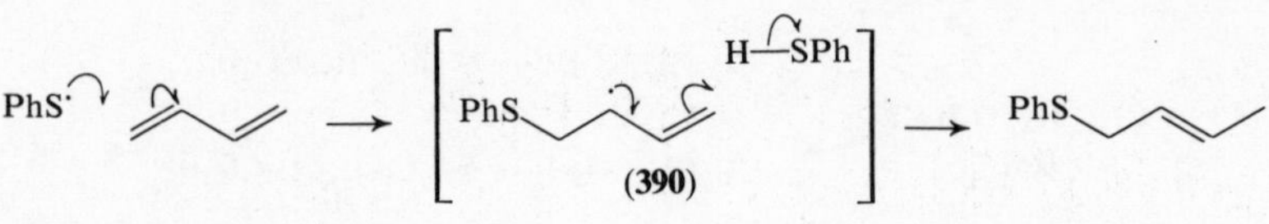

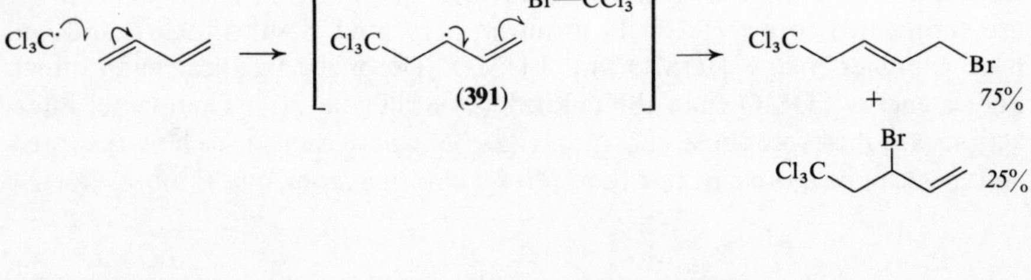

PhS· → [PhS (**390**) H—SPh] → PhS

Cl₃C· → [Cl₃C (**391**) Br—CCl₃] → Cl₃C Br 75%
+
Cl₃C Br 25%

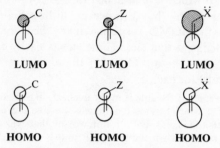

regardless of which frontier orbital is the more important: both have the higher coefficient on the carbon atom remote from the substituent.

Fig. 5-6   Frontier orbitals for C-, Z- and $\ddot{X}$-substituted olefins

With $\ddot{X}$-substituted olefins, however, the HOMO and the LUMO are polarized in opposite directions. For most $\ddot{X}$-substituted olefins, the HOMO will be closer in energy to the SOMO of the radical, because $\ddot{X}$-substituted olefins generally have high energy HOMOs and high energy LUMOs (see p. 117). This explains the usual direction of addition. For example, oxygen atoms attack but-1-ene thus:[312]

11 parts to   *1 part*

The general rule, therefore, is that radicals add to the less substituted end of an olefin to give the more stable radical. This is true of fluoroolefins such as **393** and **394**, which add trichloromethyl radicals predominantly at the less substituted carbon. However, Tedder was able to measure the activation energies for attack at both sites in **393** and **394**, and to compare them with the activation energy for attack on ethylene;[313] the values (in kcal/mole) are shown below:

$$
\begin{array}{cccc}
3\cdot2 & 3\cdot3 \quad 5\cdot4 & 4\cdot6 \quad 8\cdot3 \\
\downarrow & \downarrow \quad\quad \downarrow & \downarrow \quad\quad \downarrow \\
CH_2 = CH_2 & CH_2 = CHF & CH_2 = CF_2 \\
 & (393) & (394)
\end{array}
$$

The odd feature of these results is that attack at the carbon atom remote from the substituents is relatively unaffected by the substituents, even though the substituents are at the site which is developing into a radical. On the other hand, the rate of attack on the carbon atom carrying the substituent is affected much more, being slower when there is one fluorine than when there is none, and slower still when there are two fluorines. This is not what was expected when the only factor taken into consideration was the stability of the radical produced. Frontier orbital considerations do offer an explanation. The trichloromethyl radical has a low energy SOMO (Table 5-1) and is therefore electrophilic, interacting much more strongly with the HOMO of an olefin than with the LUMO. The effect of the fluorine (an Ẍ-substituent) on the polarization of the HOMO is to lower the coefficient on the atom to which it is bound (Fig. 5-6). It is unusual among Ẍ-substituents, however, in lowering the energy of the HOMO. Thus the effect of the fluorine is to slow down the attack on the atom to which it is attached ($3.2 \rightarrow 5.4 \rightarrow 8.3$ kcals). Attack at the other end of the double bond, however, although slowed down by the lowering in energy of the HOMO, is speeded up by the increase in the coefficient of the fluorinated olefins at that site. These effects partly cancel, and the activation energy for attack at C-2 in these olefins is rather similar ($3.2 \rightarrow 3.3 \rightarrow 4.6$ kcals) for each.

Another striking reversal of the rule is in the cyclization of hex-5-enyl radicals (**396**). Generally, these cyclize contrathermodynamically to give the primary radical (**397**). This may be for all sorts of reasons, but it is noteworthy that cylization to the secondary radical (**395**) does take place when the groups R and R' are electron-withdrawing.[315]

(**395**) (**396**) (**397**)

It is known, in fact, that the latter observation is a result of thermodynamic control:[316] the radicals (**396** and **397**, R = CN, R' = CO_2Et) are able to approach equilibrium under the reaction conditions. But we can say, at least, that the ring-closure in the sense (**396** → **395**) will be assisted when the substituents are electron-withdrawing: the radical (**396**) will then have a relatively low energy SOMO, and it will therefore be more sensitive to the polarization of the HOMO of the olefin group. In the radical (**396**, R = R' = H) without electron-withdrawing groups, the interaction with the π-bond will be more affected by the LUMO of the olefin, and this might be part of the reason for contra-thermodynamic cyclization.

The rates of radical attack on π-bonds seem also to be sensitive to the coefficients on the atoms involved in the π-bond. For instance, let us take the relative rates at which hydrogen atoms add to C-1 in the olefins shown in Table 5-2. The increase in rate as R and R' are changed from hydrogen to methyl, could be due to the increase in the energy of the HOMO of the olefins, but the coefficients at C-1 will also increase in the same order. The presence of two methyl groups on C-1 should raise the energy of the

Table 5-2 Relative rates of addition of hydrogen atoms to olefins[317]

	Compound	Relative Rates
	R = R' = H	1
	R = Me, R' = H	1·8
	R = R' = Me	4·4
		1·5

HOMO even further, and we should expect a larger value for the relative rate for this compound if the energy of the HOMO were the important factor. In fact, a smaller one is found, and we can account for this by arguing that the two methyl groups even up the polarization in the HOMO. A steric effect of this size for a hydrogen atom attack is not very likely.

5.3.2 Attack on Substituted Aromatic Rings

The rates of attack of radicals on aromatic rings correlate with ionization potential,[318] with localization energy[319] and with superdelocalizability (p. 58),[320] a picture reminiscent of the situation in aromatic electrophilic substitution. As in that field, there are evidently a number of related factors affecting reactivity. Frontier orbitals provide useful explanations for a number of observations in the field, as the following examples show.

The partial rate factors of Table 5-3 show that a phenyl radical reacts with nitrobenzene and anisole faster than it does with benzene. This can readily

Table 5-3 Some partial rate factors for radical attack on benzene rings.[321] f is the rate of attack at the site designated relative to the rate of attack at one of the carbon atoms of benzene itself

Attacking radical	Ring being attacked	f_o	f_m	f_p
$p\text{-}O_2NC_6H_4\cdot$	PhNO$_2$	0·93	0·35	1·53
Ph·		9·38	1·16	9·05
$p\text{-}O_2NC_6H_4\cdot$	PhOMe	5·17	0·84	2·30
Ph·		3·56	0·93	1·29

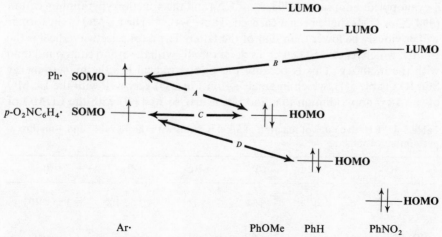

Fig. 5-7 Interactions for the attack of an aryl radical on substituted benzene rings

be explained if the energy levels come out, as they plausibly might, in the order shown in Fig. 5-7. With anisole the SOMO/HOMO interaction (*A*) is strong, and with nitrobenzene the SOMO/LUMO interaction (*B*) is strong, but with benzene neither is strong. Product development control can also explain this, since the radicals produced by attack on nitrobenzene and anisole will be more stabilized than that produced by attack on benzene. However, this cannot be the explanation for the other trend which can be seen in the table, namely that a *p*-nitrophenyl radical reacts faster with anisole and benzene than it does with nitrobenzene. But this is readily explained if the SOMO of the *p*-nitrophenyl radical is lower in energy than that of the phenyl radical, making the SOMO/HOMO interactions (*C* and *D*) strong with the former pair.

In hydrogen abstraction reactions (p. 186), alkyl radicals change, as the degree of substitution increases, from being mildly electrophilic (the methyl radical) to being mildly nucleophilic (the t-butyl radical). In addition reactions to aromatic rings, they are all relatively nucleophilic: thus they add exclusively to the 2-position of pyridinium cations (**398**). This change is reasonable,

(**398**)

because the LUMO of an aromatic ring will be lower in energy than that of a C—H bond, and the SOMO of the radical can interact more favourably with it.

Furthermore, the more substituted radicals continue to be the more nucleophilic. The *relative* rates with which the various alkyl radicals react with the 4-cyanopyridinium cation (**398**, X = CN) and the 4-methoxypyridinium cation (**398**, X = OMe) have been measured (Table 5-4).[322] The LUMO of the former will obviously be lower than that of the latter. The most selective radical is the t-butyl, which reacts 350,000 times more rapidly with the cyano compound than with the methoxy. This is because the t-butyl radical has the highest-energy SOMO (Table 5-1), which interacts (*A* on Fig. 5-8) very well with the LUMO of the 4-cyanopyridinium ion, and not nearly so well (*B*) with the LUMO of

Table 5-4 Relative rates of reaction of alkyl radicals with the 4-cyano- and 4-methoxy-pyridinium cations

R·	Me·	n-Pr·	n-Bu·	s-Bu·	t-Bu·
$\dfrac{k_{X=CN}}{k_{X=OMe}}$	46	164	203	1.3×10^4	3.5×10^5
SOMO-energy ($-$I.P. in eV)	-9.8	-8.1	-8.0	-7.4	-6.9

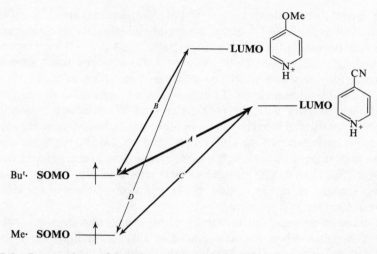

Fig. 5-8 Interactions of frontier orbitals for the reaction of alkyl radicals with substituted pyridinium cations

the 4-methoxypyridinium ion. At the other end of the scale, the methyl radical has the lowest-energy SOMO, and hence the difference between the interactions *C* and *D* on Fig. 5-8 is not so great as for the corresponding interactions (*A* and *B*) of the t-butyl radical. Therefore, it is the least selective radical, reacting only 50 times more rapidly with the cyano compound than with the methoxy. The other alkyl radicals in the table show a regular pattern, consistent with this analysis.

The most vexed subject in this field is the *site* of radical attack on substituted aromatic rings. Some react cleanly where we should expect them to. Phenyl radicals add to naphthalene (**399**), to anthracene (**400**)[323] and to thiophene

87%·Ph ·425 13% ·263 (**399**)

84% 14% ·Ph ·440 ·311 2% ·220 (**400**)

7% ·Ph ·372 93% ·602 S (**401**)

(**401**),[324] with the site-selectivity shown on the diagrams. In all three cases, the frontier orbitals are clearly in favour of this order of reactivity; we should note that, because of the symmetry in these systems, both HOMO and LUMO have the same absolute values for the coefficients, so there is no ambiguity here as to which to take.

However, there is a lot of evidence that radicals are much less selective than cations and anions. Thus, dimethylamino radicals attack toluene[325] to give 10% *ortho*-, 47% *meta*- and 43% *para*-dimethylaminotoluenes; phenyl radicals attack pyridine with little selectivity,[326] and chlorine atoms attack naphthalene completely unselectively.[327] Since almost all substituents stabilize radicals, substituted benzenes usually (but not invariably, see Table 5-3) react faster than benzene itself, and most of them, whether C-, Z- or $\ddot{X}$-substituted, show some preference for *ortho/para* attack, no doubt because attack at these sites gives the more stable intermediates. In assessing the contribution of the frontier orbitals, we are back with the problem (pp. 52–57) of how to describe the orbitals of substituted benzene rings—in other words, how to estimate not only the relative importance of the HOMO (ψ_3) and the LUMO (ψ_4^*) but also the relative importance of ψ_2 and ψ_5^*, which lie quite close in energy to the frontier orbitals. Thus the HOMO and the LUMO shown in Fig. 5-8, for example, are best thought of, not as single orbitals, but as composites of the kind discussed in Chapter 3.

One trend seems clear, and it is a trend readily explained with frontier orbitals. In an $\ddot{X}$-substituted benzene, like toluene or anisole, the proportion of *meta* attack falls as the energy of the SOMO of the attacking radical rises (Table 5-5).

Table 5-5 Product distribution in the attack of various radicals on anisole[328]

R·	%o	%m	%p	$\dfrac{\%o + 2\%p}{\%m}$	SOMO-energy (eV)[a]
Me$_3$Si·	62	31	7	2·5	−7
⬡·	67	28	5	2·8	−7·8
Ph·	69	18	13	5·3	−9·2
Me·	74	15	11	6·4	−9·8
HO$_2$CCH$_2$·	78	5	17	22·4	−10·9

R· + (MeO–C₆H₅)

[a] −I.P.

This trend is usually put down, without explanation, to the increasing 'electrophilicity' of the radicals. Because the HOMO and LUMO energies of $\ddot{X}$-substituted benzenes will be raised, we can expect that the HOMO of the aromatic ring is the more important frontier orbital. We have already seen (p. 54) how the frontier electron population is effectively higher in the *ortho* and *para* positions for an $\ddot{X}$-substituted benzene. Thus, the lower the energy of the SOMO of the radical, the better the interaction with the HOMO of the benzene ring, as we have already seen in Fig. 5-7, and hence the more *ortho* and *para* attack there is.

5.4 Ambident Radicals

5.4.1 Neutral Ambident Radicals

Some neutral radicals are ambident, but not much is known about this subject. The argument is a simple one—the site of reactivity should be largely determined by the coefficients of the SOMO.

The mono-substituted allyl radical, generated by adding a radical to a diene, usually reacts at the unsubstituted end of the allyl radical. We saw two examples of this on p. 188. This may simply be owing to product-development control, but it is also likely that the coefficient is larger at this site.

The cyclohexadienylradical (**402**) should have a higher coefficient at C-3 (see p. 45) than at C-1, and indeed it seems that this site most readily extracts a hydrogen atom from another molecule.[329] This is clearly not product-development control.

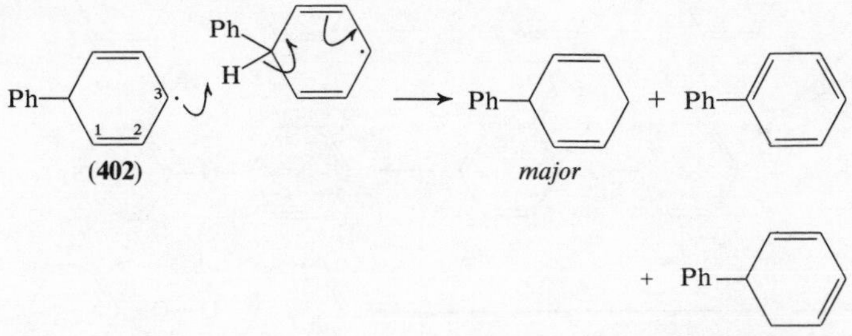

(**402**)   *major*

minor

As we have already seen (p. 182), α-carbonylmethyl radicals (e.g. **403**) react more readily at the carbon atom than at the oxygen atom. Here is another example:[330]

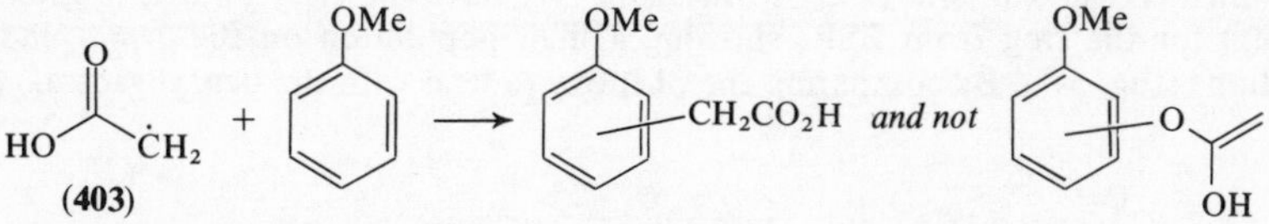

(**403**)

Cathodic electrolysis of pyridinium ions (**404**) causes an electron to be added to the ring. This electron is in an orbital very like the LUMO of pyridine (p. 68), which we know to have the largest coefficient on the 4-position. This is the site of dimerization.[331]

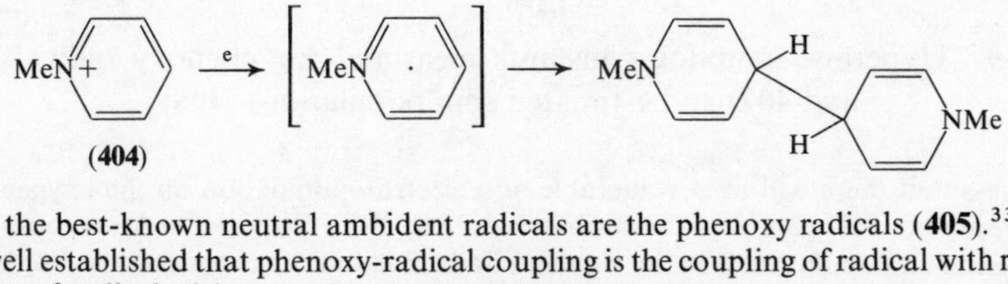

(**404**)

But the best-known neutral ambident radicals are the phenoxy radicals (**405**).[332] It is now well established that phenoxy-radical coupling is the coupling of radical with radical and not of radical with neutral molecule, although the attack of a radical on a phenate ion may occasionally be an important pathway. In either event, we can expect a large number of possible products. As it happens, all these possibilities have been observed with one compound or another. In general, *o-p*, *p-p*, and *o-o* are rather more common than *O-p* and *O-o*, and these are much more common than *O-O*. The electron-distribution in a phenoxy radical is not easy to measure. We have the electron distribution in the SOMO for the ring from ESR, showing a high population on the *ortho-* and *para-* positions (Fig. 5-9). By comparing the phenoxy radical with the benzyl radical, we can

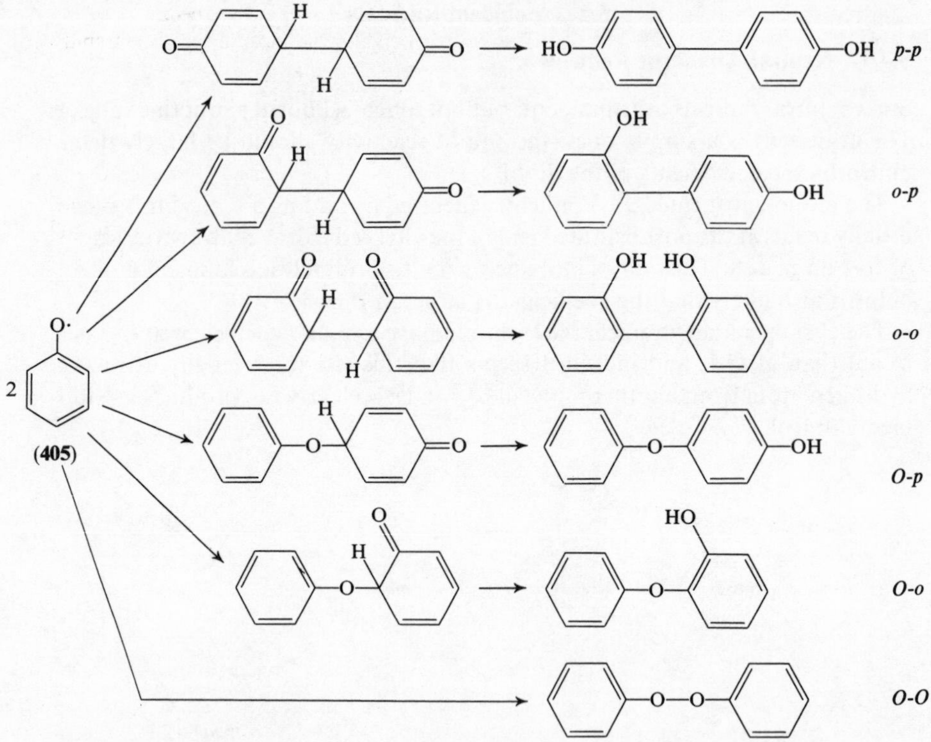

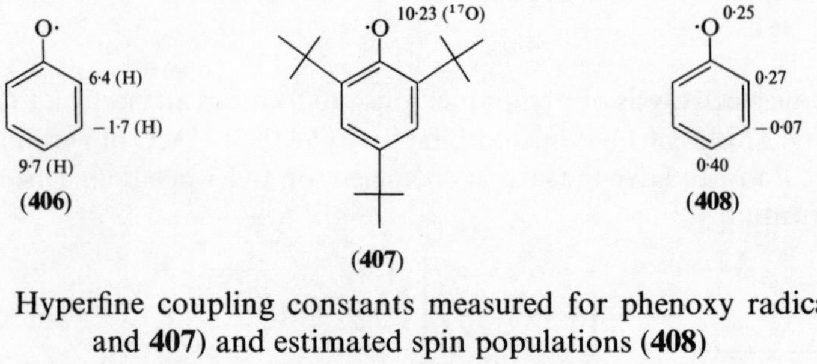

Fig. 5-9   Hyperfine coupling constants measured for phenoxy radicals (**406** and **407**) and estimated spin populations (**408**)

also guess that there will be considerable odd-electron population on the oxygen atom (a coupling of 10·23 gauss has been observed[333] to ^{17}O). We need to know how much of the electron population is in the ring and how much on the oxygen atom. For this we can use the McConnell equation (p. 22). The value of Q_{CH}^{H} is generally about -24, and the value of Q_{OC}^{O} (the coupling of an electron in a p-type orbital on oxygen to the oxygen nucleus when all the electron population is in that orbital) has been̶ to be about -40 (measured[334] on p-semiquinone). Using these numbers we get spin populations as shown on **408**, but these numbers must be regarded as orders of magnitude only—the hyperfine splittings and the Q-values were all got from quite different compounds, and the McConnell equation is far from fool proof. This electron-distribution tells us that all three positions—oxygen, *ortho*, and *para*—have quite high coefficients. We should remember that the β-value for C—O bond-formation is less than for C—C bond-formation. It is not clear from the experimental evidence whether p–p coupling really is preferred, as these numbers would suggest. The problem is that products are often obtained in low yield, and that the mass balance is usually poor. In addition there is the statistical effect of there being two *ortho* positions to one *para*. We can guess that there will not be much in it, and that does seem to be the case.

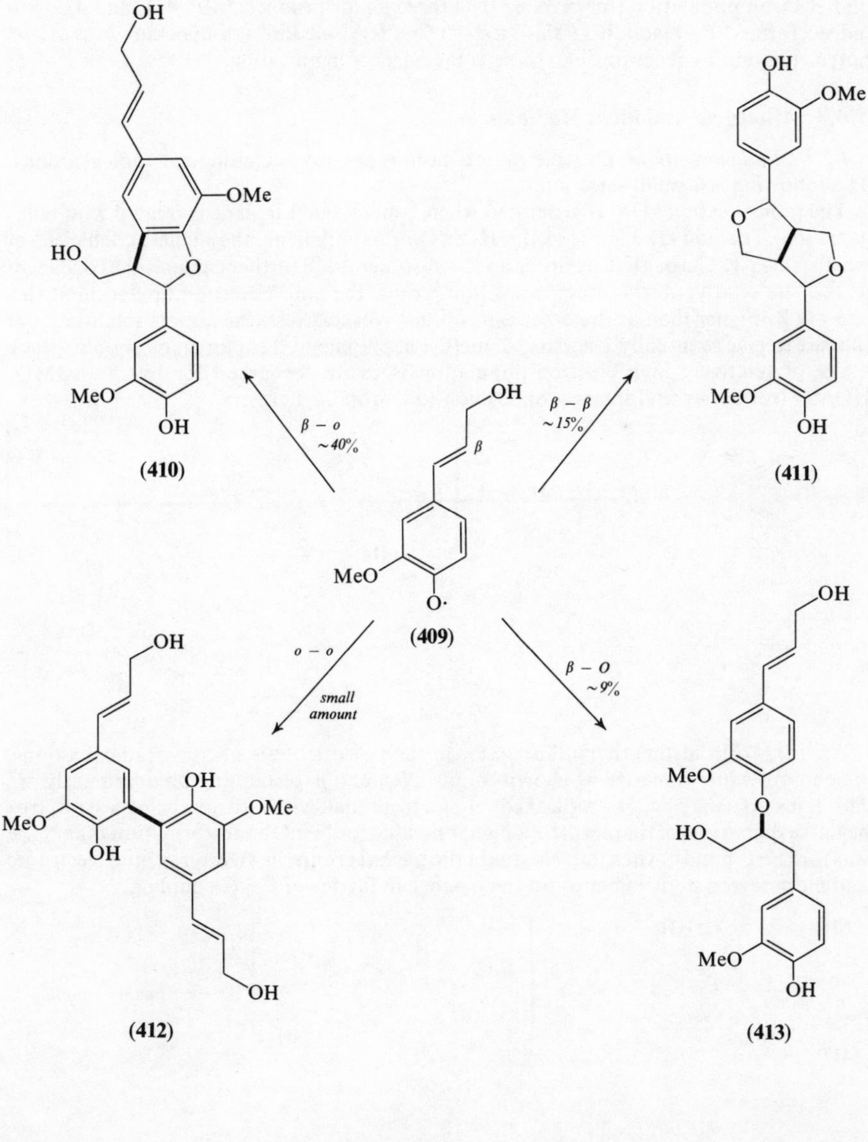

(410)

(411)

(412)

(413)

(409)

β – o
~ 40%

β – β
~ 15%

o – o
small amount

β – O
~ 9%

...*para* attack is in the Elbs persulphate oxidation
...ten as the combination of a phenoxy radical (**414**)
...it could be phenoxide ion undergoing electrophilic
...*para* attack is favoured over *ortho* attack by a factor of about

A slightly more complicated phenoxy radical is the important one (**409**) from coniferyl alcohol. Here we would expect the β-carbon of the styrene moiety to be the site of highest odd-electron population (on carbon). The three major products (**410**, **411**, and **413**) are indeed formed by reaction at this site.[336] Coniferyl alcohol is important because its polymerization by reactions like these is the basis of lignification.

5.4.2 Charged Ambient Radicals

5.4.2.1 Radical Cations. Organic radical cations are not as common as radical anions. The following is a small selection.

The radical cation (**416**) is generated when 2-methylnaphthalene is treated with manganic ion. The odd electron is in the HOMO of naphthalene, the highest coefficient of which is at C-1. The methyl group, as an $\ddot{X}$-substituent, will further enhance the coefficient at this site relative to the other α-positions; thus, the *total* electron-population at this site will be higher than at the other α-positions. Nevertheless, the acetate ion attacks at this site to give eventually 1-acetoxy-2-methyl naphthalene. That an anion should attack a site of relatively high electron-population is easily accounted for by the SOMO/HOMO frontier orbital interaction leading to a drop in energy.

Aniline (**417**) and dimethylaniline (**418**), on anodic electrolysis, also give radical cations, which apparently dimerize as shown to give *N-p* and *p-p* coupling predominantly.[337] This is just like the phenoxy radical coupling except that, with nitrogen being less electronegative than oxygen, there will be a higher β-value for N—C bond-formation than there was for O—C bond-formation. No doubt the methyl groups in **418** exert a mild electronic and more severe steric effect to tip the balance in favour of C—C coupling.

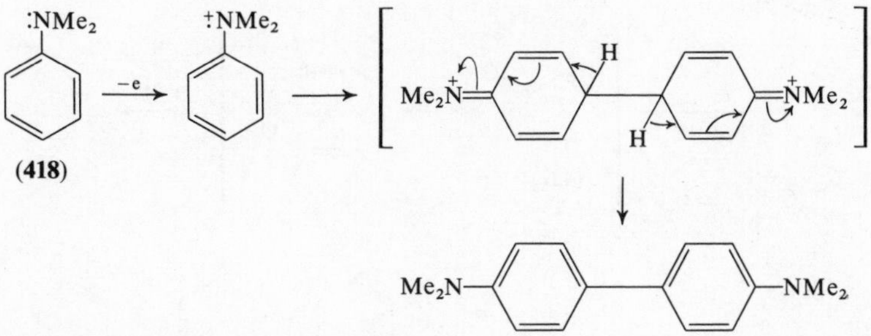

(418)

As with phenolic coupling, bonding between the heteroatoms is made unfavourable by the low β-value for such bond-formation.

The enamine **(419)** with oxygen undergoes[338] a radical-chain reaction, leading, after hydrolysis, to the 1,4-diketone **(420)**. The site of attack contrasts with that we would expect by analogy with the site of protonation of a dienolate ion (p. 46), but the heteroatom is different, and that may be decisive.

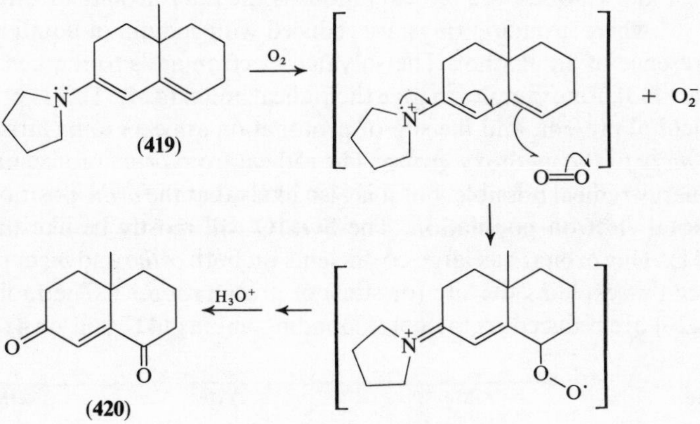

(419)

(420)

5.4.2.2 Radical Anions. Radical anions are common intermediates in organic reactions; they are easily prepared from compounds with low-enough LUMOs by the addition of an electron (from a dissolving metal or from a cathode, or the solvated electron itself). Those derived from carbonyl groups **(421)** dimerize at carbon;[339] those derived from α,β-unsaturated carbonyl compounds **(422)** dimerize at the β-position,[340] and pyridines dimerize predominantly at the 4-position.[341] In each case, the odd electron has been fed into the orbital which was the LUMO of the starting material; the site of coupling therefore should, and does, correlate with the site at which nucleophiles attack the neutral compounds.

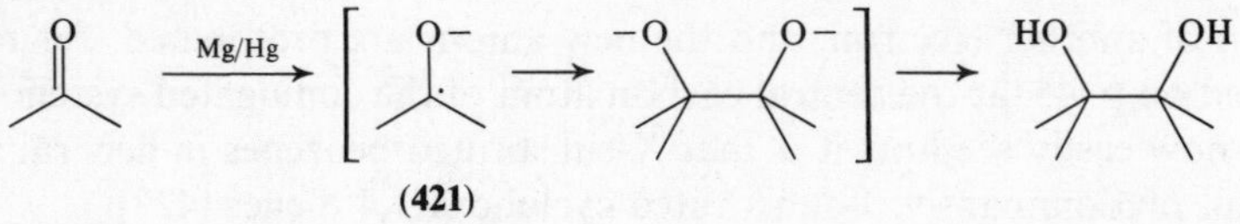

(421)

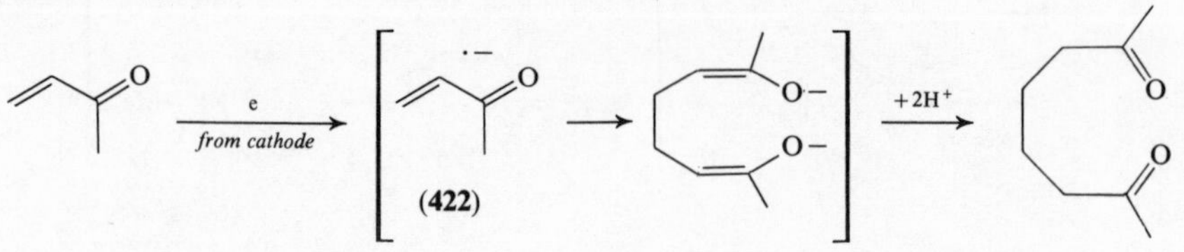

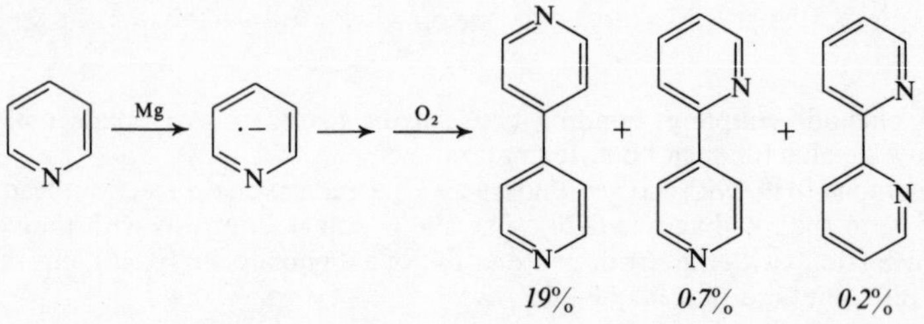

19%   0·7%   0·2%

The best known of all the radical anions is the intermediate in Birch reduction,[342, 343] where aromatic rings are reduced with sodium in liquid ammonia in the presence of an alcohol. The solvated electron adds to the benzene ring of anisole (423), for example, to give the radical anion (424). This is protonated by the alcohol present, and the site of protonation appears to be largely either *ortho* or *meta* to the methoxy group. The radical from *ortho* protonation is the lowest-energy radical possible, but it is also likely that the *ortho* position has the highest total electron population. The SOMO will mostly be like the orbital ψ_5^* (p. 61). This orbital has large coefficients on both *ortho* and *meta* positions, and hence this explains the site (or sites) of protonation.[344] The radicals (425 and/or 426) are reduced to the corresponding anions (427 and/or 428) by the

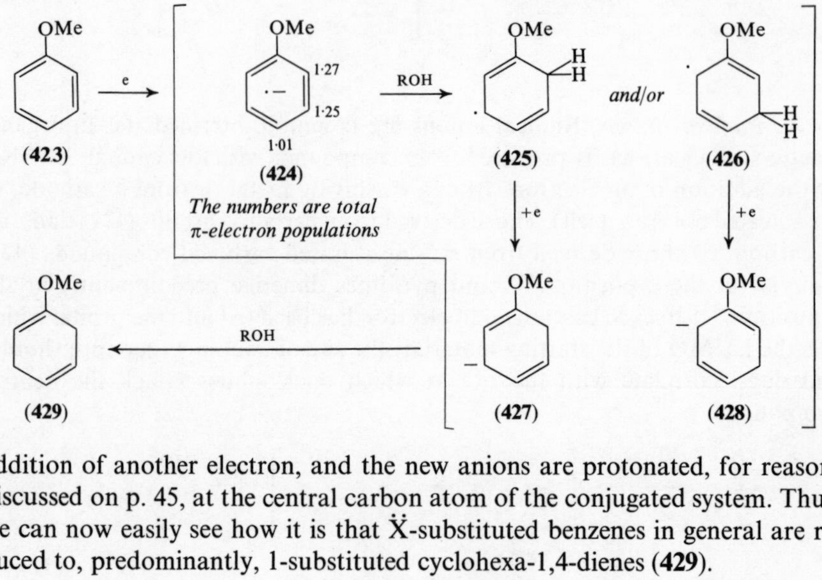

addition of another electron, and the new anions are protonated, for reasons discussed on p. 45, at the central carbon atom of the conjugated system. Thus, we can now easily see how it is that $\ddot{X}$-substituted benzenes in general are reduced to, predominantly, 1-substituted cyclohexa-1,4-dienes (429).

By contrast, C-substituted benzenes are reduced to 3-substituted cyclohexa-1,4-dienes,[343] and this too fits nicely into a frontier orbital analysis. We can use biphenyl (430) as an example. The Hückel coefficients for the SOMO of the radical anion are shown on 431. Since this molecule does not have heteroatoms

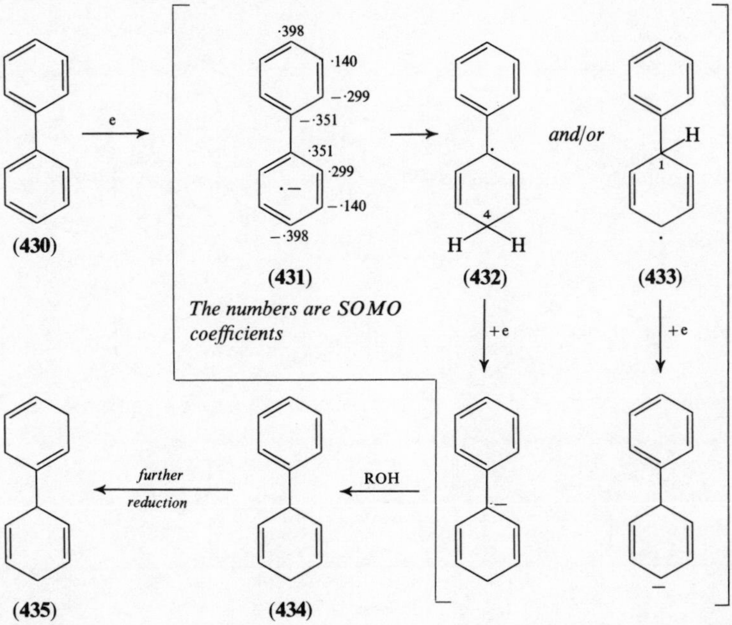

(430)

(431)

The numbers are SOMO coefficients

(432)

and/or

(433)

(434)

(435)

further reduction

ROH

+e

+e

disturbing the symmetry of the orbitals, this distribution of electron density (squared) is also the distribution of the excess charge. So, regardless of whether it is the Coulombic or the frontier orbital term that is more important, both contributions lead to protonation at C-4 or C-1 to give either 432 or 433. Reduction and protonation of these intermediates (either, or both, may be involved, but the former is much the more likely) leads then to the observed product (434).[345] It is amusing that further reduction of this molecule also takes place, but now the benzene ring is an $\ddot{X}$-substituted one. The final product, accordingly, is 435. The Birch reduction of benzoic acid is the same type as that of biphenyl, and the product is 436.[346] In the reaction medium, it will be

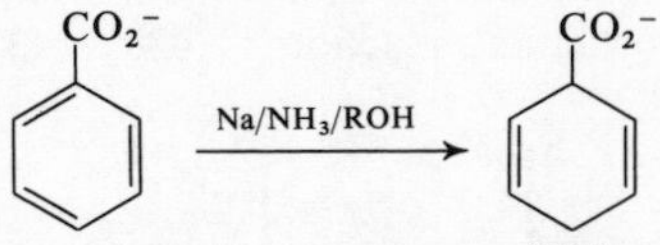

CO_2^-

Na/NH$_3$/ROH

CO_2^-

(436)

benzoate ion that is being reduced. Because of the delocalization of the negative charge in the benzoate ion, we should probably regard the carboxylate ion more as a C- than as a Z-substituent.

Here are some other compounds which are easily reduced[347-351] in a comparably explicable manner, except that, with reductions in liquid ammonia without alcohol present, two electrons must be added to the LUMO of the starting material before it is basic enough to abstract a proton from ammonia. The addition of the second electron (to the radical anion) is often so slow that dimerization of the radical-anion occurs to some extent.

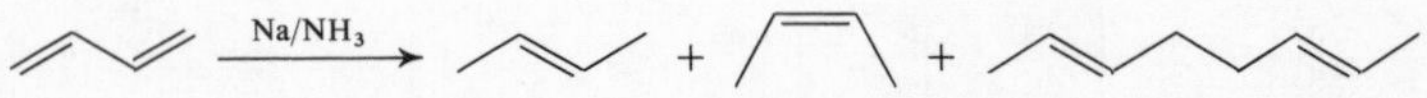

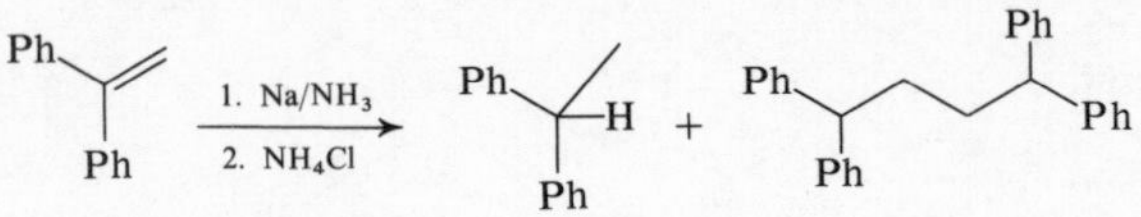

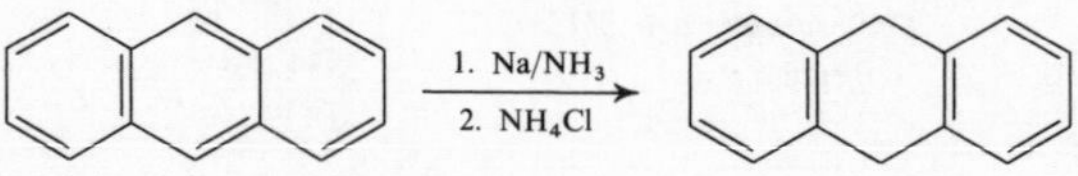

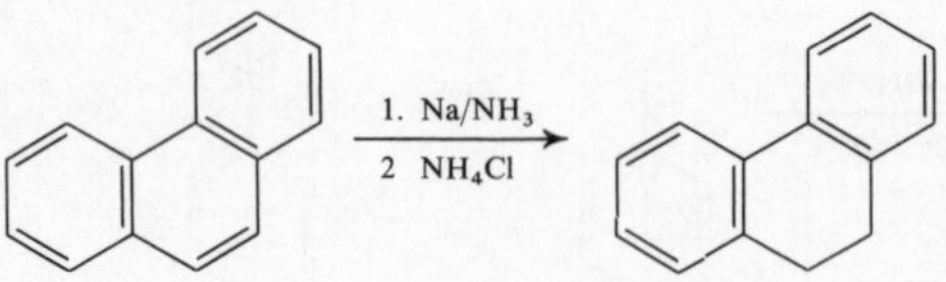

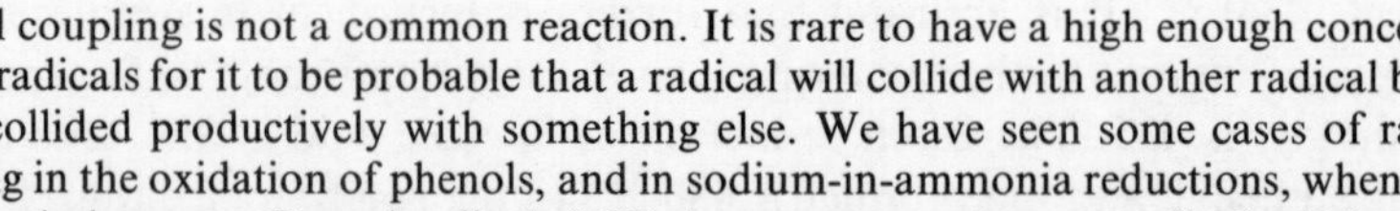

5.5 Radical Coupling

Radical coupling is not a common reaction. It is rare to have a high enough concentration of radicals for it to be probable that a radical will collide with another radical before it has collided productively with something else. We have seen some cases of radical coupling in the oxidation of phenols, and in sodium-in-ammonia reductions, when there is no alcohol present. Benzyl radicals (437), in contrast to phenoxy radicals, do have the highest odd-electron population on the exocyclic atom, and their coupling does give dibenzyl.[352] Similarly in Kornblum's reaction[353] (438 + 439 → 441), the benzyl radical

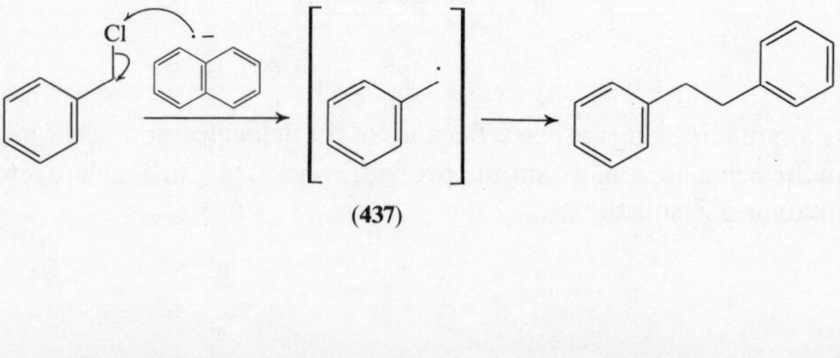

(437)

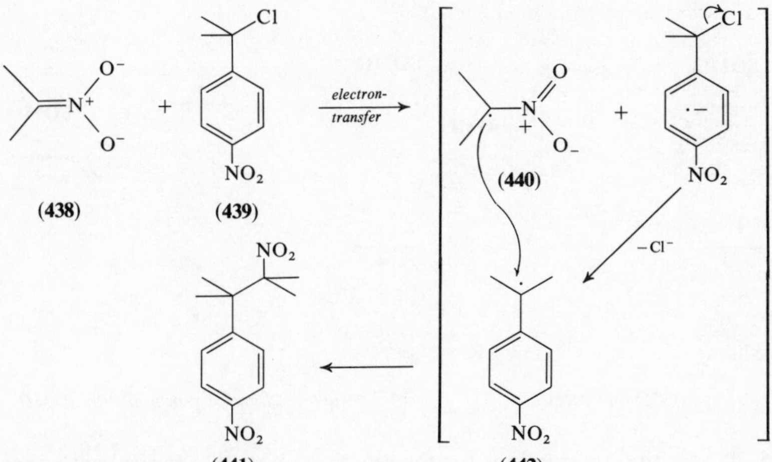

(**442**) is site-specific at the exocyclic carbon, and the α-nitromethyl radical (**440**) is also site-specific at carbon, just as we should expect from the frontier orbitals. Here are more examples of radical coupling:[354, 355, 329]

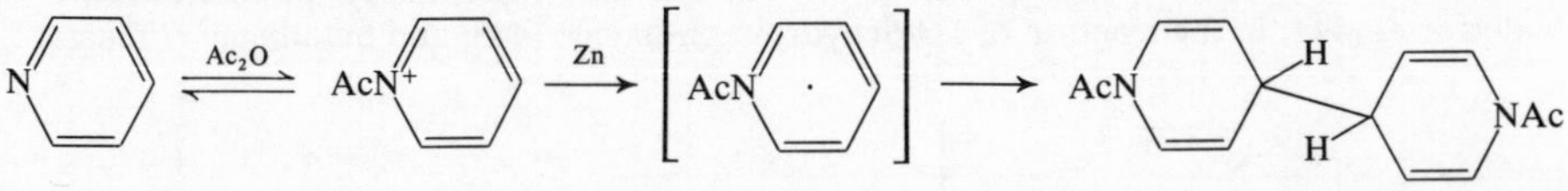

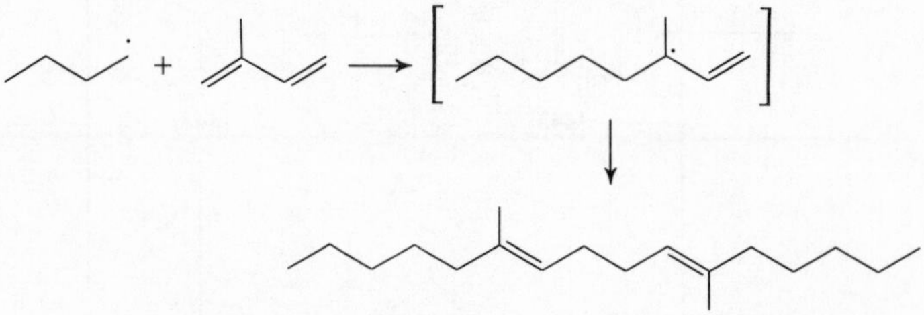

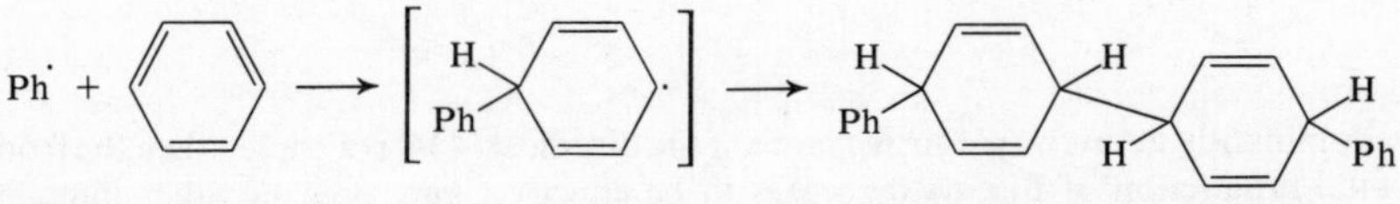

(*A minor product in the reaction seen earlier on p.* 195.)

The discussion on pp. 29–31 established the many reasons why an allyl anion and an allyl cation react with electrophiles and nucleophiles respectively at C-1 (and C-3) of the allyl system. The force of these arguments is less when they are applied to the reaction of an allyl radical with a radical. Although the frontier orbital interaction (Fig. 5-10a) will still favour attack at C-1 in the usual way, the interaction of the lowest filled π-orbital will not be negligible, especially with a radical having a low-energy SOMO (Fig. 5-10b). Since the lowest filled π-orbital has the larger coefficient on C-2, reaction at this site

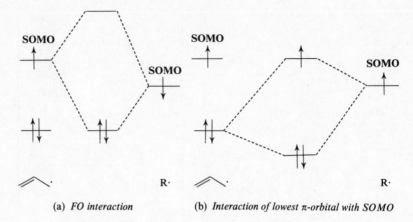

(a) *FO interaction*   (b) *Interaction of lowest π-orbital with SOMO*

Fig. 5-10 Orbital interactions for the reaction of an allyl radical with another radical

is made less unfavourable than it might at first appear; also, with radicals, there is little contribution from Coulombic forces. All this may explain why the intermediate diradical (**444**) in the reaction of 1,8-dehydronaphthalene (**443**) and butadiene[356] reacts

predominantly in this way. Furthermore, if the diradical (**444**) is a triplet, then the frontier orbital interaction of Fig. 5-10a ceases to be effective, and only the other interaction (Fig. 5-10b) is left.

5.6 Photochemically-Induced Radical Reactions

Some photochemical reactions are simply radical reactions: the light is used solely to generate radicals, and the radicals are in their electronic ground states. When this happens, the radicals usually show the same pattern of reactivity and selectivity as the corresponding radicals generated in more conventional ways. The following examples more or less belong in this category.

The Photo-Fries Rearrangement:[357]

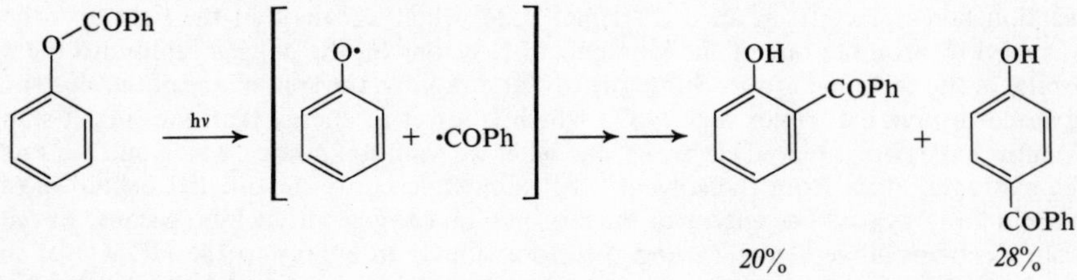

20%   28%

Radical Coupling:[358]

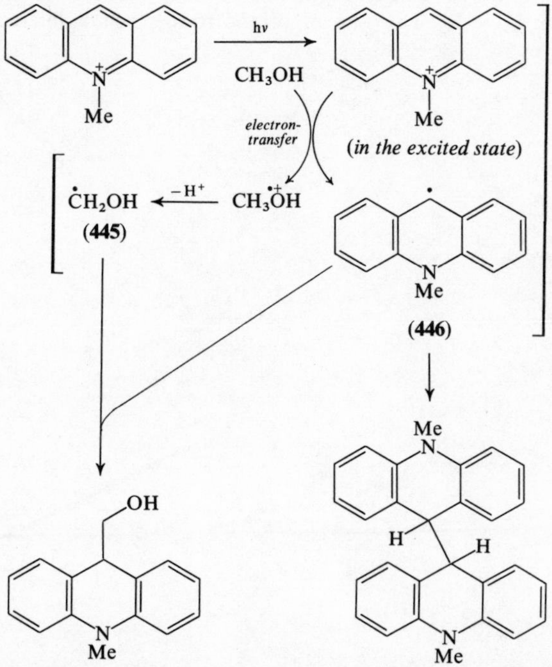

$\overset{\cdot}{C}H_2OH \xleftarrow{-H^+} CH_3\overset{\cdot +}{O}H$

(445)

electron-transfer

(in the excited state)

(446)

In this reaction, the light, by promoting an electron from the HOMO into the LUMO, has made available a low-energy orbital, the former HOMO, into which an electron can be fed, even from so poor a donor as methanol. After that, the reactions of the radicals (445 and 446) are normal.

Photopinacolization:

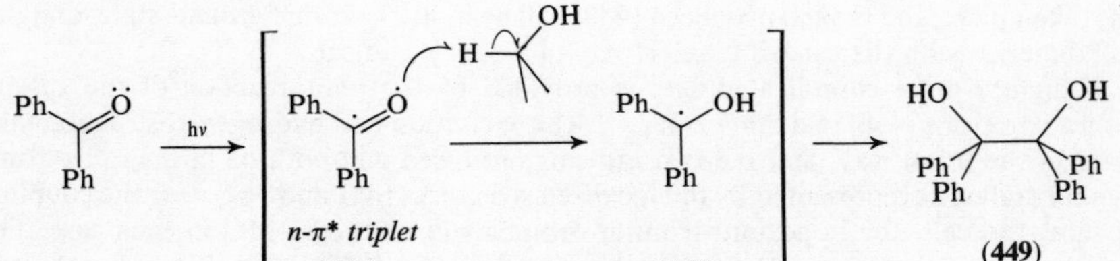

n-π* triplet

(447)   (448)   (449)

The formation of benzpinacol (**449**) by the action of light on benzophenone was almost the first organic photochemical reaction to be discovered.[359] It is now known that the reaction takes place from an n-π^* triplet state, which means that the light absorbed was used to promote one of the lone pair of electrons on the oxygen atom into the π^* orbital of the carbonyl group. Subsequently, and rapidly, the spin of one of the electrons inverted to give the triplet state (**447**), which is lower in energy than the singlet state. We now have two unpaired electrons, and what we want to know is: which one abstracts the hydrogen atom from the solvent? It is almost certainly the one left behind in the p orbital on oxygen. The energy of the p orbital on oxygen will be low (because oxygen is an electronegative element), and therefore similar in energy to the HOMO of the C—H bond of the solvent. The other odd electron will be in a carbonyl π^* orbital, which is comparatively low in energy and therefore well separated from the rather high energy of the σ^* orbital of the C—H bond. The energy levels should be something like those in Fig. 5-11, with $E_1 < E_2$. Thus the important frontier orbital interaction is the same

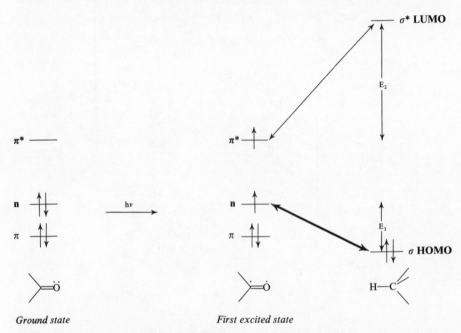

Ground state   First excited state

Fig. 5-11   Frontier orbitals for a ketone in the ground state, in the first excited triplet state and for a C—H bond

as that for most radicals abstracting a hydrogen atom (see p. 186); the n-π^* triplet state has long been recognized as an electrophilic radical. When the hydrogen abstraction has taken place, the radical produced (**448**) will be in its electronic ground-state, and this will dimerize with the usual site-selectivity of a benzyl radical.

A slightly more complicated case is provided by the photoreaction of the α,β-unsaturated ketone (**450**) in diethyl ether.[360] The excitation and hydrogen abstraction take place in the usual way, and the two radicals produced will both be in their electronic ground-states, as represented by the localized structures (**451** and **452**). For the coupling of these radicals, the important frontier orbitals will be the SOMO in each case. The radical (**452**) derived from the ether will be like the radical (**382**) on p. 184; it has the odd electron in an orbital like the LUMO of a protonated carbonyl group, and the larger coefficient will therefore be on carbon. The other radical (**451**) has the odd electron in an orbital which is like that of the LUMO of protonated acrolein (see p. 163), where the largest coefficient is on the carbonyl carbon atom. Thus the observed sites of coupling are accounted for. In ethanol, incidentally, the major product[361] is the result of self-coupling of the radical (**451**) at the same carbon atom (C-3).

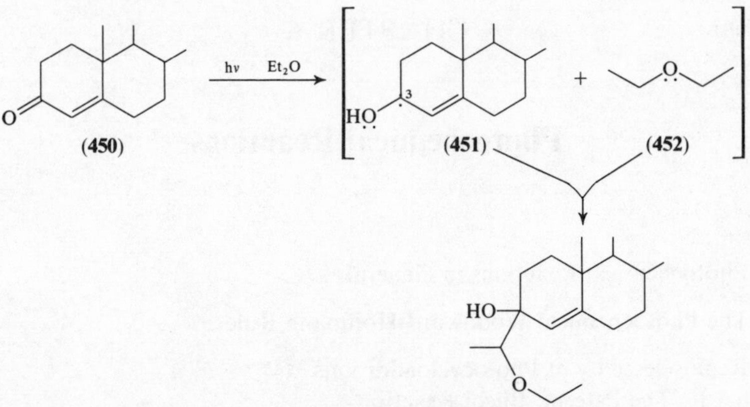

But not all photochemical additions to $\alpha\beta$-unsaturated carbonyl compounds take place at the carbonyl carbon. Thus the ketone (**453**) is attacked in the conjugate position,[362]

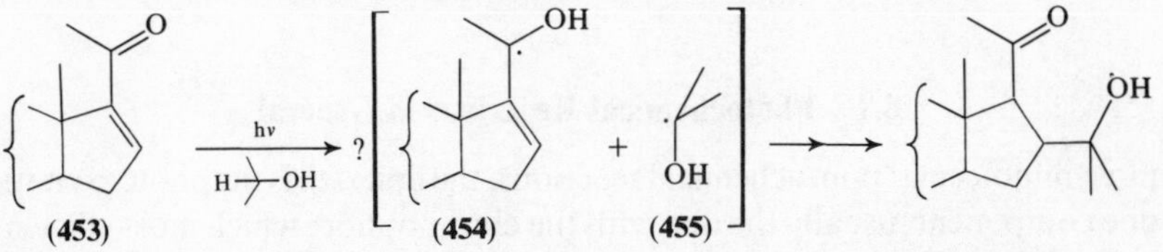

and this is quite a common and synthetically useful pattern of behaviour.[363] It is consistent with frontier orbital control only if the reaction is not a radical coupling (**454** + **455**), but attack of the radical (**455**) on a ground-state $\alpha\beta$-unsaturated ketone (**453**). This would be a chain reaction, and should have a high quantum yield; the reaction is described as being "rapid".

A word of warning. In this chapter, we have not tried any quantitative correlations. No doubt they could be made, but they will be complicated. The reason is that the interactions we have been looking at, especially the SOMO/HOMO interactions, are often between orbitals quite close in energy. Such interactions lead to *first-order* perturbations and the third term of equation 2-7 is *not* appropriate.

CHAPTER 6

Photochemical Reactions

6.1   Photochemical Reactions in General

In most bimolecular photochemical reactions, the first step is the photoexcitation of one component, usually the one with the chromophore which most efficiently absorbs the light. Typically, if a conjugated system is present in one component, it can absorb a photon of relatively long wavelength, and an electron leaves the HOMO and arrives in the LUMO as it does so. Alternatively, an electron in a non-bonding orbital, like that of the lone pair on the oxygen atom of a ketone, is promoted from this orbital to the LUMO of the carbonyl group. The excited states produced are called π-π^* and n-π^*, respectively. The second step of the reaction is between the photochemically excited molecule and another molecule, which may or may not be the same compound, in its ground-state.

For this kind of reaction, there will generally be two very energetically profitable frontier orbital interactions: (1) the interaction between the singly-occupied π^* orbital of the excited molecule and the LUMO of the molecule which is in its ground-state (shown at the top of Fig. 6-1), and (2) the interaction of the singly-occupied n or π orbital of the excited molecule and the HOMO of the molecule which is in its ground-state (shown at the bottom of Fig. 6-1).[364, 112] Both interactions will usually be strong, because the interacting orbitals are likely to be close in energy; partly for this reason, this step of a photochemical reaction is often very fast. Because they are so strong, the perturbations are first-order, and the mathematical treatment of them would not take the form of the third term of equation 2-7.

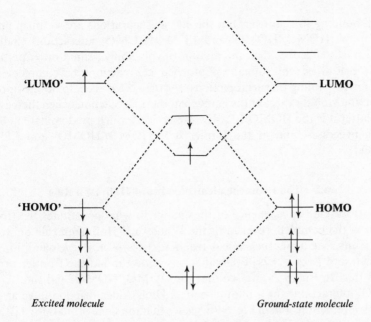

'LUMO'

LUMO

'HOMO'

HOMO

Excited molecule

Ground-state molecule

Fig. 6-1 Frontier orbital interactions between a photochemically excited molecule and a ground-state molecule

In ground-state reactions, the first-order interactions of occupied orbitals with occupied orbitals are antibonding in their overall effect, and there is therefore a large repulsion between the two components of a bimolecular reaction. The bonding interactions of occupied orbitals with unoccupied orbitals are merely second-order effects lowering the energy of the transition state. In photochemical reactions, however, the strong interactions shown in Fig. 6-1 can sometimes create a situation in which the *total* energy is lower than when the two components of the reaction were not interacting. This lower-energy state can be identified as the intermediate now well established in some photochemical reactions; this intermediate is usually called an excimer or exciplex.

Usually, in a bimolecular photochemical reaction:
The HOMO and the LUMO of one component in its ground-state interact with what were the HOMO and LUMO respectively of the other component when it was in its ground-state.

The important frontier orbitals in a photochemical reaction are therefore HOMO/'HOMO' and LUMO/'LUMO', where the inverted commas remind us that these orbitals are not the actual HOMO and LUMO at the time of the reaction, but were the HOMO and LUMO in the ground state, before the excitation took place. The HOMO/'LUMO' and LUMO/'HOMO' interactions

210

are still bonding in character, but the energy separations are so much greater than for the HOMO/'HOMO' and LUMO/'LUMO' interactions that they are much less effective: their interactions involve only second-order perturbations. We now see why so many photochemical reactions are complementary to the corresponding thermal reactions, for they often seem to do the opposite of what you would expect of the equivalent thermal reaction, when there is one. In the latter it is the HOMO/LUMO interactions which predominate in bond-making processes, and in the former it is HOMO/'HOMO' and LUMO/'LUMO'.

6.2 The Photochemical Woodward-Hoffmann Rule

The most striking consequence of the change in what constitutes the frontier orbitals is the complete reversal of the Woodward-Hoffmann rule quoted on p. 95. Thus, for a photochemical reaction, [π4s + π2s] cycloadditions are forbidden and [π2s + π2s] cycloadditions allowed, and the frontier orbitals explain this. In Fig. 6-2a, we see that the HOMO/'HOMO' and the LUMO/'LUMO' interactions for a photochemical Diels-Alder reaction have an antibonding interaction, and in Fig. 6-2b, we see that the same orbitals for a [2 + 2] cycloaddition are bonding. This is borne out in practice: there are very few

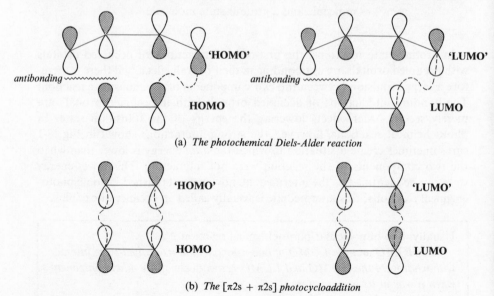

(a) *The photochemical Diels-Alder reaction*

(b) *The [π2s + π2s] photocycloaddition*

Fig. 6-2 Frontier orbitals for [4 + 2] and [2 + 2] photocycloadditions

photochemical Diels-Alder reactions (and none that are known to be concerted), and a great many [2 + 2] cycloadditions. Some of these latter (**456 → 457 + 458** and **459 → 460 + 461**, for example) have been shown to be stereospecifically suprafacial on both components.[365]

In the following examples we see several cases of photochemical reactions which are allowed, when their thermal equivalents (pp. 88–106) were forbidden. In some cases the reactions simply did not occur thermally, in others they showed different stereochemistry. In each case, the application of the appropriate frontier orbitals explains this change.

Cycloadditions:[365–367]

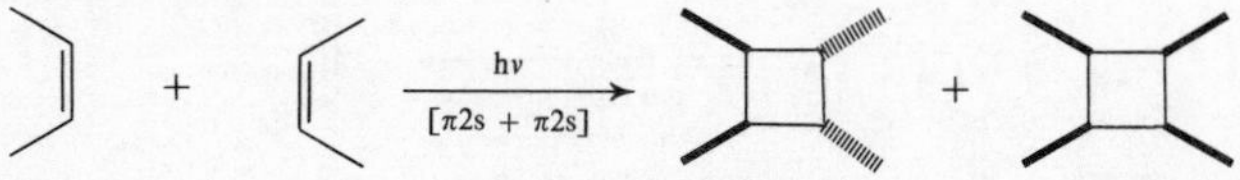

(456) + $\xrightarrow[{[\pi 2s + \pi 2s]}]{h\nu}$ (457) + (458)

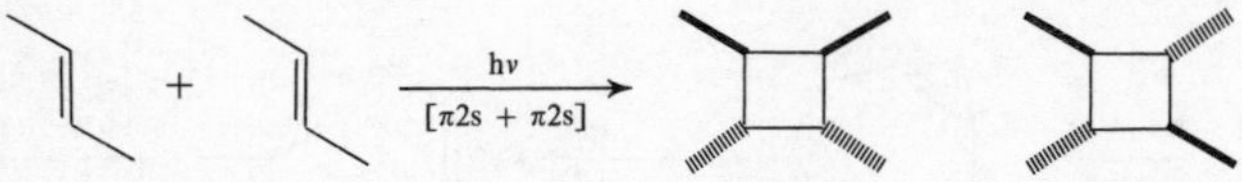

(459) + $\xrightarrow[{[\pi 2s + \pi 2s]}]{h\nu}$ (460) (461)

$\xrightarrow[{[\pi 4s + \pi 4s]}]{h\nu}$

$\xrightarrow[{[\pi 6s + \pi 6s]}]{h\nu/H^+}$

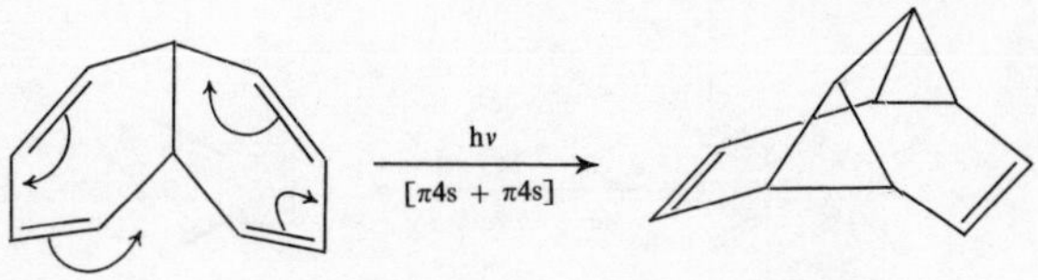

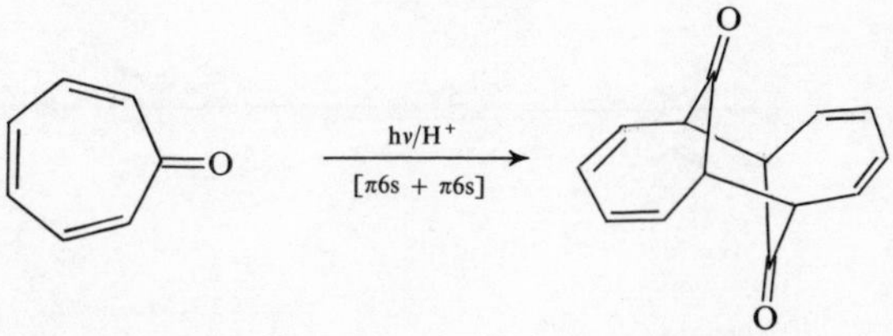

Cheletropic reactions:[368]

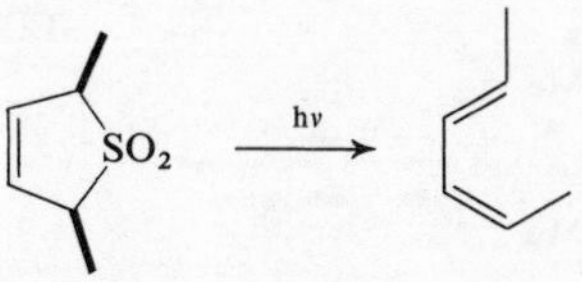

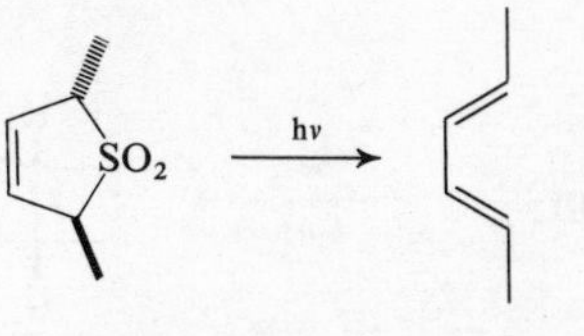

$\xrightarrow{h\nu}$

$\xrightarrow{h\nu}$

Sigmatropic rearrangements:[369–371]

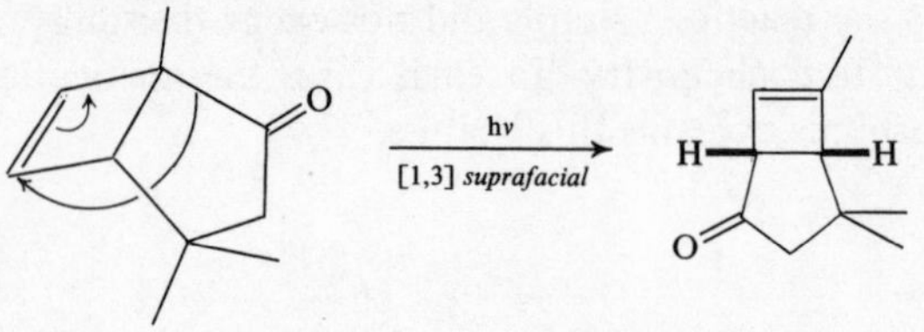

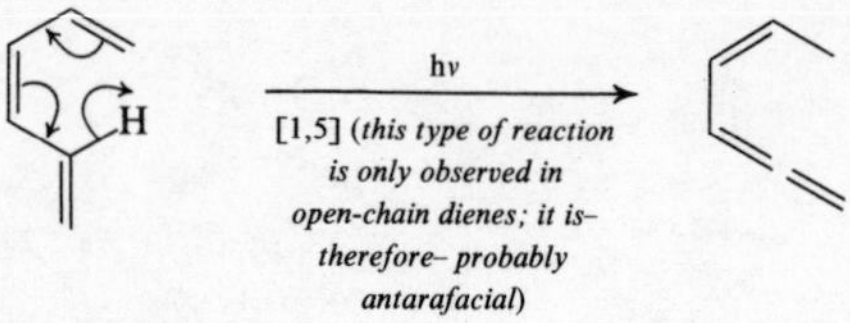

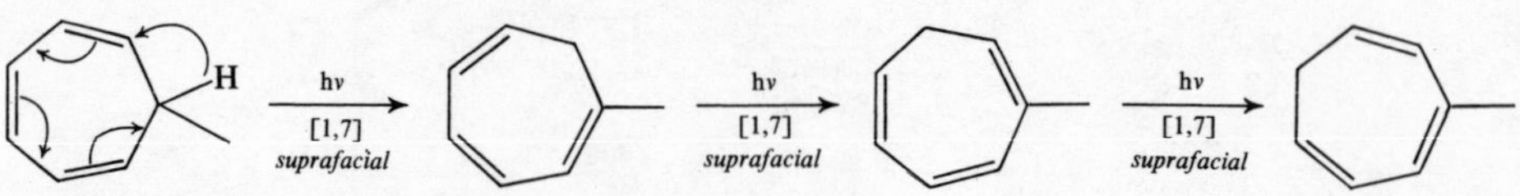

Electrocyclic reactions:[372,373,150,374]

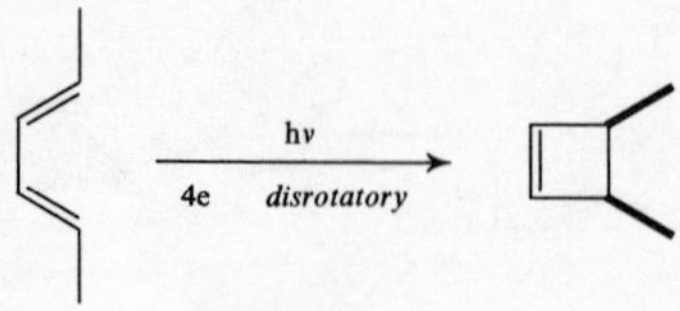

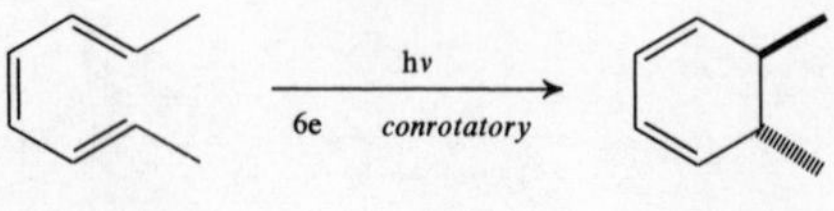

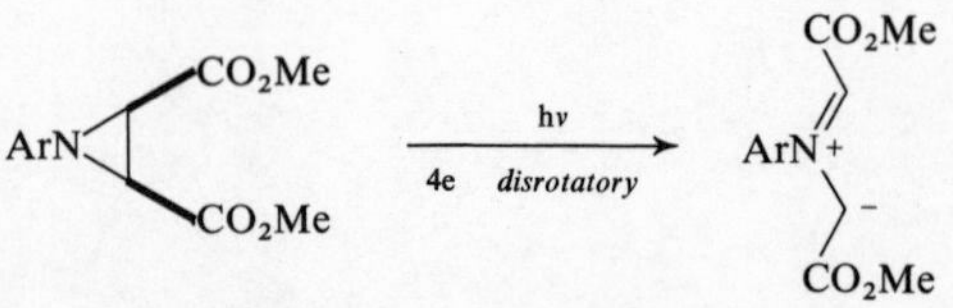

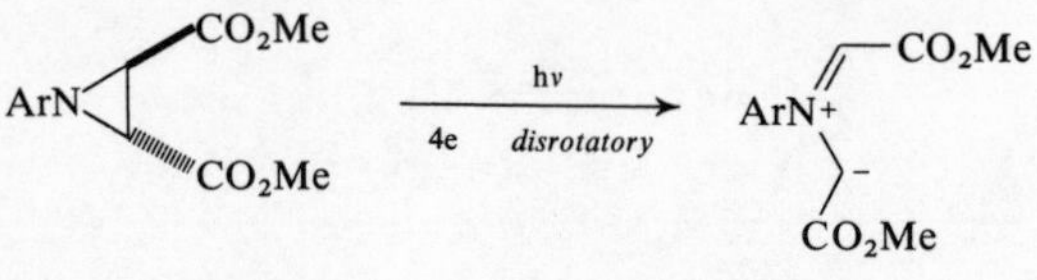

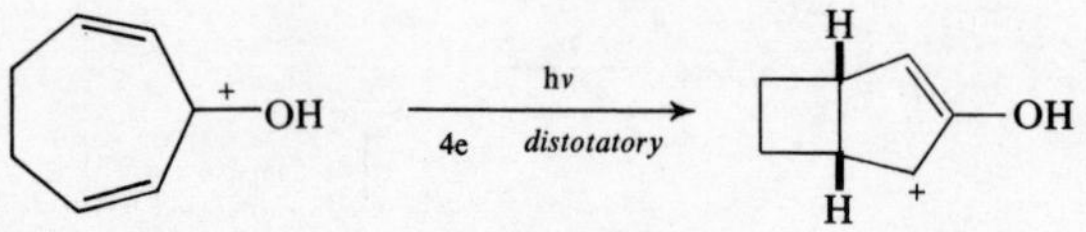

In photochemical reactions, it is even harder to *prove* the pericyclic nature of a process than in thermal reactions. Thus we should remember that some of the reactions in the examples above may not actually be pericyclic. Nevertheless, the contrast with the corresponding thermal reactions shown in Chapter 4 is striking.

6.3   Regioselectivity of Photocycloadditions

We have just seen that many photochemical reactions are complementary to the corresponding ground-state reactions, and that the frontier orbitals explain this change. As with thermal pericyclic reactions, discussed in Chapter 4, the frontier orbitals can also explain many of the finer points of these reactions, most notably the regioselectivity of photocycloadditions.[118]

6.3.1   The Paterno-Büchi Reaction

One well-known class of photocycloadditions is that of aldehydes and ketones with olefins to give oxetanes[375] (e.g. the reaction of **462** with **463** to give **464** and **465**).[376] This kind of reaction is known as the Paterno-Büchi reaction. The excited state of the ketone is the n-π* one, and it is the orbitals of this state which interact with the ground-state orbitals of the olefin. Often it is the triplet state which is involved, but occasionally the singlet state is important. The orientation usually observed is shown in the following examples, where C- and Ẍ-substituted olefins are involved.[377,376]

(462)   (463)

9 parts   :   1 part

(464)   (465)

We can explain the orientation if we assume that the major interaction is from the singly-occupied n-orbital of the ketone with the HOMO of the olefin. C- and Ẍ-substituted olefins do of course have high-energy HOMOs, which makes it reasonable that this should be the major interaction. The carbon atom with the larger coefficient in the HOMO of the olefin is then the one to which the oxygen atom becomes bonded. We can also explain the orientation—and this is the more usual explanation—by looking at the energy of the intermediates produced (**466** and **467**). Again, the lower energy diradical usually seems to be the major one: for example the radical (**466**) is lower in energy than the radical (**467**), and it is from the former that most of the product comes. This

(462) (463) (466) (467)

argument, of course, applies only to the triplet-state reaction, where an intermediate is likely to be involved. Singlet-state reactions may or may not involve the diradical.

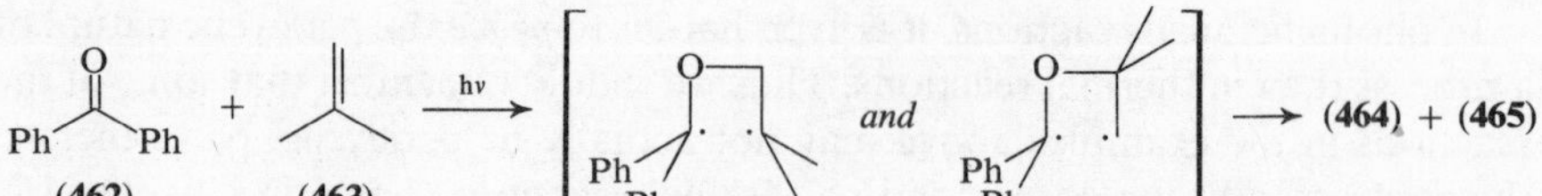

However, the photocycloaddition of ketones to Z-substituted olefins is anomalous if this argument is used; but their behaviour is entirely consistent with the frontier orbital argument. Irradiation of acrylonitrile (**469**) in acetone (**468**) solution gives the adduct (**470**), together with dimers of acrylonitrile.[378]

(468) (469) (470)

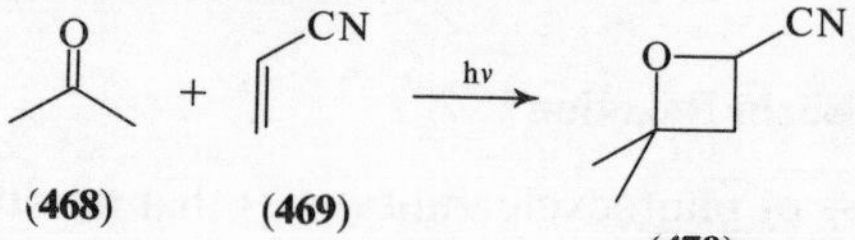

A Z-substituted olefin has a much lower-energy HOMO than the C- and $\ddot{X}$-substituted olefins, and it has, of course, a correspondingly lower LUMO. Thus the interaction of the π^* orbital of the ketone with the LUMO of the olefin (**471**) ought to be much more important for a Z-substituted olefin than it

'LUMO' (π^*) LUMO 'HOMO' (n) HOMO

(471) (472)

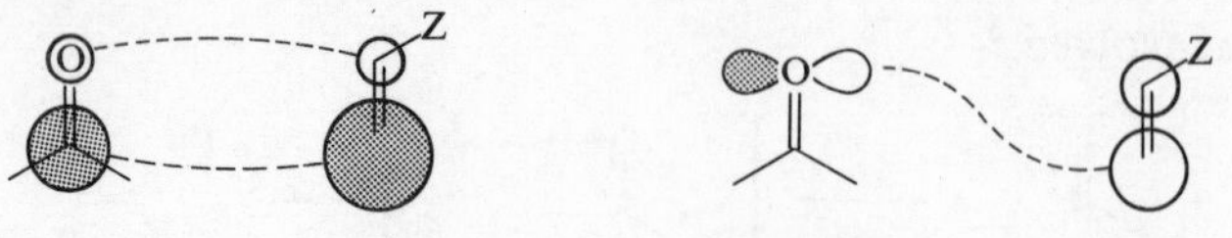

was for the other olefins. In this interaction, the two larger lobes are on the carbonyl carbon and on the β-carbon respectively, and it is these two which become bonded. Furthermore, this reaction is probably concerted. It is a singlet-state reaction, and, with β-substituted acrylonitriles, the reaction is stereospecific, with retention of configuration on the olefin component. If both bonds are being formed at the same time, the other interaction, that of the singly-occupied n-orbital with the HOMO of the acrylonitrile (**472**), must also be looked at. The HOMO of a Z-substituted olefin is polarized with the larger coefficient on the β-carbon. However, the energy-separations probably make this a less important interaction; and, furthermore, this orbital is much less polarized than the LUMO.

6.3.2 The Dimerization of Olefins

When olefins are irradiated, they often dimerize to give cyclobutanes.[379] Sometimes a singlet-state reaction is involved, but more often it is a triplet-state reaction. Concerted, stereospecific cycloadditions are rare, and appear to be common only with relatively unsubstituted olefins; but this area is fraught with difficulties in proving or disproving concertedness. Fortunately, this does not matter from the point of view of frontier orbital analysis. Regardless of whether both bonds are formed at once, or whether they are formed one at a time, the orientation should be determined by the large–large interaction in the frontier orbitals. As usual, there are the two potentially profitable interactions shown on Fig. 6-1. In a dimerization reaction, however, the set of orbitals on the left and the set on the right will have identical energies. The bonding interactions should therefore be very strong, and they should lead to the preferred formation of what are known as head-to-head (HH) dimers. For example, if we take a Z-substituted olefin, the important interactions will be those shown in the centre of Fig. 6-3. If the reaction takes place between the

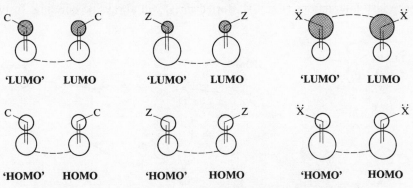

Fig. 6-3 Regioselectivity in the photodimerization of olefins

left hand component in its excited triplet-state and the right hand component in its ground-state, only the large–large bonding will actually take place, and, as it happens, the most stable of the possible diradicals will be produced.

If we were to look only at the simplest examples of each kind of olefin, we would find that this analysis seemed to be supported:

A Z-Substituted Olefin:[380]

A 'sensitized' photolysis is one in which the light is absorbed by one compound, such as acetone or benzophenone, which then interacts with the substrate (acrylonitrile in this case) in such a way that the latter is promoted to an excited state and the former reverts to the ground state. The sensitizers generally used fulfil this function from their *triplet states*; thus, almost all sensitized photolyses are triplet-state reactions.

216

A C-Substituted Olefin:[381]

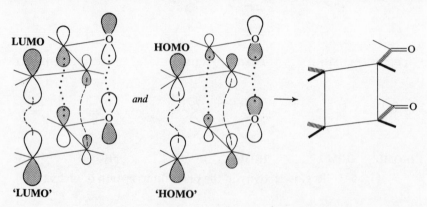

An Ẍ-Substituted Olefin:[382]

For the singlet-state reaction, we can make a further prediction: since the orbitals which are interacting are identical on each component, the *endo*-HH adduct should be preferred over the *exo*-HH adduct. This is because the secondary interactions (Fig. 6-4, dotted lines) will always be bonding. In two

LUMO   **HOMO**

'LUMO'   *and*   **'HOMO'**

Fig. 6-4 Secondary orbital interactions (dotted lines) in photocycloadditions

cases, where the singlet and triplet-state reactions have been carefully looked at and separated, this proves to be true. Thus coumarin (**473**), in the excited singlet state, dimerizes to give only the *syn*-HH dimer (**474**), but in the triplet state it gives both *syn*-HH (**474**) and *anti*-HH isomers (**475**), with only a trace of head-

(**473**)   (**474**)

to-tail (HT) products.[383] Acenaphthylene (**476**) also gives the *syn* dimer (**477**) from the singlet-state reaction and a mixture of the *syn* and *anti* dimers (**477** and **478**) from the triplet-state reaction.[384]

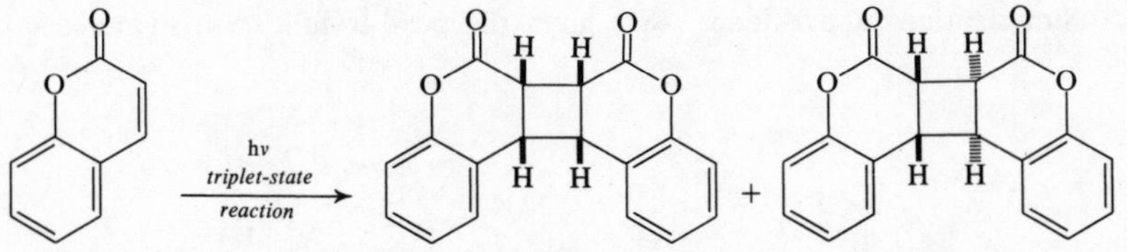

(473) (474) (475)

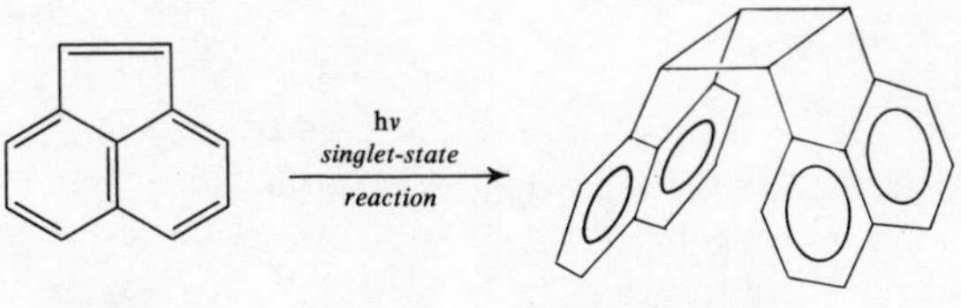

(476) (477)

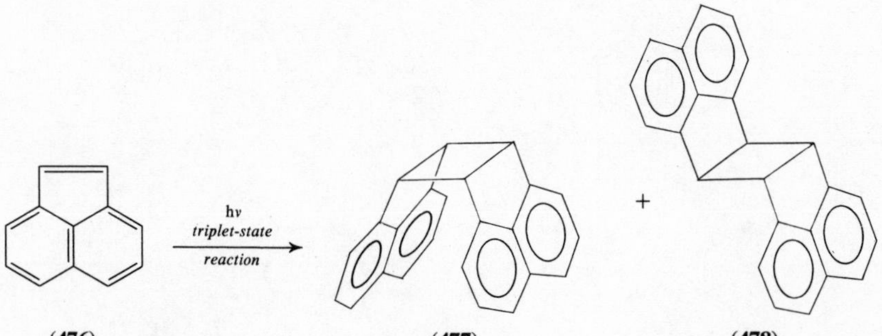

(476) (477) (478)

The whole truth, however, is not nearly as simple as this. The experimental evidence has not been collected systematically, and it is not known, in most cases, whether singlet or triplet states are involved. In the last ten years, at least sixty papers have been published in which reasonably reliable structures have been assigned to the cyclobutane dimers produced by irradiation of a great variety of unsymmetrical olefins in solution. (The stereo- and regioselectivity in *solid-state* photodimerization reactions have also received a lot of attention, but they are determined by the alignment of the monomers in the crystal lattice, not by frontier orbital effects.) For irradiation in solution, head-to-head dimers are described as major or sole products in forty of these papers, and head-to-tail dimers are described in twenty of them. In addition, almost all photodimerizations of 9-substituted anthracenes (**479**) give the head-to-head dimer (**480**),[385] whereas the

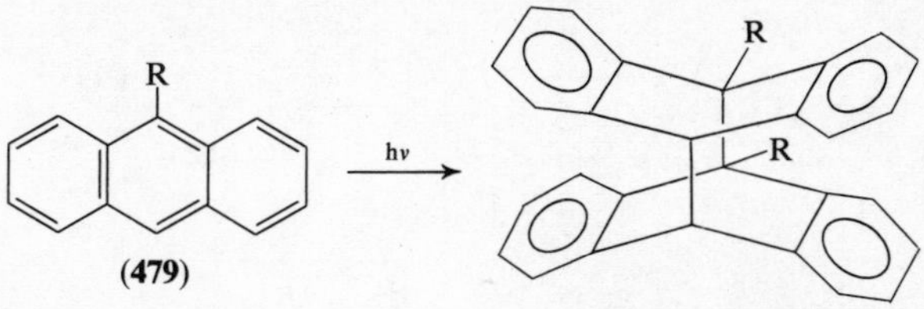

(479)

HH

R = Me, Cl, CO$_2$H, CHO
but not R = Br

(480)

photodimerization of pyridones (**481**) gives the head-to-tail, centrosymmetric dimer (**482**).[386]

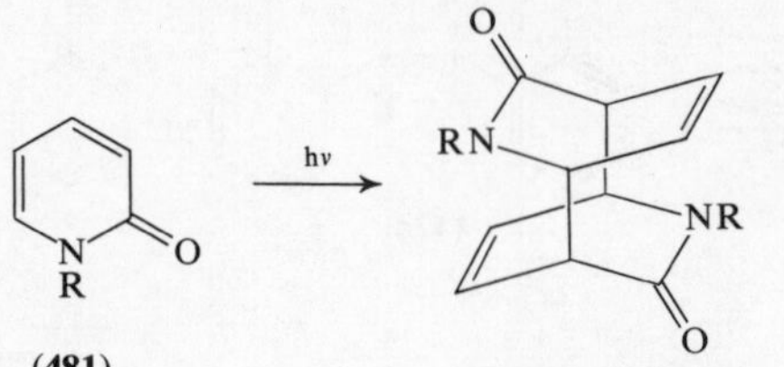

(**481**)

R = H *or* Me

(**482**)

HT

Here are a few examples from this extensive literature:

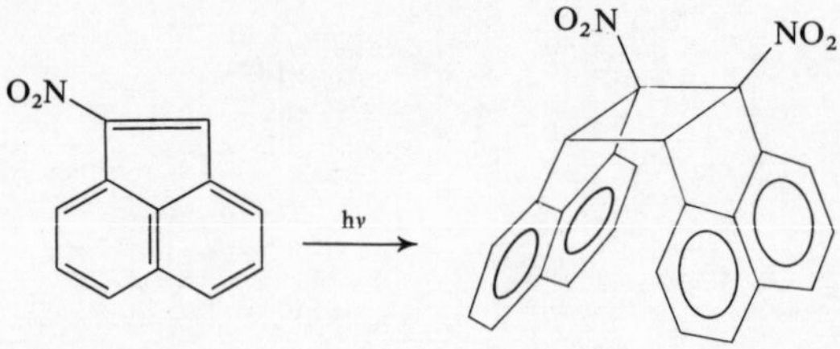

HH[387]

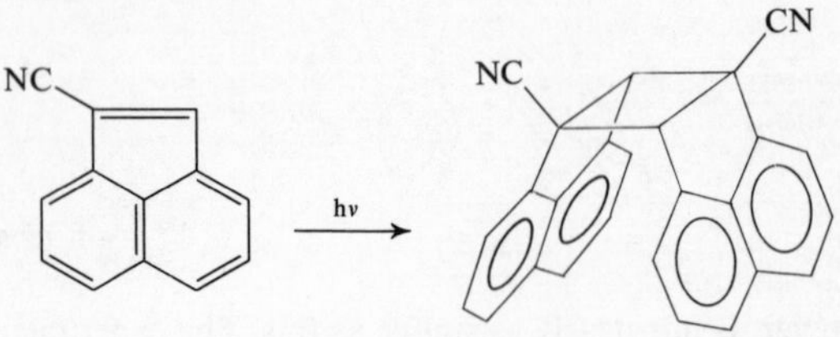

HT[388]

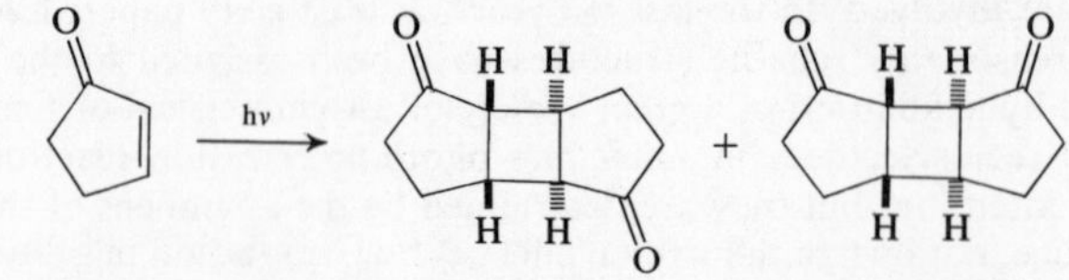

major HT   +   *minor* HH[389]

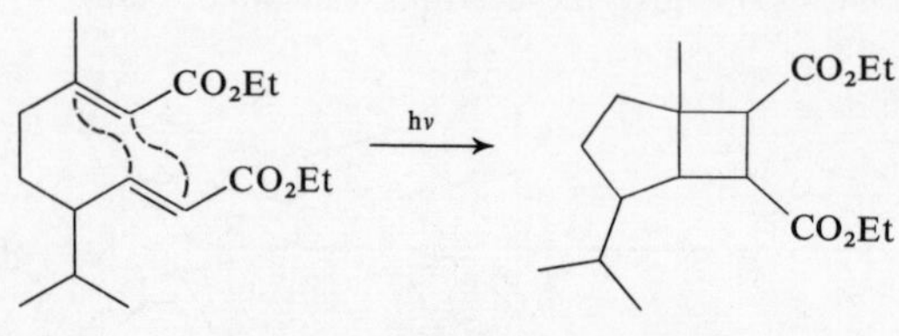

HH[390]

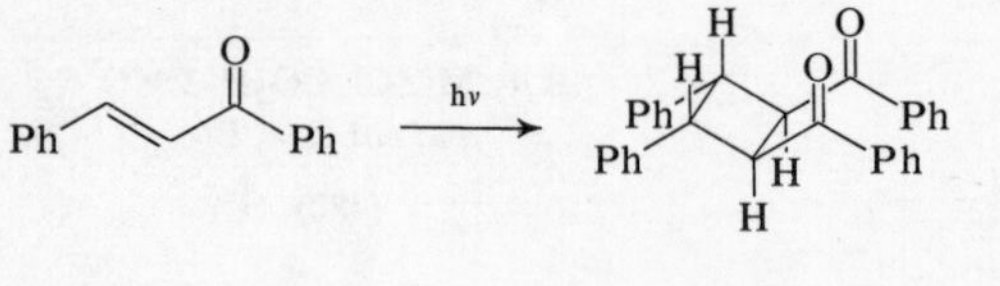

HH[391]

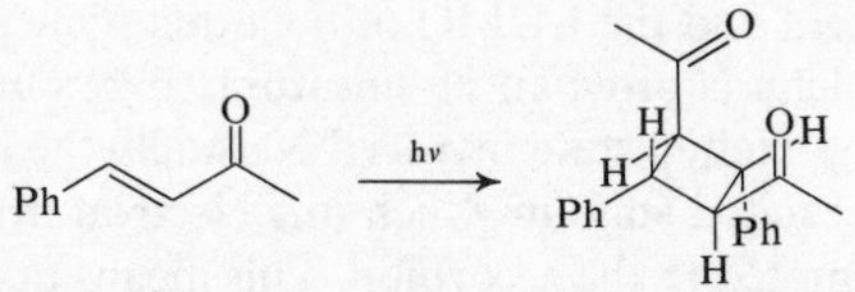

The most obvious factors which will lead molecules to adopt the HT course are dipole-dipole repulsions and steric effects.[118] That both of these factors are overridden more often than not is striking evidence for the importance of frontier orbital effects.

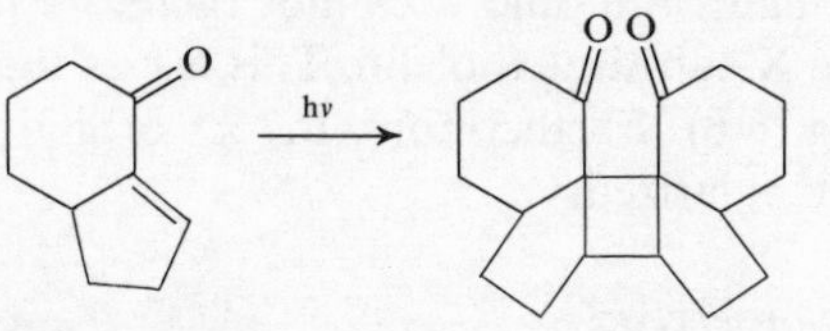

6.3.3 Cross-Coupling of Olefins

Frontier orbital effects are evidently much more dominant in cross-coupling photoreactions than they were in photodimerizations. The regioselectivity observed is almost always that predicted by frontier orbital theory. Thus there is no difficulty in predicting the orientation when a Z-substituted olefin couples with a C-substituted olefin: whether we look at the π^* orbital of the one component interacting with the LUMO of the other (Fig. 6-5a), or at the π orbital

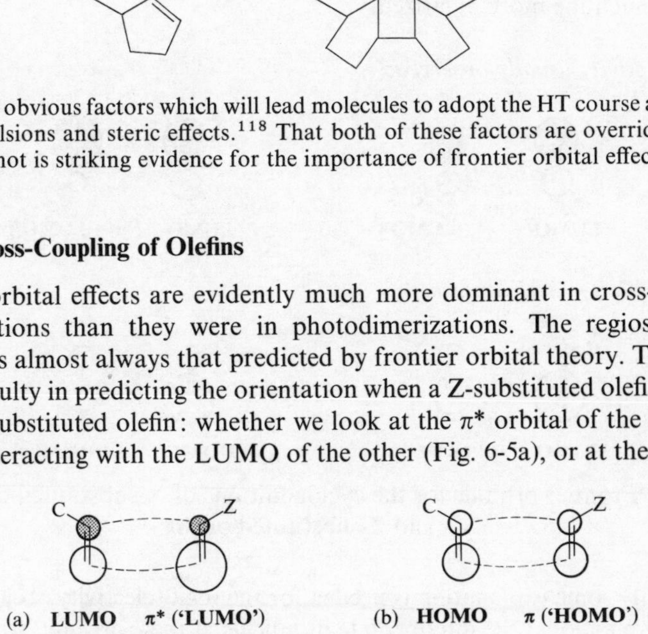

(a) **LUMO** π^* ('**LUMO**')   (b) **HOMO** π ('**HOMO**')

Fig. 6-5 Frontier orbitals for the cycloaddition of a C-substituted olefin with a Z-substituted olefin

interacting with the HOMO (Fig. 6-5b), we get the same answer. Here are two examples of this orientation:[394, 395]

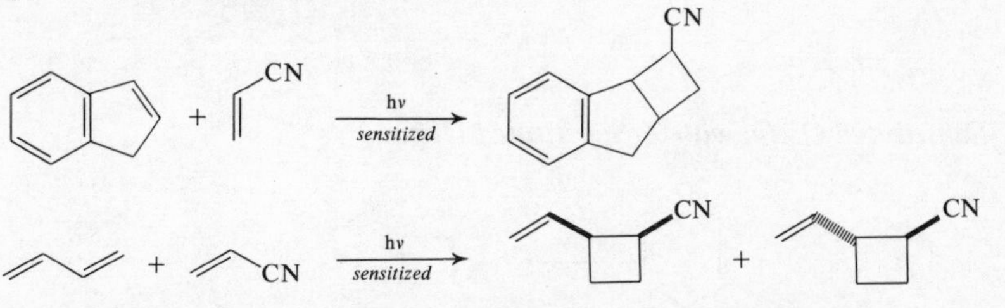

In the combinations of $\ddot{X}$-substituted olefins with C- or Z-substituted olefins, it *does* matter which frontier orbital one takes. The observed reactions are, in fact, almost always accounted for by considering only the interaction of the π^*

orbital of the one component and the LUMO of the other. One reason for this is that the Z-substituted olefin is often an αβ-unsaturated ketone or aldehyde. With these compounds, the excited state involved is usually the n-π* singlet or triplet, in other words an excited state in which one electron from a lone pair on the oxygen atom is promoted to the π* orbital. This means that the π orbital of the double bond remains full, and it cannot therefore interact favourably with the HOMO of the Ẍ-substituted olefin. This leaves the π* orbital to play the dominant role (Fig. 6-6). Furthermore, the π* orbital of a Z-substituted olefin is much the more polarized.

Regioselectivity usually observed:

| LUMO | π* ('LUMO') | | LUMO | π* ('LUMO') |

Regioselectivity rarely observed:

| HOMO | π ('HOMO') | | HOMO | π ('HOMO') |

Fig. 6-6 Frontier orbitals for the cycloadditions of Ẍ-substituted olefins with C- and Z-substituted olefins

Certainly some explanation is needed for the regioselectivity observed in the following reactions; the intermediate diradicals, if there are any, would not be the most stable of the possible diradicals.

Ẍ-Substituted Olefin with C-Substituted Olefin:[396]

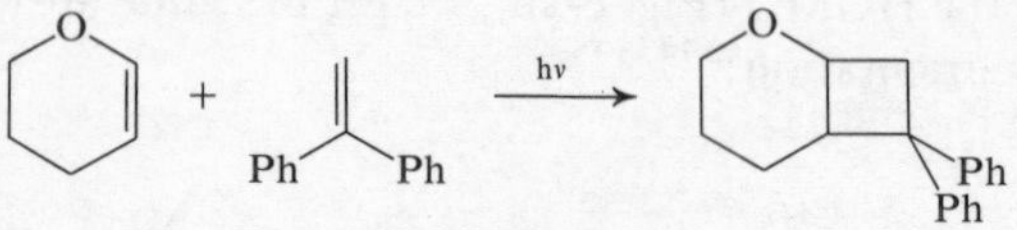

Ẍ-Substituted Olefin with Z-Substituted Olefin:[397, 398]

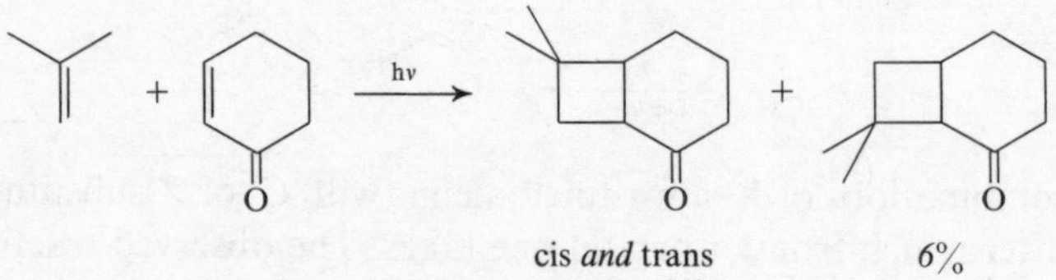

cis *and* trans
isomers: 33%   6%

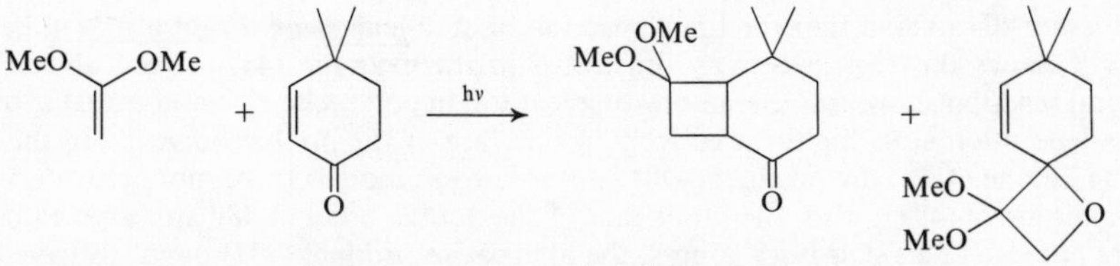

(481)   cis *and* trans *isomers*

In the ketone (481), the steric hindrance provided by the gem dimethyl group is not enough to change the regioselectivity; all it does is to cause the appearance of some Paterno-Büchi product (with the usual regioselectivity for that reaction).

The unsaturated ketone (483) undergoes a photocycloaddition reaction with isobutylene (482) to give the adduct (484).[399] This follows the usual pattern for a Z-substi-

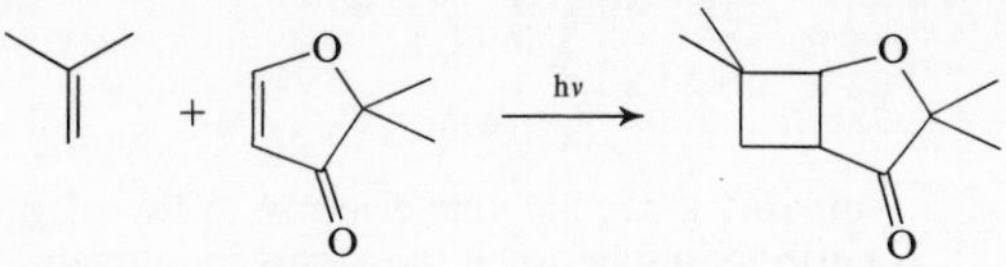

(482)   (483)   (484)

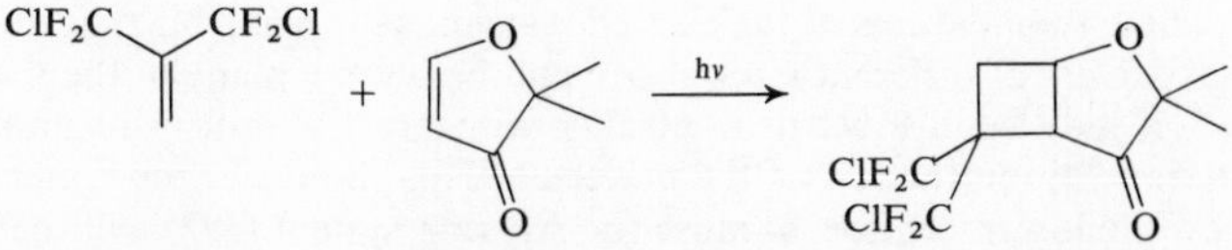

(485)   (483)   (486)

tuted olefin reacting with an Ẍ-substituted olefin. It is particularly satisfying to note that when the methyl groups of the isobutylene are replaced by chlorodifluoromethyl groups (485), the regioselectivity is inverted,[400] and the product is 486. These groups are no longer Ẍ-substituents; they are more like Z-substituents, except that they are not *conjugated* and electron-withdrawing. Without the component of conjugation, they can be expected to induce polarization in both frontier orbitals opposite to that of an Ẍ-substituted olefin.

Z-Substituted Olefins with Cumulated Double Bonds:[401]

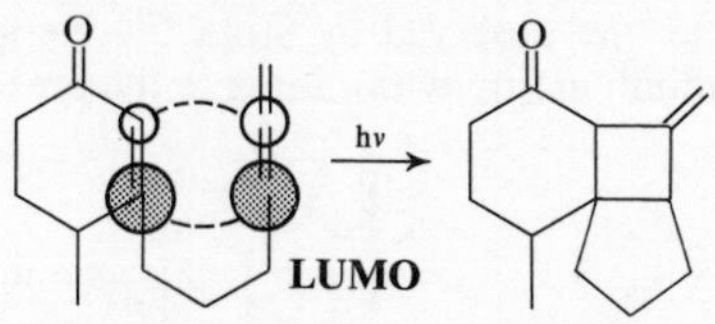

π^*
(487)   (488)

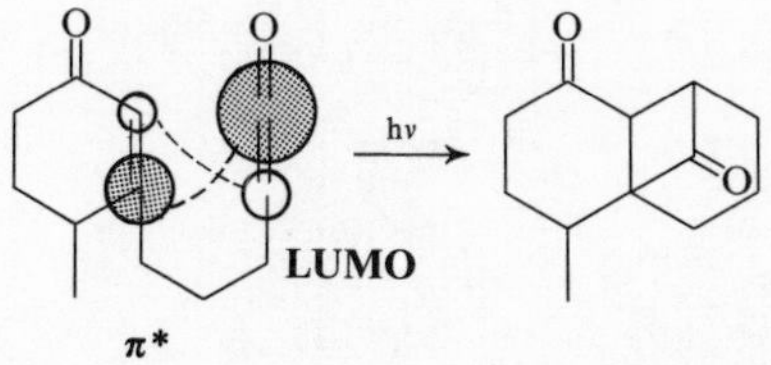

π^*

Wiesner discovered that the photoreaction of an allene to an $\alpha\beta$-unsaturated ketone always shows the regioselectivity illustrated in the example (**487** → **488**) above. The general reaction shows another interesting feature: in polycyclic $\alpha\beta$-unsaturated ketones, the allene often adds highly selectively to one face of the double bond.[402] In the case of the ketone (**489**), the product (**490**) appeared from models to be more crowded than the alternative (**491**); also, the lower face of the double bond in **489** appeared to be the more hindered face. On both counts, the alternative product (**491**) ought to have been formed and was not.

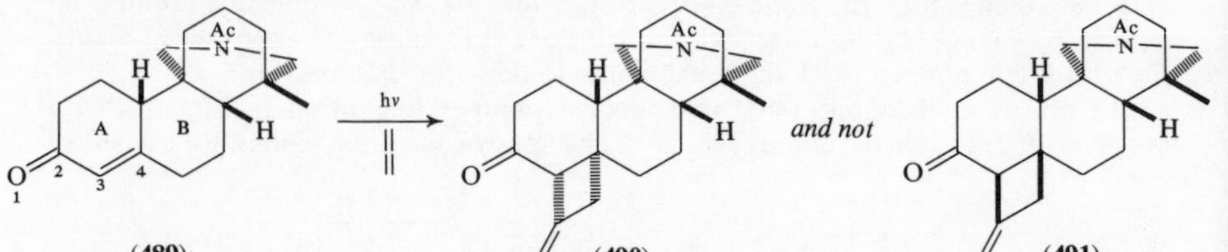

The excited state, presumably n-π^*, has one electron in the π^* orbital. This orbital, like ψ_3^* of butadiene, is antibonding between the atoms numbered 3 and 4 on structure (**489**). Thus the amount of bonding between C-3 and C-4 is reduced. Furthermore, the total electron population on C-4 will be higher than in the ground state; C-4 will be more like an anion. These two features of the excited state make it plausible that the p orbital on C-4 may have lobes of different sizes above and below the plane of the C=C bond. In other words, it may bend towards a tetrahedral geometry and, in so doing, relieve some of the steric strain in the rest of the molecule. This should be more efficient if the larger lobe is on the lower surface, because the AB ring system (**492**) will then be more like that of a *trans* decalin. If the larger lobe is down, its overlap with the allene will obviously be best when the allene approaches from the lower surface. Hence the stereo-selectivity observed.

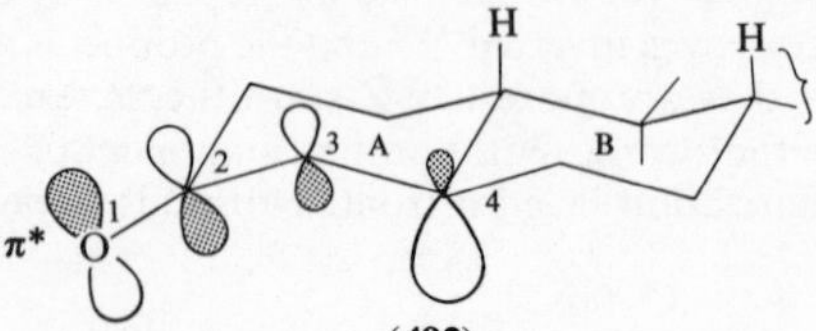

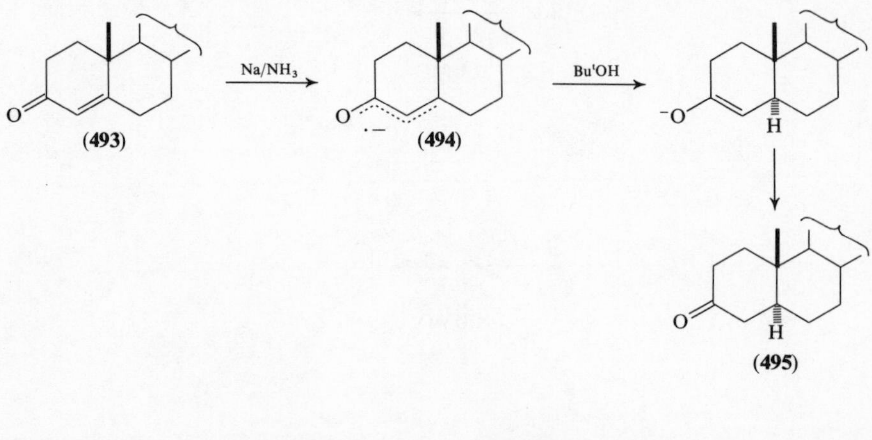

This problem is related to one observed by Stork.[403] He found that octalones like **493** could be reduced by sodium in liquid ammonia in the presence of an alcohol (Birch reduction), and that the product was exclusively the *trans*-decalone (**495**), even when that was the less stable isomer. In this reaction, an electron is fed into the LUMO of **493**, and hence the frontier orbital of the intermediate (**494**) is the same as that involved in the photochemical reaction above. Thus the unsymmetrical p orbital can again lead a reagent, a proton this time, to become attached to the lower surface, and the *trans* decalone will be the product. Photocycloaddition of allene to the unsaturated ketone (**493**) also gives the product of attack from the lower surface.

It is worth concluding by noting how often the regioselectivity observed in a photo-cycloaddition is the opposite of that which would be expected for a ground-state reaction. The corresponding ground-state reactions are not often observed, because of the symmetry-imposed barrier to the concerted reaction. But when cyclobutanes, for example, are produced in ground-state reactions, the orientation can almost always be accounted for by invoking the most stable diradical as an intermediate.[404] As we have seen, this does not always work for photocycloadditions. One particularly attractive case, in which a cycloaddition of the same general type is observed in the ground-state and the excited state, shows exactly opposite regioselectivity in each state:[405]

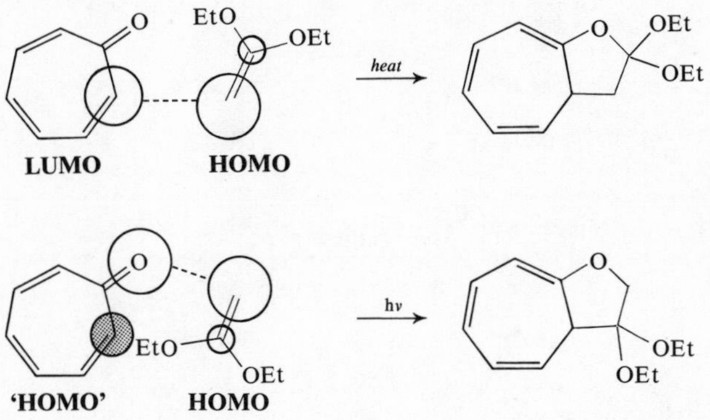

We can see from the coefficients of Fig. 4-53, and from the discussion in this chapter, how the frontier orbitals make this happen, even though neither reaction is necessarily concerted.

6.4 Photochemical Aromatic Substitution Reactions

6.4.1 Nucleophilic Substitution

In certain cases, light catalyses substitution reactions in aromatic compounds. One of the fascinating features of these reactions is an almost complete change in site-selectivity from that observed in the ground-state reactions.[406,407] When the nitrocatechol ether (**497**) is irradiated in alkali[406] or in methylamine,[408]

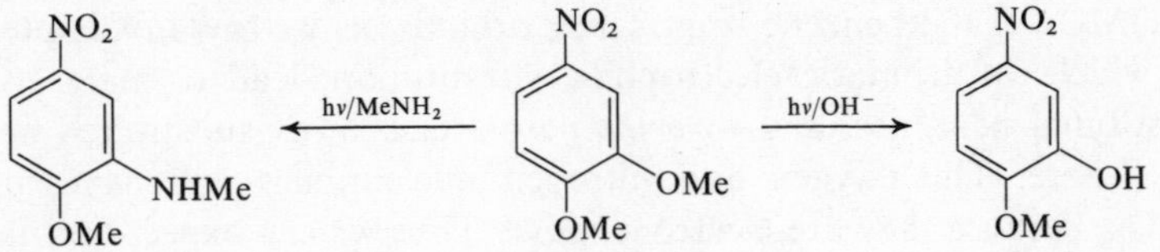

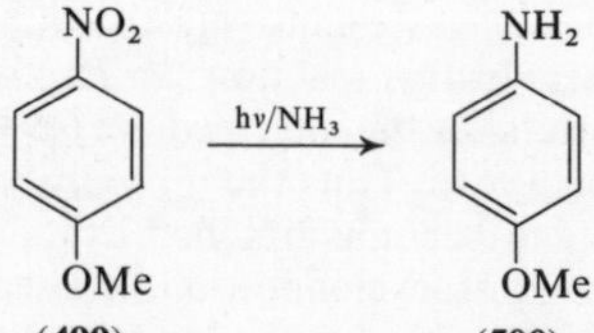

(499) → (500)

the *nucleophilic* substitution takes place *meta* to the nitro group. The nucleophilic substitution of *p*-nitroanisole (499), on the other hand, takes place *para* to the Ẍ-substituent.[406] Furthermore, with the *meta*-isomer (501), it takes place *meta* to the Z-substituent.[409] 3-Bromopyridine (503) readily gives 3-hydroxy-pyridine (504) on irradiation in aqueous alkali,[410] and the pyridinium cation (505) gives the aziridine (506),[411] where again the nucleophile has attacked the pyridine ring at C-3.

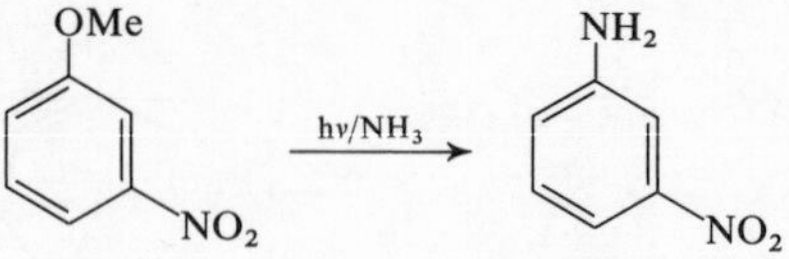

(501) → (502)

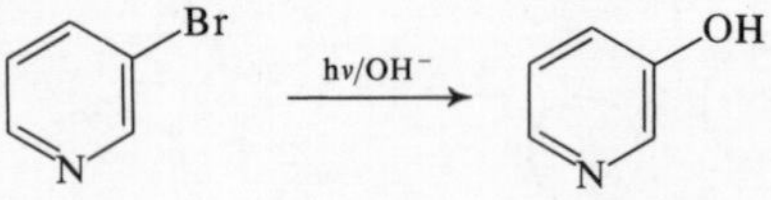

(503) → (504)

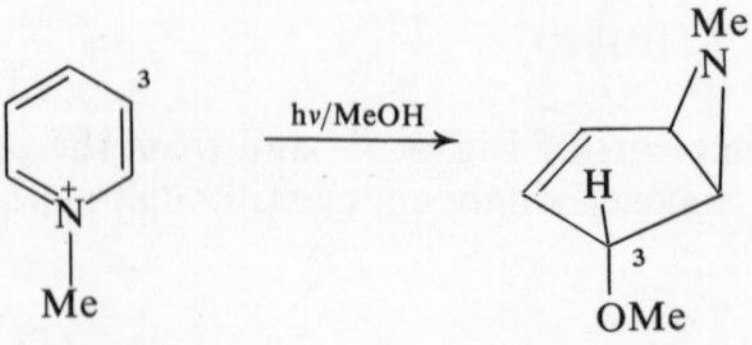

(505) → (506)

All these examples have a nucleophile attacking a photoexcited aromatic ring at a site where nucleophiles do not attack the aromatic ring in the ground state. We can now easily see why this should happen. In the ground-state reaction, the HOMO of a nucleophile interacts productively only with the LUMOs of the aromatic ring. However, in the excited-state reaction, the HOMO of the nucleophile can interact productively with what were, before excitation, the HOMOs of the benzene ring. These orbitals, as we saw in Chapter 3, are those which, in aromatic electrophilic substitution, lead to *meta* attack on Z-substituted benzenes, and to *ortho/para* attack on Ẍ-substituted benzenes. Just so here. The oxygen and nitrogen nucleophiles will have low-lying HOMOs, because they are electronegative. Thus we can expect that it will be these orbitals (essentially lone pairs) which provide the important frontier orbital for the right hand side of Fig. 6-1. The only available LUMOs in the nucleophiles will be very high in energy.

Benzene rings (**507**) without activating substituents can also be attacked by nucleophiles, provided they are in an excited state.[412] This is clearly another

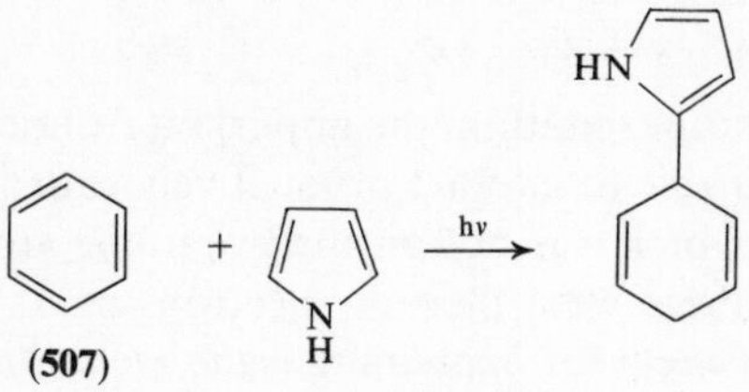

(**507**)

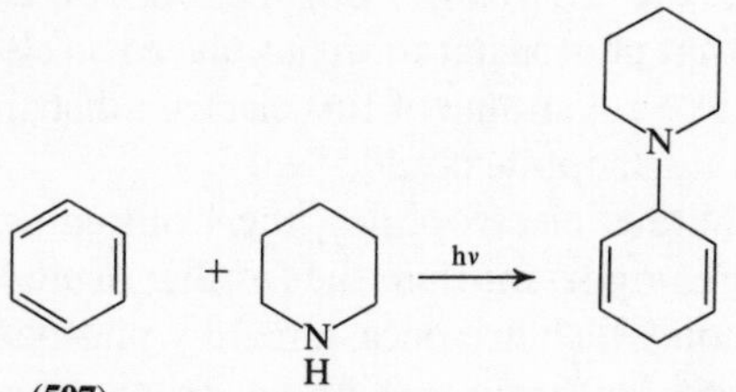

(**507**)

consequence of the ability of the HOMO of the nucleophile to interact productively with the singly-occupied *bonding* orbital of the benzene ring when the latter is in an excited state.

6.4.2 Electrophilic Substitution

A nearly complementary pattern of reactivity has been found for photochemical *electrophilic* substitution.[406, 407] Proton exchange in the photolysis of toluene (**508**) takes place most rapidly at the *meta* position.[406] In anisole (**509**), the corresponding reaction[413] is predominantly *ortho* and *meta*. Nitrobenzene (**510**), on the other hand, exchanges protons fastest at the *para*-position.[413]

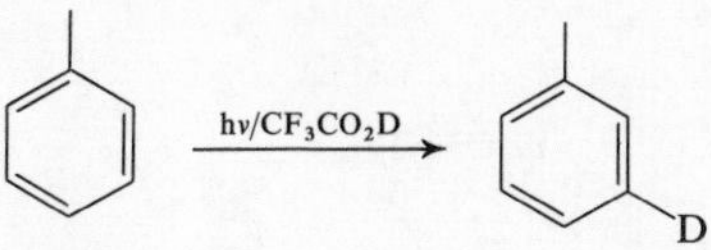

(**508**)

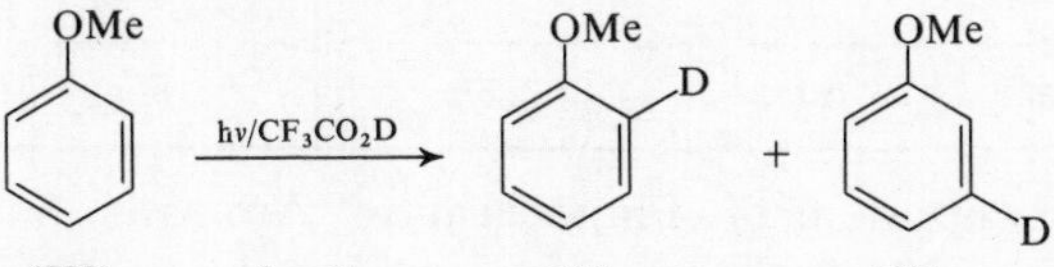

hv/CF$_3$CO$_2$D

(**509**)   *after 2h:*   11%   12%

hv/CF$_3$CO$_2$D

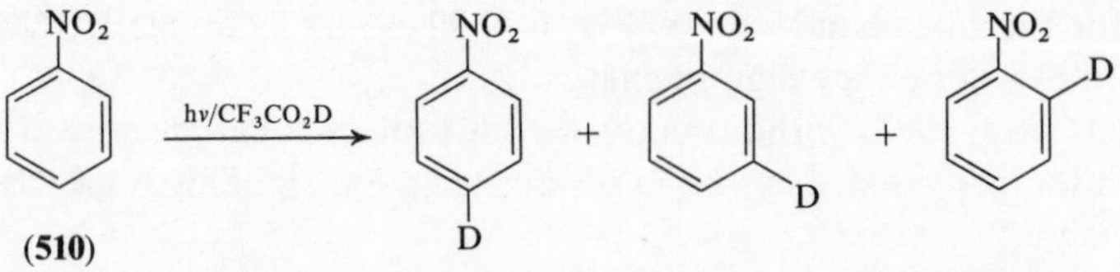

NO$_2$ hv/CF$_3$CO$_2$D   NO$_2$ +   NO$_2$ +   NO$_2$

(510)   D   D   D

after 4h: *8·6%*   *5%*   *< 1%*

Again, because of photoexcitation, the important frontier orbital of the electrophile (the LUMO) is able to interact productively with an orbital (a π^* orbital) of the benzene ring which was not productive in the ground state of any drop in energy (because there were then no electrons in it). Certainly ψ_4^* has an electron-distribution ideal for explaining *ortho/meta* attack in anisole, and, as we saw in Chapter 3, the LUMOs of nitrobenzene do lead to reactivity at the *para*-position. Now that photoexcitation has placed an electron in these orbitals, an electrophile can take advantage of this electron distribution, whereas, in the ground state, only a nucleophile could.

Of course, with charged electrophiles, the Coulombic term of equation 2-7 will probably be more important than the frontier orbital term. But the *changes* in electron distribution which are occasioned by photoexcitation all take place in frontier orbitals. So an argument based on changes in the *total* electron

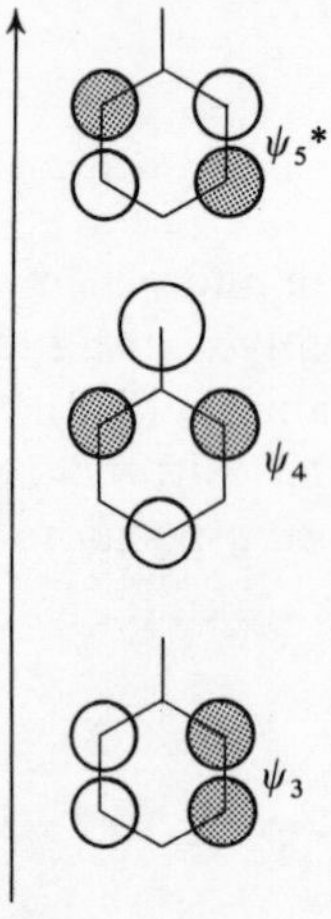

	Model for Z-substituted benzene		Model for Ẍ-substituted benzene	
	Ground-State	Excited State	Ground-State	Excited State
ψ_5^*			**LUMO**	*HSOMO*
ψ_4	**LUMO**	*HSOMO*	*HOMO*	**LSOMO**
ψ_3	*HOMO*	**LSOMO**		
Attack by nucleophile	o/p	(o)/m	(o/m)	o/p
Attack by electrophile	(o)/m	o/p	o/p	(o)/m

Energy

Fig. 6-7 Rationalization of Photochemical Aromatic Substitution (The brackets identify the less common or unobserved reaction.)

distribution is very similar to the one just given. Figure 6-7 summarizes the effect of frontier orbitals on aromatic nucleophilic and electrophilic substitution, both in the ground state and in the excited state.

As we saw in Chapter 3, the orbitals of the benzyl system can be used as models for the orbitals of Z- and Ẍ-substituted benzenes. (The more important frontier orbital for attack by a nucleophile is shown in Fig. 6-7 in bold face, and the more important frontier orbital for attack by an electrophile is shown in italics.) It is then clear how the favoured sites of attack listed at the bottom follow from the orbital picture. In those cases where *meta* attack is observed, this simple picture predicts both *ortho* and *meta* attack. As we saw in Chapter 3, we must not forget that ψ_2 is very little lower in energy than ψ_3, and $\psi_6{}^*$ is very little higher in energy than $\psi_5{}^*$. When the contribution of these orbitals is taken into account, the *meta* position, rather than the *ortho* and the *meta* position, is the favoured site of attack.

6.4.3 Radical Substitution

Photorearrangement of the compound (511) gives the isomer (513).[414] This can be understood most easily as beginning with a cyclization to the diradical (512). We may note that this diradical is probably less stable than the alternative (514), but the latter cannot be involved, because it would not lead to the observed product (513). The formation of the less stable radical (512) is easily accounted for by the high coefficient *meta* to the methoxy group in the 'LUMO'.

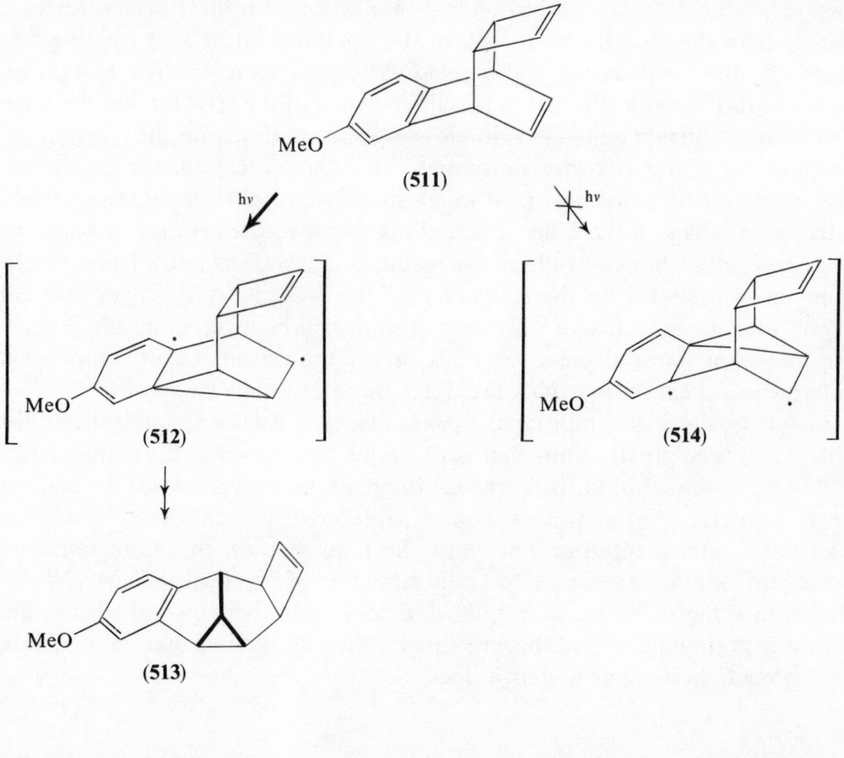

(511)

(512)

(514)

(513)

CHAPTER 7

Exceptions

There are many. Nor need we be surprised that so simple and partial a theory does not explain every feature of chemical reactivity. For the theory is not only simple: it is also clearly a simplification of the truth.

In the first place, by concentrating, in most cases, on the interactions of the HOMO of one component and the LUMO of the other, we know we are leaving out of account all the other orbital interactions, many of which are nevertheless bonding in character. We have seen that these other interactions are generally less energetically profitable than the HOMO/LUMO interaction; but there are, of course, many more of them. If other factors intervene to make the best HOMO/LUMO interaction energetically difficult to take advantage of, we can expect the interactions of lower orbitals than the HOMO (and higher orbitals than the LUMO) to become influential in determining the course or rate of a reaction.

Some of the most interesting apparent violations of the Woodward-Hoffmann rules have been explained in just this way.[415] It can happen that, although the HOMO/LUMO interaction for a particular reaction is, because of symmetry, antibonding, the corresponding interaction of a filled orbital lower in energy than the HOMO may, instead, be bonding. To be sure, the bonding gained by the involvement of this *subjacent orbital*, as it is called, will be less than the antibonding effect from the involvement of the HOMO, but the interactions may still add up to provide a lower-energy pathway for the "forbidden" reaction than for a stepwise alternative. That these exceptional reactions are also very rare demonstrates how much more important are HOMO/LUMO interactions than all the other interactions. Most of the features of reactivity described in this book have been the results of constraints much less powerful than those provided by the symmetry of the orbitals. Accordingly, we can expect that the influence of subjacent orbitals will be much more often felt in the transition states of these reactions, where the various factors involved in determining chemical reactivity are much more delicately balanced.

In the second place, molecular orbital theory is itself a simplification. Not only has it been grossly simplified in this book by taking out the mathematics, and in other ways, but fairly severe approximations must inevitably be made in order to arrive even at the most serviceable mathematical description of it. The more refinements one puts into the mathematics, the more remote it becomes from the experience and understanding of practical chemists. We are therefore obliged to use a simplified theory: it is perhaps surprising, and certainly gratifying, to find that the theory, even at the level used in this book, can be made to work as well as it does.

In the third place, we know that there are many factors of which frontier orbital theory takes little or no account. They were mentioned specifically on p. 32. In particular, arguments based on steric effects are, quite rightly, common in organic chemistry. One of the lessons to be learned from so many of the examples in this book is just how often quite powerful steric effects are over-ridden by frontier orbital effects.

Finally, the frontier orbital theory is not really about the transition state itself. It only applies to the early stages of the interactions between orbitals, and we have to trust that there is continuity from there on up to the transition state. A thorough mathematical description of the transition state is well beyond the capabilities of theoretical organic chemistry: the ground states are difficult and costly enough. Recently, semi-empirical methods (like MINDO) which use measured bond strengths and other thermodynamic data, have been very successful in estimating not only ground-state properties, but also transition-state energies.[416] Perhaps explanations of chemical reactivity based on MINDO or its successors will replace the explanations based on frontier orbital theory (and on product stability) that we have used in this book. In the meantime, frontier orbital theory needs no computer, and yet still relates to the physical picture organic chemists have of the molecules they are interested in. It is worth emphasizing that orbital interactions as a whole are almost always antibonding in their total effect. The frontier orbital interactions merely provide a small perturbation (except, of course, in photochemical and some radical reactions), which, to a greater or lesser extent, lowers the energy of the transition state. Frontier orbital theory is at its most powerful when it is used to compare two closely similar reaction pathways for which the total antibonding effect is similar. Only then is the second-order perturbation from the interaction of the frontier orbitals important enough to direct the course of the reaction.

References

1. R. B. Woodward and R. Hoffmann, *The Conservation of Orbital Symmetry*, Verlag Chemie, Weinheim, 1970.
2. C. A. Coulson and H. C. Longuet-Higgins, *Proc. Roy. Soc. A*, **192**, 16 (1947).
3. K. Fukui, *Accts. Chem. Res.*, **4**, 57 (1971).
4. For a treatment of the elementary theory leading to these equations, see A. Streitwieser, *Molecular Orbital Theory for Organic Chemists*, Wiley, New York, 1961.
5. For a more advanced discussion of molecular orbital theory, see M. J. S. Dewar, *Molecular Orbital Theory for Organic Chemists*, McGraw-Hill, New York, 1969.
6. And for a more recent and highly readable text, see W. T. Borden, *Modern Molecular Orbital Theory for Organic Chemists*, Prentice-Hall, Englewood Cliffs, New Jersey, 1975.
7. A. Streitwieser, J. I. Brauman and C. A. Coulson, *Supplemental Tables of Molecular Orbital Calculations*, Pergamon Press, Oxford, 1965.
8. D. W. Turner, *Molecular Photoelectron Spectroscopy*, Wiley, London, 1970.
9. P. B. Ayscough, *Electron Spin Resonance in Chemistry*, Methuen, London, 1967.
10. J. E. Wertz and J. R. Bolton, *Electron Spin Resonance*, McGraw-Hill, New York, 1972.
11. For a fuller account of perturbation theory, see ref. 5 and M. J. S. Dewar and R. C. Dougherty, *The PMO Theory of Organic Chemistry*, Plenum Press, New York, 1975.
12. G. S. Hammond, *J. Amer. Chem. Soc.*, **77**, 334 (1955).
13. K. Fukui, T. Yonezawa and H. Shingu, *J. Chem. Phys.*, **20**, 722 (1952).
14. G. Klopman, *J. Amer. Chem. Soc.*, **90**, 223 (1968).
15. L. Salem, *J. Amer. Chem. Soc.*, **90**, 543 and 553 (1968).
16. R. G. Pearson, *J. Chem. Educ.*, **45**, 581 and 643 (1968); B. Saville, *Angew. Chem. Internat. Edn.*, **6**, 928 (1967); J. Seyden-Penne, *Bull. Soc. chim. France*, 3871 (1968); T.-L. Ho, *Chem. Rev.*, **75**, 1 (1975).
17. R. G. Pearson, *J. Amer. Chem. Soc.*, **85**, 3533 (1963).
18. R. G. Pearson and J. Songstad, *J. Amer. Chem. Soc.*, **89**, 1827 (1967).
19. G. Klopman (Editor), *Chemical Reactivity and Reaction Paths*, Wiley, New York, 1974.
20. R. F. Hudson and G. Klopman, *Tetrahedron Letters*, 1103 (1967).
21. G. Klopman, ref. 19, Chapter 4; R. F. Hudson, ref. 19, Chapter 5.
22. G. Klopman, *J. Amer. Chem. Soc.*, **86**, 4550 (1964).
23. J. O. Edwards and R. G. Pearson, *J. Amer. Chem. Soc.*, **84**, 16 (1962).
24. J. O. Edwards, *J. Amer. Chem. Soc.*, **76**, 1540 (1954).
25. M. Arbelot, J. Metzger, M. Chanon, C. Guimon and G. Pfister-Guillouzo, *J. Amer. Chem. Soc.*, **96**, 6218 (1974).
26. R. C. Dougherty, *Tetrahedron Letters*, 385 (1975).
27. M. A. Gautier, *Ann. Chim.*, [4]**17**, 103 (1869).
28. H. L. Jackson and B. C. McKusick, *Org. Synth. Coll. Vol.* **4**, 438 (1963).
29. For a picture of the σ-framework orbitals, see W. L. Jorgensen and L. Salem, *The Organic Chemist's Book of Orbitals*, Academic Press, New York, 1973.

30. Ref. 9, p. 84.
31. P. A. Chopard, R. F. Hudson and G. Klopman, *J. Chem. Soc.*, 1379 (1965).
32. P. Sarthou, F. Guibé and G. Bram, *Chem. Comm.*, 377 (1974).
33. R. W. Fessenden and R. H. Schuler, *J. Chem. Phys.*, **38**, 773 (1963).
34. J. P. Colpa and E. de Boer, *Mol. Phys.*, **7**, 333 (1963).
35. N. A. J. Rogers and A. Sattar, *Tetrahedron Letters*, 1471 (1965).
36. P. V. Alston and R. M. Ottenbrite, *J. Org. Chem.*, **40**, 1111 (1975)
37. R. F. Hudson, *Angew. Chem. Internat. Edn.*, **12**, 36 (1973).
38. J. W. Ogilvie, J. T. Tildon and B. S. Strauch, *Biochemistry*, **3**, 754 (1964).
39. L. Radom, J. A. Pople and P. von R. Schleyer, *J. Amer. Chem. Soc.*, **94**, 5935 (1972); R. Hoffmann, L. Radom, J. A. Pople, P. von R. Schleyer, W. J. Hehre and L. Salem, *ibid.*, 6222; L. Libit and R. Hoffmann, *ibid.*, **96**, 1370 (1974).
40. K. N. Houk, *J. Amer. Chem. Soc.*, **95**, 4092 (1973).
41. G. W. Wheland, *J. Amer. Chem. Soc.*, **64**, 900 (1942).
42. J. Blyth and A. W. Hofmann, *Annalen*, **53**, 289 (1845).
43. H. Hübner and H. Lüddens, *Ber.*, **8**, 870 (1875).
44. C. A. Coulson and H. C. Longuet-Higgins, *Proc. Roy. Soc. A*, **192**, 16 (1947).
45. K. Fukui, T. Yonezawa, C. Nagata and H. Shingu, *J. Chem. Phys.*, **22**, 1433 (1954).
46. The coefficients for benzofuran (61) were calculated by Dr. A. J. Stone, personal communication, using a simple Hückel programme and Streitwieser's recommended parameters.[4]
47. C. Gassmann, *Ber.*, **29**, 1243 and 1522 (1896).
48. C. Liebermann and L. Lindemann, *Ber.*, **13**, 1584 (1880); J. Meisenheimer, *Ber.*, **33**, 3547 (1900).
49. J. Schmidt, *Ber.*, **33**, 3251 (1900); M. J. S. Dewar and E. W. T. Warford, *J. Chem. Soc.*, 3570 (1956).
50. A. G. Anderson and J. A. Nelson, *J. Amer. Chem. Soc.*, **72**, 3824 (1950).
51. W. Baker, J. W. Barton and J. F. W. McOmie, *J. Chem. Soc.*, 2666 (1958); J. W. Barton and K. E. Whitaker, *J. Chem. Soc.(C)*, 1663 (1968).
52. I. J. Rinkes, *Rec. Trav. chim.*, **53**, 1167 (1934).
53. R. Stoermer and B. Kahlert, *Ber.*, **35**, 1640 (1902).
54. G. Berti, A. Da Settimo and E. Nannipieri, *J. Chem. Soc.(C)*, 2145 (1968).
55. Nitration: E. T. Borrows, D. O. Holland and J. Kenyon, *J. Chem. Soc.*, 1077 (1946); other electrophiles: W. L. Mosby, *Heterocyclic Systems with Bridgehead Nitrogen Atoms*, Interscience, New York, 1961, Part 1, p. 254.
56. H. J. den Hertog and J. Overhoff, *Rec. Trav. chim.*, **49**, 552 (1930).
57. W. Meigen, *J. prakt. Chem.*, [2]**77**, 472 (1908); L. F. Fieser and E. B. Hershberg, *J. Amer. Chem. Soc.*, **62**, 1640 (1940).
58. P. Fortner, *Monatsh.*, **14**, 146 (1893); A. Claus and K. Hoffmann, *J. prakt. Chem.*, [2]**47**, 252 (1893); F. T. Tyson, *J. Amer. Chem. Soc.*, **61**, 183 (1939).
59. K. Fukui, T. Yonezawa and C. Nagata, *Bull. Chem. Soc. Japan*, **27**, 423 (1954).
60. See, for example, S. F. Mason, *Progress in Organic Chemistry*, **6**, 214 (1964), and ref. 37.
61. K. Fukui, T. Yonezawa and C. Nagata, *J. Chem. Phys.*, **26**, 831 (1957).
62. J. H. D. Eland and C. J. Danby, *Z. Naturforsch.*, **23A**, 355 (1968); A. D. Baker, D. P. May and D. W. Turner, *J. Chem. Soc. (B)*, 22 (1968).
63. H. E. Zimmerman, *Tetrahedron*, **16**, 169 (1961).
64. B. R. Russell, R. M. Hedges and W. R. Carper, *Mol. Phys.*, **12**, 283 (1967); K. Kumaki, S.-I. Hata, K. Mizuno and S. Tomioka, *Chem. and Parm. Bull (Japan)*, **17**, 1751 (1969); P. Lazzeretti and F. Taddei, *Org. Mag. Res.*, **3**, 283 (1971); G. R. Howe, *J. Chem. Soc. (B)*, 981 (1971).
65. M. J. S. Dewar, *J. Chem. Soc.*, 463 (1949).
66. Data from R. O. C. Norman and R. Taylor, *Electrophilic Substitution in Benzenoid Compounds*, Elsevier, Amsterdam, 1965.

232

67. H. S. Mosher and F. J. Welch, *J. Amer. Chem. Soc.*, **77**, 2902 (1955).
68. E. Ochiai, K. Arima and M. Ishikawa, *J. Pharm. Soc. (Japan)*, **63**, 79 (1943); *Chem. Abs.*, **45**, 5151 (1951).
69. M. van Ammers and H. J. den Hertog, *Rec. Trav. chim.*, **77**, 340 (1958).
70. J. A. Dixon and D. Fishman, *J. Amer. Chem. Soc.*, **85**, 1356 (1963).
71. D. J. Shaeffer and H. E. Zieger, *J. Org. Chem.*, **34**, 3958 (1969).
72. K. Hafner and H. Weldes, *Annalen*, **606**, 90 (1957).
73. K. Ziegler and W. Schäfer, *Annalen*, **511**, 101 (1934).
74. G. Biggi, F. Del Cima and F. Pietra, *J. Amer. Chem. Soc.*, **95**, 7101 (1973) and references therein.
75. H. Decker and A. Kaufmann, *J. prakt. Chem.*, [2]**84**, 436 (1911); R. E. Lyle and P. S. Anderson, *Adv. Heterocyclic Chem.*, **6**, 45 (1970); K. Ziegler and H. Zeiser, *Ber.*, **63**, 1847 (1930); M. Freund, *Ber.*, **37**, 4666 (1904).
76. E. M. Kosower, *J. Amer. Chem. Soc.*, **78**, 3497 (1956); W. von E. Doering and W. E. McEwen, *ibid.*, **73**, 2104 (1951).
77. Data from J. Miller, *Aromatic Nucleophilic Substitution*, Elsevier, Amsterdam, 1968.
78. J. Murto, *Suomen Kem.*, **B38**, 246 (1965).
79. O. Eisenstein, J.-M. Lefour, C. Minot, N. T. Anh and G. Soussan, *Compt. rend.*, **264C**, 1310 (1972).
80. W. C. Mast and C. H. Fisher, *Ind. Eng. Chem.*, **41**, 790 (1949).
81. P. L. de Benneville, L. S. Luskin and H. J. Sims, *J. Org. Chem.*, **23**, 1355 (1958).
82. D. Vorländer and A. Knötzsch, *Annalen*, **294**, 317 (1897).
83. See, however, B. Deschamps, N. T. Anh and J. Seyden-Penne, *Tetrahedron Letters*, 527 (1973), and G. Kyriakolov, M. C. Roux-Schmitt and J. Seyden-Penne, *Tetrahedron*, **31**, 1883 (1975).
84. W. P. Jencks, *Catalysis in Chemistry and Enzymology*, McGraw-Hill, New York, 1969, p. 530.
85. K. Morsch, *Monatsh.*, **63**, 220 (1933).
86. M. C. Moureu, *Ann. Chim. (France)*, [7]**2**, 145 (1894).
87. J. Bottin, O. Eisenstein, C. Minot and N. T. Anh, *Tetrahedron Letters*, 3015 (1972).
88. Data from: H. O. House, *Modern Synthetic Reactions*, Benjamin, Menlo Park, 2nd Edn., 1972, p. 93.
89. J. A. Marshall and R. D. Carroll, *J. Org. Chem.*, **30**, 2748 (1965).
90. F. G. Bordwell and D. A. Schexnayder, *J. Org. Chem.*, **33**, 3240 (1968).
91. I. Fleming and E. J. Thomas, *Tetrahedron*, **28**, 4989 (1972).
92. F. G. Bordwell and T. G. Mecca, *J. Amer. Chem. Soc.*, **94**, 5825 (1972).
93. H. J. den Hertog and H. C. van der Plas, *Adv. Heterocyclic Chem.*, **4**, 121 (1965).
94. W. Adam, A. Grimison and R. Hoffmann, *J. Amer. Chem. Soc.*, **91**, 2590 (1969).
95. G. W. J. Fleet and I. Fleming, *J. Chem. Soc. (C)*, 1758 (1969); G. W. J. Fleet, I. Fleming and D. Philippides, *J. Chem. Soc. (C)*, 3948 (1971).
96. M. J. Pieterse and H. J. den Hertog, *Rec. Trav. chim.*, **80**, 1376 (1961).
97. E. B. Biehl, E. Nie and K. C. Hsu, *J. Org. Chem.*, **34**, 3595 (1969).
98. R. Hoffmann, A. Imamura and W. J. Hehre, *J. Amer. Chem. Soc.*, **90**, 1499 (1968).
99. G. Klopman, ref. 19, p. 83.
100. D. E. Applequist and G. N. Chmurny, *J. Amer. Chem. Soc.*, **89**, 875 (1967).
101. J. Gerstein and W. P. Jencks, *J. Amer. Chem. Soc.*, **86**, 4655 (1964).
102. R. U. Lemieux, quoted in E. L. Eliel, N. L. Allinger, S. J. Angyal and G. A. Morrison, *Conformational Analysis*, Wiley, New York, 1967, p. 376.
103. C. L. Jungius, *Z. phys. Chem.*, **52**, 97 (1905).
104. S. David, O. Eisenstein, W. J. Hehre, L. Salem and R. Hoffmann, *J. Amer. Chem. Soc.*, **95**, 3806 (1973).
105. W. Adcock, S. Q. A. Rizvi and W. Kitching, *J. Amer. Chem. Soc.*, **94**, 3657 (1972).

233

106. T. G. Traylor, W. Hanstein, H. J. Berwin, N. C. Clinton and R. S. Brown, *J. Amer. Chem. Soc.*, **93**, 5714 (1971); C. Eaborn, *Chem. Comm.*, 1255 (1972).
107. But see, Z. Ardalan and E. A. C. Lucken, *Helv. Chim. Acta*, **56**, 1724 (1973).
108. E. T. McBee, I. Serfaty and T. Hodgins, *J. Amer. Chem. Soc.*, **93**, 5711 (1971).
109. K. Fukui, H. Hao and H. Fujimoto, *Bull. Chem. Soc. Japan*, **42**, 348 (1969).
110. J. W. Baker, *Hyperconjugation*, Oxford University Press, London, 1952.
111. M. J. S. Dewar, *Hyperconjugation*, The Ronald Press, New York, 1962.
112. W. C. Herndon, *Chem. Rev.*, **72**, 157 (1972), and ref. 19, Chapter 7.
113. R. Sustmann, *Pure Appl. Chem.*, **40**, 569 (1975).
114. O. Eisenstein, J.-M. Lefour and N. T. Anh, *Chem. Comm.*, 969 (1971).
115. O. Eisenstein and N. T. Anh, *Bull. Soc. chim. France*, 2721 and 2723 (1973).
116. K. N. Houk, J. Sims, C. R. Watts and L. J. Luskus, *J. Amer. Chem. Soc.*, **95**, 7301 (1973).
117. K. N. Houk, *Accts. Chem. Res.*, **8**, 361 (1975).
118. N. D. Epiotis, *J. Amer. Chem. Soc.*, **95**, 5624 (1973).
119. See, for example, K. Kraft and G. Koltzenburg, *Tetrahedron Letters*, 4357 and 4723 (1967); M. J. Goldstein and M. S. Benzon, *J. Amer. Chem. Soc.*, **94**, 5119 (1972); D. Bellus, H.-C. Mez, G. Rihs and H. Sauter, *ibid.*, **96**, 5007 (1974); J. A. Berson, E. W. Petrillo and P. Bickart, *ibid.*, 636; and the reactions of olefins with ketenes, discussed on p. 143.
120. G. C. Farrant and R. Feldmann, *Tetrahedron Letters*, 4979 (1970).
121. R. C. Cookson, B. V. Drake, J. Hudec and A. Morrison, *Chem. Comm.*, 15 (1966); S. Ito, Y. Fujise, T. Okuda and Y. Inoue, *Bull. Chem. Soc., Japan*, **39**, 135 (1966).
122. R. Eidenschink and T. Kauffmann, *Angew. Chem. Internat. Edn.*, **11**, 292 (1972).
123. H. M. R. Hoffmann, D. R. Joy and A. K. Suter, *J. Chem. Soc. (B)*, 57 (1968).
124. F. Walls, J. Padilla, P. Joseph-Nathan, F. Giral and J. Romo, *Tetrahedron Letters*, 1577 (1965).
125. R. Huisgen, *Angew. Chem. Internat. Edn.*, **2**, 565 (1963).
126. M. G. Evans, *Trans. Faraday Soc.*, **35**, 824 (1939).
127. M. J. S. Dewar, *Angew. Chem. Internat. Edn.*, **10**, 761 (1971).
128. W. von E. Doering, personal communication to R. B. Woodward, reported in ref. 1.
129. W. L. Mock, *J. Amer. Chem. Soc.*, **88**, 2857 (1966); S. D. McGregor and D. M. Lemal, *J. Amer. Chem. Soc.*, **88**, 2858 (1966).
130. W. L. Mock, *J. Amer. Chem. Soc.*, **91**, 5682 (1969).
131. W. R. Roth, J. Konig and K. Stein, *Chem. Ber.*, **103**, 426 (1970).
132. K. Alder and H.-J. Ache, *Chem. Ber.*, **95**, 503 and 511 (1962).
133. One case of a [1,3] antarafacial shift is reported by J. E. Baldwin and R. H. Fleming, *J. Amer. Chem. Soc.*, **94**, 2140 (1972).
134. F. Näf, R. Decorzant, W. Thommen, B. Willhalm and G. Ohloff, *Helv. Chim. Acta*, **58**, 1016 (1975).
135. A. P. ter Borg, E. Razenberg and H. Kloosterziel, *Rec. Trav. chim.*, **84**, 1230 (1965); K. W. Egger, *J. Amer. Chem. Soc.*, **89**, 3688 (1967).
136. E. Vogel, K.-H. Ott and K. Gajek, *Annalen*, **644**, 172 (1961).
137. L. Claisen and E. Tietze, *Ber.*, **58**, 275 (1925).
138. P. Radlick and S. Winstein, *J. Amer. Chem. Soc.*, **85**, 344 (1963).
139. J. E. Baldwin, R. E. Hackler and D. P. Kelly, *Chem. Comm.*, 537 (1968); G. M. Blackburn, W. D. Ollis, J. D. Placket, C. Smith and I. O. Sutherland, *ibid.*, 186; R. B. Bates and D. Feld, *Tetrahedron Letters*, 417 (1968); B. M. Trost and R. LaRochelle, *ibid.*, 5029.
140. D. R. Rayner, E. G. Miller, P. Bickart, A. J. Gordon and K. Mislow, *J. Amer. Chem. Soc.*, **88**, 3138 (1966).
141. U. Schöllkopf and I. Hoppe, *Annalen*, **765**, 153 (1972).
142. H.-J. Hansen, B. Sutter and H. Schmid, *Helv. Chim. Acta*, **51**, 828 (1968).

234

143. J. A. Berson and G. L. Nelson, *J. Amer. Chem. Soc.*, **89**, 5303 (1967).
144. F. T. Bond and L. Scerbo, *Tetrahedron Letters*, 2789 (1968); W. R. Roth and A. Friedrich, *ibid.*, 2607 (1969).
145. H. E. Zimmerman, D. S. Crumrine, D. Dopp and P. S. Huyffer, *J. Amer. Chem. Soc.*, **91**, 434 (1969); T. M. Brennan and R. K. Hill, *ibid.*, **90**, 5614 (1968).
146. R. E. K. Winter, *Tetrahedron Letters*, 1207 (1965); for a more complete study of this reaction, including the measurement of the minute amount of forbidden reaction taking place, see J. I. Brauman and W. C. Archie, *J. Amer. Chem. Soc.*, **94**, 4262 (1972).
147. E. Vogel, W. Grimme and E. Dinné, *Tetrahedron Letters*, 391 (1965); E. N. Marvell, G. Caple and B. Schatz, *ibid.*, 385.
148. E. N. Marvell and J. Seubert, *J. Amer. Chem. Soc.*, **89**, 3377 (1967); R. Huisgen, A. Dahmen and H. Huber, *ibid.*, 7130.
149. P. von R. Schleyer, T. M. Su, M. Saunders and J. C. Rosenfeld, *J. Amer. Chem. Soc.*, **91**, 5174 (1969).
150. R. Huisgen, W. Scheer and H. Huber, *J. Amer. Chem. Soc.*, **89**, 1753 (1967).
151. N. W. K. Chiu and T. S. Sorensen, *Canad. J. Chem.*, **51**, 2776 (1973).
152. C. W. Shoppee and G. N. Henderson, *J. Chem. Soc. Perkin I*, 765 (1975). For other examples of dienyl anions going to cyclopentenyl anions, see P. R. Stapp and R. F. Kleinschmidt, *J. Org. Chem.*, **30**, 3006 (1965); L. H. Slaugh, *ibid.*, **32**, 108 (1967); R. B. Bates and D. A. McCombs, *Tetrahedron Letters*, 977 (1969).
153. H. C. Longuet-Higgins and E. W. Abrahamson, *J. Amer. Chem. Soc.*, **87**, 2045 (1965), and ref. 1.
154. A. C. Day, *J. Amer. Chem. Soc.*, **97**, 2431 (1975).
155. R. Hoffmann and R. B. Woodward, *J. Amer. Chem. Soc.*, **87**, 4388 and 4389 (1965).
156. W. von E. Doering and W. R. Roth, *Tetrahedron*, **18**, 67 (1962), and *Angew. Chem. Internat. Edn.*, **2**, 115 (1963). For an estimate of the difference in energy of these two transition states, see M. J. Goldstein and M. S. Benzon, *J. Amer. Chem. Soc.*, **94**, 7147 (1972).
157. H. M. R. Hoffmann and D. R. Joy, *J. Chem. Soc. (B)*, 1182 (1968).
158. R. Gree and R. Carrie, *Bull. Soc. chim. France*, 1319 (1975); R. Gree, F. Tonnard and R. Carrie, *ibid.*, 1325.
159. (a) E. H. Farmer and F. L. Warren, *J. Chem. Soc.*, 897 (1929); (b) G. M. Whitman, British Patent 616671 (1949); *Chem. Abs.*, **43**, 5418 (1949).
160. J. Sauer and H. Wiest, *Angew. Chem. Internat. Edn.*, **1**, 268 (1962).
161. D. L. Fields, T. H. Regan and J. C. Dignan, *J. Org. Chem.*, **33**, 390 (1968); see also, N. A. Porter, I. J. Westerman, T. G. Wallis and C. K. Bradsher, *J. Amer. Chem. Soc.*, **96**, 5104 (1974).
162. A. I. Konovalov and B. N. Solomonov, *Doklady Akad. Nauk S.S.S.R. Ser. Kim.*, **211**, 1115 (1973).
163. R. Sustmann and R. Schubert, *Angew. Chem. Internat. Edn.*, **11**, 840 (1972).
164. R. Sustmann and H. Trill, *Angew. Chem. Internat. Edn.*, **11**, 838 (1972).
165. R. C. Cookson, S. S. H. Gilani and I. D. R. Stevens, *J. Chem. Soc. (C)*, 1905 (1967).
166. O. Wichterle, *Coll. Czech. Chem. Comm.*, **10**, 497 (1938).
167. R. B. Woodward and T. J. Katz, *Tetrahedron*, **5**, 70 (1959).
168. K. Alder, M. Schumacher and O. Wolff, *Annalen*, **564**, 79 (1949).
169. K. Alder and K. Heimbach, *Chem. Ber.*, **86**, 1312 (1953).
170. P. V. Alston, R. M. Ottenbrite and D. D. Shillady, *J. Org. Chem.*, **38**, 4075 (1973); P. V. Alston and R. M. Ottenbrite, *ibid.*, **39**, 1584 (1974); P. V. Alston and D. D. Shillady, *ibid.*, 3402.
171. For a large compilation of Diels-Alder reactions, see A. S. Onishchenko, *Diene Synthesis*, Israel Program for Scientific Translations, Jerusalem, 1964; *Dienovyi Sintez*, Izdatel'stvo Akademii Nauk, Moskva, 1963.
172. Y. A. Titov and A. I. Kuznetsova, *Doklady Akad. Nauk S.S.S.R.*, **126**, 586 (1959).

173. M. S. Kharasch and E. Sternfeld, *J. Amer. Chem. Soc.*, **61**, 2318 (1939).
174. Y. A. Titov and A. I. Kuznetsova, *Doklady Akad. Nauk S.S.S.R., Otdel. Khim. Nauk*, 1810 and 1815 (1960).
175. H. R. Snyder and G. I. Poos, *J. Amer. Chem. Soc.*, **71**, 1057 (1949).
176. I. N. Nazarov, A. I. Kuznetsova and N. V. Kuznetsov, *Zhur. obschei Khim.*, **25**, 88 (1955).
177. K. Alder and J. Haydn, *Annalen*, **570**, 201 (1950).
178. J. S. Meek, R. T. Merrow, D. E. Ramney and S. J. Cristol, *J. Amer. Chem. Soc.*, **73**, 5563 (1951).
179. C. S. Marvel and N. O. Brace, *J. Amer. Chem. Soc.*, **71**, 37 (1949); C. C. Price, G. A. Cypher and I. V. Krishnamurti, *ibid.*, **74**, 2987 (1952); C. C. J. Culvenor and T. A. Geissman, *Chem. and Ind.*, 366 (1959).
180. I. N. Nazarov, G. P. Verkholetova and L. D. Bergel'son, *Izvest. Akad. Nauk S.S.S.R., Otdel Khim. Nauk*, 511 (1948).
181. E. Lehmann and W. Paasche, *Ber.*, **68**, 1146 (1935); G. Blumenfeld, *ibid.*, **74**, 524 (1941); K. Alder, H. Vagt and W. Vogt, *Annalen*, **565**, 135 (1949); G. A. Ropp and E. C. Coyner, *J. Amer. Chem. Soc.*, **71**, 1832 (1949); J. S. Meek, F. J. Lorenzi and S. J. Cristol, *ibid.*, 1830.
182. H. R. Snyder and G. I. Poos, *J. Amer. Chem. Soc.*, **72**, 4104 (1950).
183. K. Alder and W. Vogt, *Annalen*, **564**, 120 (1949).
184. G. L. Dunn and J. K. Donohue, *Tetrahedron Letters*, 3485 (1968) and references therein.
185. A. A. Petrov and N. P. Sopov, *Zhur. obschei Khim.*, **26**, 2452 (1956).
186. K. Alder, J. Haydn and B. Krüger, *Chem. Ber.*, **86**, 1372 (1953).
187. S. Danishefsky and R. Cunningham, *J. Org. Chem.*, **30**, 3676 (1965); G. A. Berchtold, J. Ciabattoni and A. A. Tunick, *ibid.*, 3679. For another example, see D. A. Evans, C. A. Bryan and C. L. Sims, *J. Amer. Chem. Soc.*, **94**, 2891 (1972).
188. See, however, N. D. Epiotis, *J. Amer. Chem. Soc.*, **94**, 1924 (1972).
189. I. Fleming, F. L. Gianni and T. Mah, *Tetrahedron Letters*, 881 (1976).
190. R. A. Firestone, *J. Org. Chem.*, **37**, 2181 (1972).
191. T. Uyehara and Y. Kitahara, *Chem. and Ind.*, 354 (1971).
192. S. Ito, H. Takeshita and Y. Shoji, *Tetrahedron Letters*, 1815 (1969).
193. K. Alder and R. Schmitz-Josten, *Annalen*, **595**, 1 (1955).
194. W. E. Bachmann and N. C. Deno, *J. Amer. Chem. Soc.*, **71**, 3062 (1949).
195. W. E. Bachmann and L. B. Scott, *J. Amer. Chem. Soc.*, **70**, 1462 (1948).
196. T. L. Gresham and T. R. Steadman, *J. Amer. Chem. Soc.*, **71**, 737 (1949).
197. E. J. Corey and S. W. Walinsky, *J. Amer. Chem. Soc.*, **94**, 8932 (1972).
198. G. Kresze and J. Firl, *Tetrahedron Letters*, 1163 (1965); J. Firl and G. Kresze, *Chem. Ber.*, **99**, 3695 (1966).
199. G. Desimoni and G. Tacconi, *Chem. Rev.*, **75**, 651 (1975).
200. K. Alder and E. Rüden, *Ber.*, **74**, 920 (1941); S. M. Sherlin, A. Y. Berlin, T. A. Serebrennikova and F. E. Rabinovich, *Zhur. obschei Khim.*, **8**, 22 (1938).
201. A. Devaquet and L. Salem, *J. Amer. Chem. Soc.*, **91**, 3793 (1969).
202. K. N. Houk and R. W. Strozier, *J. Amer. Chem. Soc.*, **95**, 4094 (1973).
203. K. Alder, H. Offermans and E. Rüden, *Ber.*, **74**, 905 (1941).
204. C. W. Smith, D. G. Norton and S. A. Ballard, *J. Amer. Chem. Soc.*, **73**, 5273 (1951).
205. I. Fleming and M. H. Karger, *J. Chem. Soc. (C)*, 226 (1967).
206. M. Fischer and F. Wagner, *Chem. Ber.*, **102**, 3486 (1969).
207. H. C. Stevens, D. A. Reich, D. R. Brandt, K. R. Fountain and E. J. Gaughan, *J. Amer. Chem. Soc.*, **87**, 5257 (1965).
208. R. Huisgen and L. A. Feiler, *Chem. Ber.*, **102**, 3391 (1969); R. Huisgen, L. A. Feiler and P. Otto, *ibid.*, 3405; L. A. Feiler and R. Huisgen, *ibid.*, 3428; R. Huisgen, L. A. Feiler and P. Otto, *ibid.*, 3444; R. Huisgen, L. A. Feiler and G. Binsch, *ibid.*, 3460; R. Huisgen and P. Otto, *ibid.*, 3475; W. T. Brady and H. R. O'Neal,

J. Org. Chem., **32**, 612 (1967); J. C. Martin, V. W. Goodlett and R. D. Burpitt, *ibid.*, **30**, 4309 (1965); R. Montaigne and L. Ghosez, *Angew. Chem. Internat. Edn.*, **7**, 221 (1968). See, however, H. U. Wagner and R. Gompper, *Tetrahedron Letters*, 2819 (1970).

209. K. N. Houk, R. W. Strozier and J. A. Hall, *Tetrahedron Letters*, 897 (1974); R. Sustmann, A. Ansmann and F. Vahrenholt, *J. Amer. Chem. Soc.*, **94**, 8099 (1972); S. Inagaki, T. Minato, S. Yamabe, H. Fujimoto and K. Fukui, *Tetrahedron*, **30**, 2151 (1974).

210. N. Campbell and H. G. Heller, *J. Chem. Soc.*, 5473 (1965).

211. R. H. Fish, *J. Org. Chem.*, **34**, 1127 (1969).

212. L. A. Feiler, R. Huisgen and P. Koppitz, *J. Amer. Chem. Soc.*, **96**, 2270 (1974).

213. C. D. Hurd and R. D. Kimbrough, *J. Amer. Chem. Soc.*, **82**, 1373 (1960).

214. J. F. Arens, *Angew. Chem.*, **70**, 631 (1958).

215. R. Huisgen and P. Otto, *J. Amer. Chem. Soc.*, **91**, 5922 (1969).

216. H. Staudinger and S. Bereza, *Annalen*, **380**, 243 (1911).

217. H. J. Hagemeyer, *Ind. and Eng. Chem.*, **41**, 765 (1949).

218. H. Staudinger and H. W. Klever, *Ber.*, **40**, 1149 (1907).

219. R. Scarpati, *Rend. Acad. Sci. fis. mat. (Napoli)*, 25 (1958); *Chem. Abs.*, **55**, 11423 (1961).

220. H. Bergreen, *Ber.*, **21**, 337 (1888).

221. G. Wittig and G. Geissler, *Annalen*, **580**, 44 (1953).

222. C. K. Ingold and H. A. Piggott, *J. Chem. Soc.*, 2793 (1922); *ibid.*, 2745 (1923).

223. F. Chick and N. T. M. Wilsmore, *J. Chem. Soc.*, 946 (1908); H. Staudinger and St. Bereza, *Ber.*, **42**, 4908 (1909); correct structure: A. L. Wilson, quoted by A. B. Boese, *Ind. and Eng. Chem.*, **32**, 16 (1940), and proved by L. Katz and W. N. Lipscomb, *Acta Cryst.*, **5**, 313 (1952).

224. H. Staudinger and H. W. Klever, *Ber.*, **39**, 968 (1906).

225. W. E. Truce and J. R. Norell, *J. Amer. Chem. Soc.*, **85**, 3231 (1963).

226. W. A. Sheppard and J. Dieckmann, *J. Amer. Chem. Soc.*, **86**, 1891 (1964).

227. J. P. Snyder, *J. Org. Chem.*, **38**, 3965 (1973).

228. G. Opitz and H. R. Mohl, *Angew. Chem. Internat. Edn.*, **8**, 73 (1969).

229. J. Strating, L. Thijs and B. Zwanenberg, *Rec. Trav. chim.*, **83**, 631 (1964).

230. K. von Auwers and E. Cauer, *Annalen*, **470**, 284 (1929); K. von Auwers and F. König, *ibid.*, **496**, 252 (1932).

231. K. N. Houk, J. Sims, R. E. Duke, R. W. Strozier and J. K. George, *J. Amer. Chem. Soc.*, **95**, 7287 (1973); K. N. Houk, J. Sims, C. R. Watts and L. J. Luskus, *ibid.*, 7301. See also, J. Bastide, N. E. Ghandour and O. Henri-Rousseau, *Bull. Soc. chim. France*, 2290 and 2294 (1973), and J. Bastide and O. Henri-Rousseau, *ibid.*, 1037 (1974).

232. R. A. Firestone, *J. Org. Chem.*, **33**, 2285 (1968).

233. R. Huisgen, *J. Org. Chem.*, **33**, 2291 (1968).

234. (a) R. Huttel, *Ber.*, **74**, 1680 (1941); (b) K. Bowden and E. R. H. Jones, *J. Chem. Soc.*, 953 (1946).

235. D. E. McGrear, W. Wai and G. Carmichael, *Canad. J. Chem.*, **38**, 2410 (1960).

236. P. K. Kadaba and T. F. Colturi, *J. Heterocyclic Chem.*, **6**, 829 (1969).

237. R. A. Firestone, *J. Org. Chem.* **41**, 2212 (1976); see, however S. H. Groen and J. F. Arens, *Rec. Trav. chim.*, **80**, 879 (1961).

238. A. Ledwith and D. Parry, *J. Chem. Soc. (C)*, 1408 (1966).

239. F. Piozzi, A. Umani-Ronchi and L. Merlini, *Gazzetta*, **95**, 814 (1965).

240. R. Huisgen, L. Möbius and G. Szeimies, *Chem. Ber.*, **98**, 1138 (1965).

241. G. D. Buckley, *J. Chem. Soc.*, 1850 (1954).

242. W. Kirmse and L. Horner, *Annalen*, **614**, 1 (1958).

243. R. Huisgen, G. Szeimies and L. Möbius, *Chem. Ber.*, **99**, 475 (1966).

244. W. Oppolzer, *Tetrahedron Letters*, 2199 (1970).

245. H. Gotthardt and R. Huisgen, *Chem. Ber.*, **101**, 552 (1968); R. Huisgen, R. Grashey and H. Gotthardt, *ibid.* 829.
246. R. Huisgen, H. Gotthardt and R. Grashey, *Chem. Ber.*, **101**, 536 (1968).
247. R. Grashey and K. Adelsberger, *Angew. Chem. Internat. Edn.*, **1**, 267 (1962).
248. R. Criegee, G. Blust and H. Zinke, *Chem. Ber.*, **87**, 766 (1954).
249. E. J. Corey, N. M. Weinshenker, T. K. Schaaf and W. Huber, *J. Amer. Chem. Soc.*, **91**, 5675 (1969).
250. T. Inukai and T. Kojima, *J. Org. Chem.*, **32**, 869 and 872 (1967).
251. W. Kreiser, W. Haumesser and A. F. Thomas, *Helv. Chim. Acta*, **57**, 164 (1974).
252. J. Sauer and J. Kredel, *Tetrahedron Letters*, 731 (1966).
253. N. T. Anh and J. Seyden-Penne, *Tetrahedron*, **29**, 3259 (1973).
254. B. B. Snider, *J. Org. Chem.*, **39**, 255 (1974).
255. R. A. Dickinson, R. Kubela, G. A. MacAlpine, Z. Stojanac, and Z. Valenta, *Canad. J. Chem.*, **50**, 2377 (1972). See also, Z. Stojanac, R. A. Dickinson, N. Stojanac, R. J. Woznow and Z. Valenta, *ibid.*, **53**, 617 (1975).
256. M. F. Ansell, B. W. Nash and D. A. Wilson, *J. Chem. Soc.*, 3012 (1963).
257. O. Diels and K. Alder, *Annalen*, **486**, 191 (1931).
258. M. S. Kharasch and E. Sternfeld, *J. Amer. Chem. Soc.*, **61**, 2318 (1939).
259. K. Alder and H. von Brachel, *Annalen*, **608**, 195 (1957); H. Fleischaker and G. F. Woods, *J. Amer. Chem. Soc.*, **78**, 3436 (1956).
260. H. R. Snyder and G. I. Poos, *J. Amer. Chem. Soc.*, **72**, 4096 (1950).
261. G. Spiteller, G. Schmidt, H. Budzikiewicz and F. Wessely, *Monatsh.*, **91**, 129 (1960).
262. C. von der Heide, *Ber.*, **37**, 2101 (1904).
263. R. Huisgen and P. Otto, *Chem. Ber.*, **102**, 3475 (1969).
264. A. P. ter Borg and A. F. Bickel, *Rec. Trav. chim.*, **80**, 1217 (1961).
265. L. A. Nakhapetyan, I. L. Safonova and B. A. Kazanskii, *Izv. Akad. Nauk S.S.S.R., Otd. khim. Nauk*, 902 (1962).
266. J. P. Gouesnard, *Tetrahedron*, **30**, 3113 (1974), and ref. 263.
267. E. E. Waali and W. M. Jones, *J. Amer. Chem. Soc.*, **95**, 8114 (1973).
268. F. D. Lewis, C. E. Hoyle and D. E. Johnson, *J. Amer. Chem. Soc.*, **97**, 3267 (1975).
269. J. W. Cook and R. Schoental, *J. Chem. Soc.*, 170 (1948).
270. G. M. Badger, J. W. Cook and A. R. M. Gibb, *J. Chem. Soc.*, 3456 (1951).
271. M. G. Sturrock and R. A. Dunn, *J. Org. Chem.*, **33**, 2149 (1968).
272. E. Clar, *Polycyclic Hydrocarbons*, Academic Press, London, 1964.
273. J. W. Cook, J. D. Loudon and W. F. Williamson, *J. Chem. Soc.*, 911 (1950).
274. J. W. Cook and R. Schoental, *Nature*, **161**, 237 (1948).
275. R. Huisgen, A. Dahmen and H. Huber, *J. Amer. Chem. Soc.*, **89**, 7130 (1967).
276. R. B. Bates, W. H. Deines, D. A. McCombs and D. E. Potter, *J. Amer. Chem. Soc.*, **91**, 4608 (1969).
277. A. Galbraith, T. Small, R. A. Barnes and V. Boekelheide, *J. Amer. Chem. Soc.*, **83**, 453 (1961).
278. W. von E. Doering and D. W. Wiley, *Tetrahedron*, **11**, 183 (1960).
279. L. A. Paquette and J. H. Barrett, *J. Amer. Chem. Soc.*, **88**, 2590 (1966); A. L. Johnson and H. E. Simmons, *ibid.*, **89**, 3191 (1967).
280. G. C. Farrant and R. Feldman, *Tetrahedron Letters*, 4979 (1970).
281. W. L. Mock, *J. Amer. Chem. Soc.*, **89**, 1281 (1967) and **91**, 5682 (1969). See also, R. M. Dodson and J. P. Nelson, *Chem. Comm.*, 1159 (1969).
282. J. L. M. A. Schlatmann, J. Pot and E. Havinga, *Rec. Trav. chim.*, **83**, 1173 (1964); M. Akhtar and C. J. Gibbons, *Tetrahedron Letters*, 509 (1965).
283. E. E. Schweizer, D. M. Crouse and D. L. Dalrymple, *Chem. Comm.*, 354 (1969).
284. J. C. Gilbert, K. R. Smith, G. W. Klumpp and M. Schakel, *Tetrahedron Letters*, 125 (1972); J. C. Gilbert and K. R. Smith, *J. Org. Chem.*, 1976, **41**, 3883.
285. G. Frater and H. Schmid, *Helv. Chim. Acta*, **51**, 190 (1968).

238

286. M. J. Goldstein and S.-H. Dai, *J. Amer. Chem. Soc.*, **95**, 933 (1973).
287. T. Laird and W. D. Ollis, *Chem. Comm.*, 658 (1973).
288. Ref. 1, p. 167.
289. I. Fleming, *J. Chem. Soc. Perkin I*, 1019 (1973).
290. H. Prinzbach, D. Seip, L. Knothe and W. Faisst, *Annalen*, **698**, 34 (1966).
291. H. Prinzbach and H. Knöfel, *Angew. Chem. Internat. Edn.*, **8**, 881 (1969).
292. R. E. Davies, W. Henslee, A. Garza, H. Knöfel and H. Prinzbach, *Tetrahedron Letters*, 2823 (1974), and ref. 291.
293. K. N. Houk, J. K. George and R. E. Duke, *Tetrahedron*, **30**, 523 (1974).
294. K. N. Houk and L. J. Luskus, *J. Org. Chem.*, **38**, 3836 (1972).
295. P. Caramella, P. Frattini and P. Grünanger, *Tetrahedron Letters*, 3817 (1971).
296. T. Asao, T. Machiguchi, T. Kitamura and Y. Kitahara, *Chem. Comm.*, 89 (1970); R. E. Harmon, W. D. Barta, S. K. Gupta and G. Slomp, *J. Chem. Soc. (C)*, 3645 (1971).
297. M. N. Paddon-Row and R. N. Warrener, *Tetrahedron Letters*, 3797 (1974).
298. K. N. Houk and L. J. Luskus, *Tetrahedron Letters*, 4029 (1970).
299. M. N. Paddon-Row, K. Gell and R. N. Warrener, *Tetrahedron Letters*, 1975 (1975).
300. H. Takeshita, A. Mori, S. Sano and Y. Fujise, *Bull. Chem. Soc. Japan*, **48**, 1661 (1975).
301. K. Fukui, ref. 19, p. 48; G. Klopman, ref. 19, p. 111.
302. K. J. Saunders, *Organic Polymer Chemistry*, Chapman and Hall, London, 1973, p. 30.
303. P. I. Abell, in J. K. Kochi (Editor), *Free Radicals*, Wiley, New York, 1973, Vol. II, p. 96.
304. G. A. Russell, in J. K. Kochi (Editor), *Free Radicals*, Wiley, New York, 1973, Vol. I, p. 295; R. W. Henderson, *J. Amer. Chem. Soc.*, **97**, 213 (1975).
305. K. U. Ingold, *Pure Appl. Chem.*, **15**, 49 (1967).
306. K. Fukui, H. Kato and T. Yonezawa, *Bull. Chem. Soc. Japan*, **33**, 1197 (1960) and **35**, 1475 (1962).
307. H. Magritte and A. Bruylants, *Ind. chim. belge*, **22**, 547 (1957).
308. J. Fossey, Ph.D. Thesis, Paris, 1974.
309. A. A. Oswald, K. Griesebaum, W. A. Thaler and B. E. Hudson, *J. Amer. Chem. Soc.*, **84**, 3897 (1962).
310. C. S. Hsia Chen and R. F. Stamm, *J. Org. Chem.*, **28**, 1580 (1963).
311. G. W. Kabalka, H. C. Brown, A. Suzuki, S. Honma, A. Arase and M. Itoh, *J. Amer. Chem. Soc.*, **92**, 710 (1970).
312. R. J. Cvetanovic, *Canad. J. Chem.*, **36**, 623 (1958).
313. J. M. Tedder and J. C. Walton, *Trans. Faraday Soc.*, **62**, 1859 (1966).
314. D. P. Johari, H. W. Sidebottom, J. M. Tedder and J. C. Walton, *J. Chem. Soc. (B)*, 95 (1971).
315. M. Julia, *Pure Appl. Chem.*, **15**, 167 (1967).
316. M. Julia, M. Maumy and L. Mion, *Bull. Soc. chim. France*, 2641 (1967); M. Julia and M. Maumy, *ibid.*, 2427 (1969).
317. K. R. Jennings and R. J. Cvetanovic, *J. Chem. Phys.*, **35**, 1233 (1961); there are more data in R. J. Cvetanovic and R. S. Irwin, *ibid.*, **46**, 1694 (1967).
318. I. M. Whittemore, A. P. Stefani and M. Swarc, *J. Amer. Chem. Soc.*, **84**, 3799 (1962).
319. M. Swarc and J. H. Binks, in *Theoretical Organic Chemistry, The Kekule Symposium*, Butterworths, London, 1959, p. 262.
320. K. Fukui, T. Yonezawa and C. Nagata, *J. Chem. Phys.*, **27**, 1247 (1957).
321. M. J. Perkins, in J. K. Kochi (Editor), *Free Radicals*, Wiley, New York, 1973, Vol. II, p. 231.
322. F. Minisci, R. Mondelli, G. R. Gardini and O. Porta, *Tetrahedron*, **28**, 2403 (1972).
323. S. C. Dickerman and G. B. Vermont, *J. Amer. Chem. Soc.*, **84**, 4150 (1962).

324. C. M. Camaggi, G. De Luca and A. Tundo, *J. Chem. Soc. Perkin I*, 412 (1972).
325. F. Minisci, R. Galli, R. Bernardi and A. Galli, *Chim. Ind. (Milan)*, **49**, 386 (1967).
326. D. H. Hey, C. J. M. Stirling and G. H. Williams, *J. Chem. Soc.*, 3963 (1955).
327. E. C. Kooyman, *Adv. Free Radical Chemistry*, **1**, 145 (1965).
328. H. Sakurai, in J. K. Kochi, *Free Radicals*, Wiley, New York, 1973, Vol. II, p. 787.
329. D. F. DeTar and R. A. J. Long, *J. Amer. Chem. Soc.*, **80**, 4742 (1958).
330. E. I. Heiba, R. M. Dessau and W. J. Koehl, *J. Amer. Chem. Soc.*, **91**, 138 (1969).
331. B. Emmert, *Ber.*, **42**, 1997 (1909).
332. W. I. Taylor and A. R. Battersby (Editors), *Oxidative Coupling of Phenols*, Dekker, New York, 1967.
333. A. Rieker and K. Scheffler, *Tetrahedron Letters*, 1337 (1965).
334. M. Broze, Z. Luz and B. L. Silver, *J. Chem. Phys.*, **46**, 4891 (1967).
335. J. Forrest and V. Petrow, *J. Chem. Soc.*, 2340 (1950).
336. K. Freudenberg and H. H. Hübner, *Chem. Ber.*, **85**, 1181 (1952); K. Freudenberg and D. Rasenack, *ibid.*, **86**, 755 (1953); K. Freudenberg and H. Schlütter, *ibid.*, **88**, 617 (1955); K. Freudenberg and K.-C. Renner, *ibid.*, **98**, 1879 (1965).
337. D. M. Mohilnar, R. N. Adams and W. J. Argersinger, *J. Amer. Chem. Soc.*, **84**, 3618 (1962); T. Mizoguchi and R. N. Adams, *ibid.*, 2058.
338. S. K. Malhotra, J. J. Hostynek and A. F. Lundin, *J. Amer. Chem. Soc.*, **90**, 6565 (1968).
339. R. Adams and E. W. Adams, *Org. Synth.*, *Collective Vol.*, **1**, 459 (1941).
340. M. M. Bazier and J. P. Petrovich, *Progr. Phys. Org. Chem.*, **7**, 188 (1970).
341. I.C.I. Belgian Patent 620,301 (1963); *Chem. Abs.*, **59**, 8712f (1963).
342. A. J. Birch, *J. Chem. Soc.*, 430 (1944).
343. M. Smith, in R. L. Augustine (Editor), *Reduction*, Dekker, New York, 1968, p. 121; H. Smith, *Organic Reactions in Liquid Ammonia*, Interscience, New York, 1963; but, for dimethylaniline, see A. J. Birch, E. G. Hutchinson and G. Subba Rao, *J. Chem. Soc. (C)*, 637 (1971).
344. H. E. Zimmerman, in P. de Mayo (Editor), *Molecular Rearrangements*, Interscience, New York, 1963, Vol. I, p. 345.
345. W. Hückel and R. Schwen, *Chem. Ber.*, **89**, 150 (1956).
346. H. Plieninger and G. Ege, *Angew. Chem.*, **70**, 505 (1958); M. E. Kuehne and B. F. Lambert, *J. Amer. Chem. Soc.*, **81**, 4278 (1959).
347. N. L. Bauld, *J. Amer. Chem. Soc.*, **84**, 4347 (1962), and ref. 350.
348. H. Gilman and J. C. Bailie, *J. Amer. Chem. Soc.*, **65**, 267 (1943).
349. H. F. Miller and G. B. Bachman, *J. Amer. Chem. Soc.*, **57**, 766 (1935).
350. W. Hückel and H. Bretschneider, *Annalen*, **540**, 157 (1939).
351. A. J. Birch, A. R. Murray, and H. Smith, *J. Chem. Soc.*, 1945 (1951); W. Hückel and H. Schlee, *Chem. Ber.*, **88**, 346 (1955).
352. H. Güsten and L. Horner, *Angew. Chem. Internat. Edn.*, **1**, 455 (1962).
353. N. Kornblum, R. T. Swiger, G. W. Earl, H. W. Pinnick and F. W. Stuchal, *J. Amer. Chem. Soc.*, **92**, 5513 (1970).
354. O. Dimroth and F. Frister, *Ber.*, **55**, 1223 (1922).
355. M. S. Kharasch, P. G. Holton and W. Nudenberg, *J. Org. Chem.*, **19**, 1600 (1954).
356. J. Meinwald, L. V. Dunkerton and G. W. Gruber, *J. Amer. Chem. Soc.*, **97**, 681 (1975).
357. J. C. Anderson and C. B. Reese, *J. Chem. Soc.*, 1781 (1963); reviewed by V. I. Stenberg, *Org. Photochem.*, **1**, 127 (1967), and by D. Bellus, *Adv. Photochem.*, 109 (1971).
358. F. Mader and V. Zanker, *Chem. Ber.*, **97**, 2418 (1964).
359. G. Ciamician and P. Silber, *Ber.*, **33**, 2911 (1900).
360. B. Nann, D. Gravel, R. Schorta, H. Wehrli, K. Schaffner, and O. Jeger, *Helv. Chim. Acta*, **46**, 2473 (1963).
361. A. Butenandt and L. Poschmann, *Ber.*, **73**, 893 (1940).

240

362. I. A. Williams and P. Bladon, *Tetrahedron Letters*, 257 (1964).
363. B. Fraser-Reid, D. R. Hicks, D. L. Walker, D. E. Iley, M. B. Yunker and J. Saunders, *Tetrahedron Letters*, 297 (1975); D. L. Walker and B. Fraser-Reid, *J. Amer. Chem. Soc.*, **97**, 6251 (1975).
364. K. Fukui, ref. 19, p. 49.
365. H. Yamazaki and R. J. Cvetanovic, *J. Amer. Chem. Soc.*, **91**, 520 (1969).
366. S. Masamune and N. Darby, *Accts. Chem. Res.*, **5**, 272 (1972).
367. T. Mukai, T. Tezuka and Y. Akasaki, *J. Amer. Chem. Soc.*, **88**, 5025 (1966).
368. J. Saltiel and L. Metts, *J. Amer. Chem. Soc.*, **88**, 2232 (1967).
369. G. Büchi and E. M. Burgess, *J. Amer. Chem. Soc.*, **82**, 4333 (1960).
370. R. Srinivasan, *J. Chem. Phys.*, **38**, 1039 (1963).
371. A. P. ter Borg and H. Kloosterziel, *Rec. Trav. chim.*, **88**, 266 (1969).
372. R. Srinivasan, *J. Amer. Chem. Soc.*, **90**, 4498 (1968).
373. G. J. Fonken, *Tetrahedron Letters*, 549 (1962).
374. H. Nozaki, M. Kurita and R. Noyori, *Tetrahedron Letters*, 3635 (1968).
375. D. R. Arnold, *Adv. Photochem.*, **6**, 301 (1968); for a theoretical discussion of this reaction, see L. Salem, *J. Amer. Chem. Soc.*, **96**, 3486 (1974).
376. D. R. Arnold, R. L. Hinman and A. H. Glick, *Tetrahedron Letters*, 1425 (1964).
377. J. Saltiel, R. M. Coates and W. G. Dauben, *J. Amer. Chem. Soc.*, **88**, 2745 (1966).
378. J. A. Barltrop and H. A. J. Carless, *J. Amer. Chem. Soc.*, **94**, 1951 (1972).
379. D. J. Trecker, *Org. Photochem.*, **2**, 63 (1969).
380. S. Hosaka and S. Wakamatsu, *Tetrahedron Letters*, 219 (1968).
381. G. S. Hammond, N. J. Turro and R. S. H. Liu, *J. Org. Chem.*, **28**, 3297 (1963).
382. S. Kuwata, Y. Shigemitsu and Y. Odaira, *Chem. Comm.*, 2 (1972).
383. R. Hoffman, P. Wells and H. Morrison, *J. Org. Chem.*, **36**, 102 (1971).
384. D. O. Cowan and R. L. E. Drisko, *J. Amer. Chem. Soc.*, **92**, 6286 (1970).
385. F. D. Greene, S. L. Misrock and J. R. Wolfe, *J. Amer. Chem. Soc.*, **77**, 3852 (1955); D. E. Applequist, R. L. Little, E. C. Friedrich and R. E. Wall, *ibid.*, **81**, 452 (1959).
386. E. C. Taylor, R. O. Kan and W. W. Paudler, *J. Amer. Chem. Soc.*, **83**, 4484 (1961); G. Slomp, F. A. MacKellar and L. A. Paquette, *ibid.*, 4472.
387. T. S. Cantrell and H. Schechter, *J. Org. Chem.*, **33**, 114 (1968).
388. H. Bouas-Laurent, A. Castellan, J. P. Desvergne, G. Dumartin, C. Courseille, J. Gaultier and C. Hauw, *Chem. Comm.*, 1267 (1972).
389. P. E. Eaton, *J. Amer. Chem. Soc.*, **84**, 2344 (1962); J. L. Ruhlen and P. A. Leermakers, *ibid.*, **89**, 4944 (1967).
390. M. Brown, *Chem. Comm.*, 340 (1965).
391. G. Montaudo and S. Caccamese, *J. Org. Chem.*, **38**, 710 (1973).
392. N. Sugiyama, K. Yamada, Y. Watari and T. Koyama, *J. Chem. Soc. Japan*, **87**, 737 (1966).
393. F. Weisbuch, P. Scribe and C. Provelenghiou, *Tetrahedron Letters*, 3441 (1973).
394. J. J. McCullough and C. W. Huang, *Chem. Comm.*, 815 (1967).
395. W. L. Dilling and R. D. Kroenig, *Tetrahedron Letters*, 695 (1970).
396. P. Servé, H. M. Rosenberg and R. Rondeau, *Canad. J. Chem.*, **47**, 4295 (1969).
397. E. J. Corey, J. D. Bass, R. LeMaheu and R. B. Mitra, *J. Amer. Chem. Soc.*, **86**, 5570 (1964).
398. O. L. Chapman, T. H. Koch, F. Kelin, P. J. Nelson and E. L. Brown, *J. Amer. Chem. Soc.*, **90**, 1657 (1968).
399. P. Margaretha, *Tetrahedron*, **29**, 1317 (1973).
400. P. Margaretha, *Helv. Chim. Acta*, **57**, 1866 (1974).
401. D. Becker, Z. Harel and D. Birnbaum, *Chem. Comm.*, 377 (1975).
402. K. Wiesner, *Tetrahedron*, **31**, 1655 (1975).
403. G. Stork and S. D. Darling, *J. Amer. Chem. Soc.*, **86**, 1761 (1964).
404. J. D. Roberts and C. M. Sharts, *Org. Reactions*, **12**, 1 (1962); P. D. Bartlett, *Quart. Rev.*, **24**, 473 (1970).

405. T. S. Cantrell, *Tetrahedron Letters*, 907 (1975).
406. E. Havinga and M. E. Kronenberg, *Pure Appl. Chem.*, **16**, 137 (1968).
407. J. Cornelisse, *Pure Appl. Chem.*, **41**, 433 (1975); J. Cornelisse and E. Havinga, *Chem. Rev.*, **75**, 353 (1975); J. Cornelisse, G. P. de Gunst and E. Havinga, *Adv. Phys. Org. Chem.*, **11**, 225 (1975).
408. M. E. Kronenberg, V. van der Heyden and E. Havinga, *Rec. Trav. chim.*, **86**, 254 (1967).
409. A. van Vliet, M. E. Kronenberg, J. Cornelisse and E. Havinga, *Tetrahedron*, **26**, 1061 (1970).
410. G. H. D. van der Stegen, E. J. Poziomek, M. E. Kronenberg and E. Havinga, *Tetrahedron Letters*, 6371 (1966).
411. L. Kaplan, J. W. Pavlick and K. E. Wilzbach, *J. Amer. Chem. Soc.*, **94**, 3283 (1972).
412. M. Bellas, D. Bryce-Smith and A. Gilbert, *Chem. Comm.*, 263 and 862 (1967).
413. B. A. de Bie and E. Havinga, *Tetrahedron*, **21**, 2359 (1965).
414. L. A. Paquette, D. M. Cottrell, R. A. Snow, K. B. Gifkins and J. Clardy, *J. Amer. Chem. Soc.*, **97**, 3275 (1975). For some other photochemical rearrangements which can be accounted for by similar arguments[3], see L. B. Jones and V. K. Jones, *J. Amer. Chem. Soc.*, **89**, 1880 (1967); S. Boué and R. Srinivasan, *ibid.*, **92**, 3226 (1970); W. G. Dauben, K. Koch, S. L. Smith and O. L. Chapman, *ibid.*, **85**, 2616 (1963) and O. L. Chapman and D. J. Pasto, *ibid.*, **82**, 3642 (1960).
415. J. A. Berson and L. Salem, *J. Amer. Chem. Soc.*, **94**, 8917 (1972); J. A. Berson, *Accts. Chem. Res.*, **5**, 406 (1972); H. E. Zimmerman, *ibid.*, 393; J. E. Baldwin, A. H. Andrist and R. K. Pinschmidt, *ibid.*, 402.
416. M. J. S. Dewar, *Chem. in Britain*, **11**, 97 (1975).

Index

248